Springer Proceedings in Business and Economics

More information about this series at http://www.springer.com/series/11960

Hui Yang · Robin Qiu
Editors

Advances in Service Science

Proceedings of the 2018 INFORMS International Conference on Service Science

Springer

Editors
Hui Yang
Department of Industrial Engineering
Pennsylvania State University
University Park, PA, USA

Robin Qiu
Division of Engineering and
Information Science
Pennsylvania State University
Malvern, PA, USA

ISSN 2198-7246 ISSN 2198-7254 (electronic)
Springer Proceedings in Business and Economics
ISBN 978-3-030-04725-2 ISBN 978-3-030-04726-9 (eBook)
https://doi.org/10.1007/978-3-030-04726-9

Library of Congress Control Number: 2018962378

This Springer imprint is published by the registered company Springer Nature Switzerland AG
The registered company address is: Gewerbestrasse 11, 6330 Cham, Switzerland

Preface

This proceeding records the papers submitted and presented in the 2018 INFORMS International Conference Service Science (CSS2018), held in Phoenix, AZ on November 3, 2018, which includes diverse participants sharing their vision, knowledge, and experience in Service Science Research, Education, and Applications. INFORMS CSS 2018 is held right before the INFORMS annual meeting 2018. The conference proceeding is published by Springer.

Service Science related research and education have fueled increasing interests to both academic researchers and industry practitioners. Service operations and management is critical to the national and international economic growth. The objective of CSS 2018 is to disseminate the emerging research results, technology and applications in service science, and to serve as a forum for researchers, professionals, and academic of the profession to network with each other.

This year we had 40 submissions from around the world. All submissions were carefully reviewed by at least two reviewers. Paper review and selection were undertaken electronically via the easy chair system. After the rigorous review and revision process, 29 papers were finally accepted to be included in this proceeding. The major areas covered at the conference and presented in this proceeding include:

- Service theories and development
- Service research, education, and practice
- Service management, operations, engineering, design, and marketing
- Service system, modeling, and simulation
- Smart cities and public services
- Big data, machine learning, and artificial intelligence in service
- Service analytics and applications in healthcare, education, energy, finance, information technology, transportation, sports, logistics, and public services.

In addition to the accepted research papers and invited talks, INFOMRS CSS 2018 provides an opportunity for high-level exchanges between academia and industry. Dr. Richard Larson from MIT was invited to present a keynote talk at the conference, i.e., “The Services Industries: Some Insights Provided by Operations Research”. There is also a Panel discussion on “Machine Learning Meets Service

Dominant Logic", moderated by Prof. Paul R. Messinger from University of Alberta and with panelists, Dr. Mary Jo Bitner from Arizona State University, Dr. Peter I. Frazier from Cornell University and Uber, Dr. Aly Megahed from IBM, and Dr. Xin (Shane) Wang from Western University.

Finally, we would like to thank all authors for submitting their high-quality works in the field of service science, and the Program Committee members, listed on the following page, for their tireless efforts and time spent on reviewing submissions. Special appreciation is extended to the Springer Editors, Matthew Amboy and Faith Su, who have contributed tremendously to the success of the INFORMS CSS 2018 conference proceedings.

Co-editors—Proceedings of 2018 INFORMS Conference on Service Science.

University Park, USA Hui Yang
Malvern, USA Robin Qiu

Contents

The Inmate Transportation Problem and Its Application in the PA Department of Corrections

Anshul Sharma, Mohammad Shahabsafa and Tamás Terlaky

Abstract The Inmate Transportation Problem (ITP) is a common complex problem in any correctional system. We develop a weighted multi-objective mixed integer linear optimization (MILO) model for the ITP. The MILO model optimizes the transportation of the inmates within a correctional system, while considering all legal restrictions and best business practices. We test the performance of the MILO model with real datasets from the Pennsylvania Department of Corrections (PADoC) and demonstrate that the inmate transportation process at the PADoC can significantly be improved by using operations research methodologies.

1 Introduction

According to the International Centre for Prison Studies, the U.S. incarcerates 698 people for every 100,000 of its population. Having approximately 4.5% of the world's population, the U.S. has 21.4% of the world's incarcerated population [12].

Population management of the inmates is one of the most critical operations within a correctional system involving the inmate assignment to Correctional Institutions (CIs) and transportation between CIs. Transportation expenditures in a correctional systems include labor cost, maintenance and fuel costs, fixed cost for using a vehicle, etc. Efficient management of inmate transportation gives substantial savings.

A. Sharma · M. Shahabsafa · T. Terlaky (✉)
Department of Industrial and Systems Engineering, Lehigh University,
Bethlehem, PA 18015, USA
e-mail: terlaky@lehigh.edu

M. Shahabsafa
e-mail: mos313@lehigh.edu

A. Sharma
e-mail: ansel5@lehigh.edu

H. Yang and R. Qiu (eds.), *Advances in Service Science*, Springer Proceedings
in Business and Economics, https://doi.org/10.1007/978-3-030-04726-9_1

Security of the personnel and inmates is another important aspect of the transportation process. In particular, we want to curtail the total transportation cost without compromising security, while considering all the regulations and business practices. Here we study and formalize the inmate transportation process at the PADoC and develop a mathematical optimization model for the Inmate Transportation Problem (ITP).

Conventionally, inmate transportation planning has been a manual and subjective process at the PADoC, where a staff member creates trips and assigns inmates to those trips considering the transportation criteria and policies. While the general guidelines are known, the huge number of possible routes, and the complexity of the transportation problem makes it extremely difficult, if not impossible, to manually determine optimal routes for a fleet of vehicles.

In this paper, we formulate a multi-objective mixed integer linear optimization (MILO) model for the ITP. The model is validated by solving various datasets from the PADoC. The goal is to optimize the inmate transportation process to achieve the following objectives:

- reduce the number of inmates not transported in a given time period,
- reduce the total number of seats used for the inmate transportation.

2 Literature Review

The traveling salesman problem (TSP) was considered mathematically already in the 1930s, e.g., by Flood who was looking to solve a school bus routing problem [5]. He later formalized the problem in 1956 in his paper "Traveling-Salesman Problem" [7]. Dantzig and Ramser [4] formulated a generalization of the TSP as a Vehicle Routing Problem (VRP). For more information about the VRP see e.g., Crainic and Laporte [3]. A lot of work has been done on solving the TSP and the VRP [1, 2, 6, 10].

Li et al. [9] first studied the inmate assignment problem in a correctional system. They developed a decision tree based model which gives a ranked order of CIs for an inmate considering all the business rules of the inmate assignment process. Shahabsafa et al. [11] further studied the inmate assignment and scheduling problem in the PADoC, and developed the Inmate Assignment Decision Support System (IADSS) to assist the PADoC with the assignment of inmates to CIs. The core of the IADSS is a multi-objective MILO model which makes the simultaneous assignment of the inmates to the CIs and schedules their rehabilitation programs.

3 Problem Description

The Office of Population Management (OPM) is responsible for the transportation of the inmates at the PADoC. There are 25 CIs at the PADoC. On average, 35,000 transportations are scheduled annually, yielding about 650 transportations each week. Conventionally, a staff member of OPM with his experience and judgment manually makes the decisions about the transportation of inmates. The decisions are made in two main steps. First, the routes are specified for the vehicles, and then inmates are assigned to the vehicles based on their origin and destination CIs. One of the critical restrictions of the manual assignment is that there is a small set of predefined routes, and the trips are currently scheduled based only on those predefined routes. The limited number of predefined routes in the current policy significantly limits the flexibility of the transportation decisions. This manual way of planning for the transportation is clearly not efficient.

Next, we define the ITP. Given a time horizon, the set of inmates who need to be transported are identified. For each inmate the origin and the destination is predefined. In other words, the decision about the assignment of an inmate to a CI is made prior to deciding on his/her transportation. In the ITP, we decide on the vehicles used at each transportation day, their routes, and the number of inmates that are going to be assigned to the vehicles at each day.

Vehicles visit a sequence of CIs, and need to return to their starting CI, because the vehicles are maintained by the respective CIs, and the drivers need to return home at the end of the day. Trips should be scheduled in the time window [7 a.m., 7 p.m.]. This means that every route should start and finish at the same CI, and transport inmates within the given 12 h time window. Considering the travel time limit, there are a few pairs of CIs which can not be visited in a single trip. In order to be able to transport inmates between any two arbitrary CIs, the PADoC has one transfer hub, which is located at the central region of the state. Additionally, the hub helps to significantly reduce transportation costs.

The time horizon adds another level of complexity to the problem. Right now the time horizon considered for the trips is a week. The actual time horizon depends on the frequency of transportation days and the number of inmates which need to be transported. The MILO model allows to consider longer time horizon.

4 Model Development

In this section, we introduce the MILO mathematical model. Specifically, the model constructs the optimal routes for a fleet of vehicles and minimizes the total number of the allocated seats, while ensuring that the maximum number of inmates are assigned to routes in the given week. Here we define the terms and assumptions we have used to develop the MILO model.

Definition 1 A **route** is a sequence of CIs which starts and ends at the same CI. The starting CI of a route is the **origin** of the route, and two consecutive CIs of the route form a **leg**.

Definition 2 A **trip** is specified with a vehicle along with its capacity and location at a given CI, a given transportation day, and a route. The given CI is the **origin** and the final destination of the trip.

Definition 3 A **potential trip** is a trip where the vehicle with its capacity, the origin CI, and the transportation day is specified, but the route is not specified.

In ITP, we define the set of all potential trips. One of the main decisions to be made is to assign a route—if any—to potential trips and use those trips for inmate transportation.

Due to various policy restrictions and business practices we limit the set of possible routes. We use Google Maps API to calculate the pessimistic travel time between the facilities and create the distance matrix, which is then further used to create routes. In order to comply with the business practices as mentioned in Sect. 3, we make the following assumptions in generating the set of possible routes:

- We allocate a predefined time duration for getting on and off the vehicle at each CI, except for the route origin.
- The hub may only be visited at most once in a route.
- No consecutive pairs of CIs should be visited more than once.
- Only the legs that are currently used by PADoC are considered in generating the set of the routes. In this case the vehicles will travel only on the paths that are approved by the PADoC.

We do not consider special cases of inmate transportation, such as medical transports, since such requests form a small percentage of the total transportation requests, and are handled by special vehicles. We also do not consider over-night stay for an inmate during the transportation, i.e., all the inmates assigned to a trip will reach their destination at the same day.

One hub is currently used for inmate transportation in PA. The hub is necessary, because considering all the route assumptions there are no acceptable routes between some CI pairs. Furthermore, using the hub helps to reduce the cost of transportation.

We have two main objectives. We aim to minimize the number of the allocated seats and minimize the number of inmates not assigned to a trip.

4.1 Mathematical Model

In this section, we present the multi-objective MILO model for the ITP. In Table 1 the sets, the decision variables, and the parameters of the model are presented.

Table 1 The sets, decision variables, and parameters of the MILO model

Sets	
$\mathcal{C}$	Set of all CIs
$\mathcal{R}$	Set of all possible routes
$\mathcal{T}$	Set of days of the transportation
$\mathcal{P}_t$	Set of the potential trips on day t
$\mathcal{P}$	Set of the all the potential trips ($\mathcal{P} = \bigcup\limits_{t \in \mathcal{T}} \mathcal{P}_t$)
$\mathcal{K}_{ri}$	Set of the stops corresponding to CI i on route r
Variables	
x_{pr}	1, if route r is assigned to potential trip p; 0, otherwise
y_{ijp}	Number of inmates moving directly (without going to hub) from CI i to CI j on trip p
$u_{prn_1n_2}$	Number of inmates directly going from the n_1-th CI to the n_2-th CI of route r on trip p
$\overline{v}_{prn_1j}$	Number of inmates on trip p going from the n_1-th CI of route r to the hub with final destination j
$\overline{\overline{v}}_{prn_2i}$	Number of inmates on trip p going from the hub to the n_2-th CI of route r with origin i
g_{prn}	Number of inmates on the vehicle at the n-th CI of route r on trip p
$\overline{N}_{ij}$	Number of inmates that need to move from CI i to CI j, but not assigned to any trip
Parameters	
N_{ij}	Number of inmates that need to move from CI i to CI j
S_p	Number of seats of the vehicle of trip p
$S^{\max}$	Maximum number of available seats among all the vehicles
η_r	Number of stops (CIs) on route r
η_r^h	Stop number of the hub on route r if the route visits the hub; ∞, otherwise
ω_{ijr}	1, if CI i is before CI j on route r; 0, otherwise

As mentioned earlier, we have three main decisions to make. We need to allocate the trips for transportation, assign routes to the allocated trips, and specify the number of inmates that are going to be transported on each trip. The MILO model is as follows:

$$
\begin{aligned}
&\min \alpha \sum_{p\in\mathcal{P}} \sum_{r\in\mathcal{R}} S_p x_{pr} + \sum_{i,j\in\mathcal{C}|i\neq j} \overline{N}_{ij} \\
&\text{subject to} \\
&\sum_{r\in\mathcal{R}} x_{pr} \leq 1 && \forall p \in \mathcal{P}, \\
&y_{ijp} = \sum_{r\in\mathcal{R}} \sum_{n_1\in\mathcal{K}_{ri}} \sum_{n_2\in\mathcal{K}_{rj}} u_{prn_1n_2} && \forall i,j \in \mathcal{C}, p \in \mathcal{P}, i \neq j, \\
&g_{pr0} = \sum_{n=1}^{\eta_r} u_{pr0n} + \sum_{i\in\mathcal{C}} \overline{v}_{pr0i} && \forall p \in \mathcal{P}, r \in \mathcal{R}, \\
&g_{prn} = g_{pr,n-1} + \sum_{n_2>n} u_{prnn_2} - \sum_{n_1<n} u_{prn_1n} + \sum_{i\in\mathcal{C}} \overline{v}_{prni} && \forall p \in \mathcal{P}, r \in \mathcal{R}, n < \eta_r^h, \\
&g_{prn} = g_{pr,n-1} + \sum_{n_2>n} u_{prnn_2} - \sum_{n_1<n} u_{prn_1n} - \sum_{i\in\mathcal{C}} \sum_{n_1<\eta_r^h} \overline{v}_{prn_1i} + \sum_{i\in\mathcal{C}} \sum_{n_2>\eta_r^h} \overline{\overline{v}}_{prn_2i} && \forall p \in \mathcal{P}, r \in \mathcal{R}, n = \eta_r^h, \\
&g_{prn} = g_{pr,n-1} + \sum_{n_2>n} u_{prnn_2} - \sum_{n_1<n} u_{prn_1n} - \sum_{i\in\mathcal{C}} \overline{\overline{v}}_{prni} && \forall p \in \mathcal{P}, r \in \mathcal{R}, n > \eta_r^h, \\
&g_{prn} \leq S_p x_{pr} && \forall p \in \mathcal{P}, r \in \mathcal{R}, n \leq \eta_r, \\
&u_{prn_1n_2} \leq S_p x_{pr} && \forall p \in \mathcal{P}, r \in \mathcal{R}, 1 \leq n_1 < n_2 \leq \eta_r, \\
&\sum_{p\in\mathcal{P}_t} \sum_{r\in\mathcal{R}} \sum_{n_1\in\mathcal{K}_{ri}} \overline{v}_{prn_1j} = \sum_{p\in\mathcal{P}_t} \sum_{r\in\mathcal{R}} \sum_{n_2=\mathcal{K}_{rj}} \overline{\overline{v}}_{prn_2i} && \forall i,j \in \mathcal{C}, t \in \mathcal{T}, i \neq j, \\
&\overline{v}_{prn_1i} \leq S_p x_{pr} && \forall p \in \mathcal{P}, r \in \mathcal{R}, 1 \leq n_1 \leq \eta_r^h, i \in \mathcal{C}, \\
&\overline{\overline{v}}_{prn_2i} \leq S_p x_{pr} && \forall p \in \mathcal{P}, r \in \mathcal{R}, \eta_r^h \leq n_2 \leq \eta_r, i \in \mathcal{C}, \\
&N_{ij} = \sum_{p\in\mathcal{P}} y_{ijp} + \sum_{p\in\mathcal{P}} \sum_{r\in\mathcal{R}} \sum_{n_1\in\mathcal{K}_{ri}} \overline{v}_{prn_1j} + \overline{N}_{ij} && \forall i,j \in \mathcal{C}, i \neq j, \\
&y_{ijp} \leq S^{\max} \sum_{r\in\mathcal{R}} \omega_{ijr} x_{pr} && \forall i,j \in \mathcal{C}, p \in \mathcal{P}, i \neq j, \\
&x_{pr} = \{0,1\} && \forall p \in \mathcal{P}, r \in \mathcal{R}, \\
&z_p = \{0,1\} && \forall p \in \mathcal{P}, \\
&y_{ijp}, g_{prn}, \overline{v}_{prnj}, \overline{\overline{v}}_{prni}, u_{prn_1n_2}, \overline{N}_{ij} \in \mathbb{Z} && \forall i,j \in \mathcal{C}, p \in \mathcal{P}, 1 \leq n_1 < n_2 \leq \eta_r, 1 \leq n \leq \eta_r.
\end{aligned}
$$

The ITP is a multi-objective optimization problem. We had to specify and fine-tune the weights of the objectives and ensure robustness of the model in assigning inmates to trips for various datasets. In the MILO model, the parameter α is the weight of the total number of seats used for the transportation.

5 Computational Results

In this section, we discuss the output of the MILO model and compare the computational results of the model with that of the manual transportation process. For testing the model we use a dataset of 550 inmates which were transported in the first week of April 2018.

For computational experiments a computer with Dual Intel Xeon® CPU E5-2630 @ 2.20 GHz (20 cores) and 64 GB of RAM is used. Gurobi [8] is used to solve the MILO model with its default parameters and is set to use 10 threads. The solution time limit of Gurobi is set to either 1800 or 43,200 s.

There are two vehicle types, buses and vans, available at the CIs. The capacities of these buses and vans are different depending on their make and model. The capacities of buses are generally more than those of the vans. Since, we minimize the total number of seats allocated for transportation, the model tends to minimize the number of allocated trips with buses as opposed to vans.

Table 2 Output received from PADoC database

Trips	Seats used	Buses	Vans	Inmates moved	% moved (with hub)	Utilization ratio	
						Without hub	With hub
42	948	21	21	550	58	0.58	0.93

The results of the manual allocation of the trips and the assignment of the inmates to the trips in the first week of April 2018 is presented in Table 2. In that week, 550 inmates were transported in 42 trips, out of these 21 trips were made by buses. The total seats of the allocated vehicles for transportation was 948 and out of the total inmates transported 58% went through the hub.

Seat utilization ratio is the ratio of the total inmates moved to the total number of seats used in trips for the transportation. The seat utilization ratio can be greater than one, since multiple inmates can occupy the same seat in a trip, as they get on and get off at different stops. We consider two types of seat utilization ratio: "without hub" represents the utilization ratio when we consider the inmates moving through hub as taking one seat; "with hub" represents the ratio when the inmate who is going through the hub is considered to take two seats instead of a single seat. Seat utilization with hub and without hub was equal to 0.58 and 0.93, respectively in the manual transportation during the first week of April, 2018.

In Table 3, the results of the MILO model with 1800 s time limit is presented. The parameter α is the coefficient used in the objective function to penalize the allocation of the vehicles for the transportation. As α increases, the penalty associated with allocating a vehicle for transportation increases. Thus, the number of the allocated trips and more importantly the number of allocated buses for transportation decreases as α increases. There is a trade-off between the two objectives of the model: minimize the number of the inmates not transported and minimize the number of the allocated seats. The relative penalty of not assigning inmates to trips decreases as α increases. Thus, the number of inmates that are not assigned to a trip increases as α increases. Additionally, the number of inmates assigned to a trip increases, thus the utilization ratio increases.

In Table 4, the results of the MILO model with 43,200 s (12 h) time limit is presented. As we can see in Tables 3 and 4, none of the instances are solved to global optimality. The gap has decreased for all the instances with different values of α when the solution time limit increases from 1800 to 43,200 s. However, the improvements differ from an instance to another. When $\alpha = 0.1$, we have the biggest improvements and $\alpha = 1$ has the smallest improvement. As the decisions about the inmate transportation is currently made once a week we can let the solver run longer (e.g., 12 h) to obtain a better solution. Considering the optimality gap at 12 h, little improvement is expected if we run the model for longer.

One important decision to make is to specify the value of α. We reviewed the results of the MILO model with the PADoC, and we evaluated the trade-off between

Table 3 Output when Gurobi time-limit is set to 1800 s

α	Trips	Seats used	Buses	Vans	Inmates not moved	Inmates moved	% moved with hub	Seat utilization ratio		Opt. gap %
								Without hub	With hub	
0.10	27	557	12	15	2	548	43	0.98	1.41	33.40
0.30	27	529	12	15	7	543	35	1.03	1.38	30.10
0.50	24	437	9	15	24	526	39	1.20	1.68	20.80
0.75	22	404	8	14	30	520	36	1.29	1.75	13.20
1.00	18	225	3	15	179	371	16	1.65	1.91	7.72

Table 4 Output when Gurobi time-limit is set to 43,200 s (12 h)

α	Trips	Seats used	Buses	Vans	Inmates not moved	Inmates moved	% moved with hub	Seat utilization ratio		Opt. gap %
								Without hub	With hub	
0.10	25	444	9	16	1	549	44	1.24	1.78	13.60
0.30	23	430	9	14	1	549	40	1.28	1.79	9.54
0.50	23	430	9	14	1	549	41	1.28	1.80	9.11
0.75	22	404	8	14	14	536	39	1.33	1.85	7.49
1.00	19	265	4	15	129	421	19	1.59	1.89	3.92

the two objectives for different values of α. The most appropriate value of α was determined to be equal to 0.5, since only nine buses are used for the transportation of the inmates in that week, and only one inmate is not transported. This inmate can be transported in the following week.

6 Benefits and Impact

In this section we quantify the expected savings of using the MILO model for the inmate transportation process. We have identified two main saving areas that can be achieved by optimizing the process. In order to compute the savings, we compare the results of the manual transportation, presented in Table 2, with that of the MILO model for $\alpha = 0.5$, presented in Table 4.

Gas and Maintenance: Using the MILO model, the number of the buses decreased from 21 to 9. It was reported by the PADoC in 2013 that the total gas and maintenance cost for 21 buses was \$500,000. The model reduces the number of buses used by 12. Thus, the savings from gas and maintenance is projected to be \$285,700 annually.

Table 5 The projected quantified savings of optimizing the inmate transportation process

Savings	One year ($)	Five years ($)
Gas and maintenance	285,700	1,428,500
Salary	1,350,000	6,750,000
Sum	1,635,700	8,178,500

Salary: There is a reduction of 12 bus-trips and 7 van-trips. Each bus and van, used for the transportation of the inmates, need three and two correctional officers, respectively. This would result in a saving of 50 man-day which can then translate to 10 full-time correctional officer positions. The average salary and benefits of a correctional officer is $135,000. Thus, the saving from the salary would be $1,350,000 annually.

The projected quantified savings in one year and over five years are summarized in Table 5.

7 Summary

In this paper, we studied the inmate transportation process as a proof of concept at the PADoC as it is done manually, and suggest an alternative to optimize the process system-wide. We developed a multi-objective MILO model to optimize the ITP. Numerical results demonstrate that significant savings can be achieved by using the model for the ITP. Our MILO model can be advanced further to incorporate other business rules and constraints of the inmate transportation process, and can be adapted to other jurisdictions.

References

1. Applegate DL, Bixby RE, Chvátal V, Cook WJ. The traveling salesman problem: a computational study. In: Princeton series in applied mathematics; 2007.
2. Cook WJ. In pursuit of the traveling salesman: mathematics at the limits of computation. Princeton: Princeton University Press; 2014.
3. Crainic TG, Laporte G, editors. Fleet management and logistics. Springer; 1998.
4. Dantzig GB, Ramser JH. The truck dispatching problem. Manag Sci. 1959;6(1):80–91.
5. Dantzig GB, Fulkerson R, Johnson S. Solution of a large-scale traveling-salesman problem. J Oper Res Soc Am. 1954;2(4):393–410.
6. Fernández E, Laporte G, Rodríguez-Pereira J. A branch-and-cut algorithm for the multidepot rural postman problem. Transp Sci. 2017;52(2):353–69.
7. Flood MM. The traveling-salesman problem. Oper Res. 1956;4(1):61–75.
8. Gurobi Optimization Inc: Gurobi optimizer reference manual. http://www.gurobi.com; 2016.
9. Li D, Plebani LJ, Terlaky T, Wilson GR, Bucklen KB. Inmate classification: decision support tool gives help to Pennsylvania Department of Corrections. Ind Eng. 2014;46(7).

10. Padberg M, Rinaldi G. A branch-and-cut algorithm for the resolution of large-scale symmetric traveling salesman problems. SIAM Rev. 1991;33(1):60–100.
11. Shahabsafa M, Terlaky T, Gudapati C, Sharma A, Plebani LJ, Wilson GR, Bucklen KB. The inmate assignment and scheduling problem and its application in the Pennsylvania Department of Corrections. 2018;48(5):467–483.
12. Walmsley R (2015) World prison population list. http://www.prisonstudies.org/sites/default/files/resources/downloads/world_prison_population_list_11th_edition_0.pdf. Accessed 27 June 2018.

Robust Modality Selection in Radiotherapy

Sevnaz Nourollahi, Archis Ghate and Minsun Kim

Abstract External beam radiotherapy attempts to maximize tumor-damage while limiting toxicity on healthy tissue. Although several modalities with distinctive biological and physical properties are available, none is dominant. A mathematical formulation for optimal modality selection with uncertainty in these properties is presented. Uncertainty is incorporated via a robust approach. The problem decomposes into finitely many subproblems with analytically solvable Karush-Kuhn-Tucker conditions. Numerical experiments demonstrate how uncertainty affects optimal solutions even when clinical intuition is not readily available.

1 Introduction

In external beam radiotherapy (EBRT), high-energy radiation is passed through the patient's body to kill tumors. This also damages nearby healthy anatomies and organs-at-risk, collectively termed normal tissue. The objective is to maximize tumor-damage and limit toxic effects on normal tissue.

Several modalities such as photon beam x-rays, protons, and neutrons, with distinctive biological and physical properties, are available [1]. Neutrons, for example, have a higher biological cell-kill power than photons. They are thus more toxic to both the tumor and the normal tissue. Depth dose deposition profile is a crucial physical property. Photons deposit a high radiation dose near the entry point, which then decreases with distance. Protons deposit a dose that increases for some distance. It then abruptly rises to a so-called Bragg peak, and falls sharply after. A large dose differential between the tumor and normal tissue can thus be attained if the Bragg peak is positioned exactly in the tumor. While this is favorable to a patient, uncertainty about the location of the Bragg peak makes protons less desirable. Such trade-offs

S. Nourollahi · A. Ghate
Industrial & Systems Engineering, University of Washington, Seattle, WA 98195, USA

M. Kim (✉)
Radiation Oncology, University of Washington, Seattle, WA 98195, USA
e-mail: mk688@uw.edu

H. Yang and R. Qiu (eds.), *Advances in Service Science*, Springer Proceedings in Business and Economics, https://doi.org/10.1007/978-3-030-04726-9_2

render the choice of an optimal modality difficult. No single universally dominant modality has emerged [3].

The challenge is further compounded because EBRT is delivered in multiple sessions. This is called fractionation. Normal tissue possesses better damage-repair capabilities than tumors. Fractionation thus gives the normal tissue some time to recover between sessions. This hints at the benefits of long courses. However, tumors proliferate, and hence a shorter course may be desirable. Mathematical methods for optimizing the number of sessions with a single modality based on the linear-quadratic (LQ) model of dose-response are reviewed in [5]. The question of how to split the sessions into different modalities further confounds trade-offs.

Consider the following formulation of the problem with two modalities, based on the LQ model:

$$(P0) \quad \min_{N_1,d_1,N_2,d_2} \quad e^{-N_1\alpha_1^\tau d_1 - N_1\beta_1^\tau (d_1)^2 - N_2\alpha_2^\tau d_2 - N_2\beta_2^\tau (d_2)^2 + \gamma(N_1+N_2)} \tag{1}$$

$$N_1 s_{1k}\alpha_{1k}^\phi d_1 + N_1\beta_{1k}^\phi (s_{1k}d_1)^2 + N_2\alpha_{2k}^\phi d_2 + N_2\beta_{2k}^\phi (s_{2k}d_2)^2 \le B_k,\ k \in \mathcal{K}, \tag{2}$$

$$d_1, d_2 \ge 0,\ N_1 + N_2 \le N_{\max},\ N_1, N_2 \ge 0, \text{ integers.} \tag{3}$$

We solved a special case of this with a single normal tissue via Karush-Kuhn-Tucker (KKT) conditions in [4]. In (P0), modalities are indexed by subscripts $i = 1, 2$. Thus, N_i is the number of sessions with modality i, and d_i is the dose in each of these sessions. Modality 1 is assumed to be the conventional modality (photons). The objective function minimizes the fraction of surviving cells per the LQ model. Here, α_i^τ and β_i^τ are the modality-specific parameters of the linear and quadratic components of the tumor's response, respectively; $\gamma(N_1 + N_2)$ is the tumor proliferation term, which depends on the total number of treatment sessions [2]. The set of normal tissue is $\mathcal{K} = \{1, 2, \ldots, K\}$, and is indexed by k. For normal tissue k, α_{ik}^ϕ and β_{ik}^ϕ are the linear and quadratic parameters of dose-response for modality i. Also, s_{ik} is the sparing factor for normal tissue k. That is, if a dose of d_i is delivered to the tumor by modality i, then a dose $s_{ik}d_i$ is delivered to normal tissue k [[5] describes how sparing factors are obtained]. Thus, the left hand side (LHS) of constraints (2) equals the total biological effect (BE) [2] on normal tissue k. The right hand side (RHS) of constraints (2) is defined as $B_k = N_{\text{conv}}\alpha_{1k}^\phi d_{\text{conv},k} + N_{\text{conv}}\beta_{1k}^\phi (d_{\text{conv},k})^2$. Here, $d_{\text{conv},k}$ is the dose in each session that normal tissue k is known to tolerate if administered in N_{conv} sessions. This, in turn, can be rewritten as $B_k = \alpha_{1k}D_k + \beta_{1k}D_k^2/N_{\text{conv}}$ by letting $D_k = N_{\text{conv}}d_{\text{conv},k}$ for brevity. The RHS equals the BE of the conventional treatment schedule that normal tissue k is known to tolerate. Constraints (2) ensure that the BE of the selected treatment schedules is no more than that of a conventional one.

(P0) ignores uncertainty in biological and physical properties that further complicates modality selection. For protons for instance, the exact location of Bragg peak is unknown [1]. This can lead to the undesirable consequence that a large dose is delivered to the normal tissue. This can be modeled as the sparing factors s being uncertain. Similarly, the exact biological powers of different modalities are unknown, and this can be modeled as the parameters α, β being uncertain. Our objective is to propose a robust approach and solution method to address this, and to derive clinical insights.

2 Problem Formulation and Exact Solution Method

2.1 Robust Formulation Under Interval Uncertainty

First, (P0) is converted into an equivalent maximization problem after taking the natural logarithm of the objective. It suffices to solve the resulting problem with (N_1, N_2) fixed. Further, the proliferation term $\gamma(N_1 + N_2)$ becomes a constant and hence is dropped. We thus refer to the following as the nominal problem:

$$\text{(P)} \max_{d_1,d_2} \; N_1\alpha_1^\tau d_1 + N_1\beta_1^\tau (d_1)^2 + N_2\alpha_2^\tau d_2 + N_2\beta_2^\tau (d_2)^2$$

$$\text{subject to } N_1 s_{1k}\alpha_{1k}^\phi d_1 + N_1\beta_{1k}^\phi (s_{1k}d_1)^2 + N_2\alpha_{2k}^\phi d_2 + N_2\beta_{2k}^\phi (s_{2k}d_2)^2 \le B_k,\; k \in \mathcal{K},$$

$$d_1, d_2 \ge 0.$$

This section presents its robust counterpart using an interval model of uncertainty.

The robust counterpart tackles uncertainty about sparing factors s_{1k}, s_{2k}, and uncertainty about the dose-response parameters $\alpha_{1k}^\phi, \beta_{1k}^\phi, \alpha_{2k}^\phi, \beta_{2k}^\phi$. The planner assumes that these parameters belong to the intervals $R_{s_{1k}} = [s_{1k}^{\min}, s_{1k}^{\max}]$, $R_{s_{2k}} = [s_{2k}^{\min}, s_{2k}^{\max}]$ $R_{\alpha_{1k}} = [\alpha_{1k}^{\min}, \alpha_{1k}^{\max}]$, $R_{\beta_{1k}} = [\beta_{1k}^{\min}, \beta_{1k}^{\max}]$, $R_{\alpha_{2k}} = [\alpha_{2k}^{\min}, \alpha_{2k}^{\max}]$, and $R_{\beta_{2k}} = [\beta_{2k}^{\min}, \beta_{2k}^{\max}]$. The planner computes the best solution that will remain feasible irrespective of the values of these parameters from these intervals.

This yields the following robust counterpart of the nominal problem (P):

$$\max_{d_1,d_2} \; N_1\alpha_1^\tau d_1 + N_1\beta_1^\tau (d_1)^2 + N_2\alpha_2^\tau d_2 + N_2\beta_2^\tau (d_2)^2$$

$$\text{subject to } N_1\alpha_{1k}^\phi s_{1k}d_1 + N_1\beta_{1k}^\phi (s_{1k}d_1)^2 + N_2\alpha_{2k}^\phi s_{2k}d_2 + N_2\beta_{2k}^\phi (s_{2k}d_2)^2 \le \alpha_{1k}^\phi D_k + \beta_{1k}^\phi D_k^2/N_{\text{conv}}$$

$$\text{for } s_{1k} \in R_{s_{1k}},\; s_{2k} \in R_{s_{2k}},\; \alpha_{1k} \in R_{\alpha_{1k}},\; \beta_{1k} \in R_{\beta_{1k}},\; \alpha_{2k} \in R_{\alpha_{2k}},\; \beta_{2k} \in R_{\beta_{2k}},\; k \in \mathcal{K},$$

$$d_1, d_2 \ge 0.$$

After algebraic simplification, the main functional constraint above can be rewritten as

$$\alpha_{1k}^\phi (N_1 s_{1k} d_1 - D_k) + \beta_{1k}^\phi (N_1 s_{1k}^2 d_1^2 - D_k^2/N_{\text{conv}}) + N_2\alpha_{2k}^\phi s_{2k} d_2 + N_2\beta_{2k}^\phi s_{2k}^2 d_2^2 \le 0,$$

for $s_{1k} \in R_{s_{1k}}$, $s_{2k} \in R_{s_{2k}}$, $\alpha_{1k} \in R_{\alpha_{1k}}$, $\beta_{1k} \in R_{\beta_{1k}}$, $\alpha_{2k} \in R_{\alpha_{2k}}$, $\beta_{2k} \in R_{\beta_{2k}}$, $k \in \mathcal{K}$. This includes an uncountably infinite number of constraints. We next show how to decompose this into a finite set of subproblems, each with a finite number of constraints.

2.2 Decomposition into Subproblems Solved via KKT Conditions

The initial step is to identify bottleneck parameter values on the LHS of the constraints. This is easy to do for the third and the fourth terms, $N_2\alpha_{2k}s_{2k}d_2$ and $N_2\beta_{2k}s_{2k}^2d_2^2$, because the bottleneck parameter values equal their largest values. That is, $\alpha_{2k}^{\max}$, $\beta_{2k}^{\max}$ and $s_{2k}^{\max}$, for α_{2k}^{ϕ}, β_{2k}^{ϕ}, and s_{2k}. Similarly, the bottleneck value for s_{1k} is $s_{1k}^{\max}$. This does not work for the first and the second terms because the multipliers $(N_1s_{1k}^{\max}d_1 - D_k)$ and $(N_1(s_{1k}^{\max})^2d_1^2 - D_k^2/N_{\text{conv}})$ may be positive or negative. We develop an alternative approach to handle these terms.

Note that the sign of the multiplier $(N_1s_{1k}^{\max}d_1 - D_k)$ is determined by whether or not $d_1 \geq (D_k/N_1s_{1k}^{\max})$. If the sign is negative, then the largest value of $\alpha_{1k}^{\phi}(N_1s_{1k}^{\max}d_1 - D_k)$ on the LHS of the constraint is attained when $\alpha_{1k}^{\phi} = \alpha_{1k}^{\min}$. On the other hand, if the sign is positive, then the largest value of this term is attained when $\alpha_{1k}^{\phi} = \alpha_{1k}^{\max}$. Similarly, the sign of the multiplier $(N_1(s_{1k}^{\max})^2d_1^2 - D_k^2/N_{\text{conv}})$ is determined by whether or not $d_1^2 \geq (D_k^2/N_{\text{conv}}N_1(s_{1k}^{\max})^2)$. If the sign is negative, then the largest value of the second term $\beta_{1k}^{\phi}(N_1(s_{1k}^{\max})^2d_1^2 - D_k^2/N_{\text{conv}})$ is attained when $\beta_{1k}^{\phi} = \beta_{1k}^{\min}$. On the other hand, if the sign is positive, then the largest value of the second term is attained when $\beta_{1k}^{\phi} = \beta_{1k}^{\max}$. We sort $\mu_k = \frac{D_k}{N_1s_{1k}^{\max}}$ and $\nu_k = \frac{D_k^2}{N_{\text{conv}}N_1(s_{1k}^{\max})^2}$ in increasing order. The sorted indices k are stored in sequences L and Q. Suppose that, for any $i = 1, 2, \ldots, K$, L_i denotes the ith normal tissue index in the sorted sequence L. Then $\mu_{L_i} \leq \mu_{L_{i+1}}$, for $i = 1, 2, \ldots, K-1$. Similarly, suppose that, for any $j = 1, 2, \ldots, K$, Q_j denotes the jth normal tissue index in the sorted sequence Q. Then $\nu_{Q_j} \leq \nu_{Q_{j+1}}$, for $j = 1, 2, \ldots, K-1$. We use this notation to partition feasible dose values $d_1 \geq 0$ into different subsets. These subsets are indexed by pairs (ℓ, q), for $\ell \in \{0, 1, \ldots, K\}$ and $q \in \{0, 1, \ldots, K\}$. The (ℓ, q)th subset is characterized by

$$d_1 \geq \mu_k \text{ for } k \in \{L_1, L_2, \ldots, L_\ell\} \text{ and } d_1 \leq \mu_k \text{ for } k \in \{L_{\ell+1}, L_{\ell+2}, \ldots, L_K\}; \text{ and}$$
$$d_1^2 \geq \nu_k \text{ for } k \in \{Q_1, Q_2, \ldots, Q_q\} \text{ and } d_1^2 \leq \nu_k \text{ for } k \in \{Q_{q+1}, Q_{q+2}, \ldots, Q_K\}.$$

We tackle the robust problem by solving subproblems indexed by (ℓ, q), for $\ell \in \{0, 1, \ldots, K\}$ and $q \in \{0, 1, \ldots, K\}$, and then by identifying the (ℓ, q) pair and the corresponding (d_1, d_2) doses that yield the best objective. The (ℓ, q)th subproblem from this group is given by

$$\max_{d_1,d_2} \; N_1\alpha_1^\intercal d_1 + N_1\beta_1^\intercal (d_1)^2 + N_2\alpha_2^\intercal d_2 + N_2\beta_2^\intercal (d_2)^2 \tag{4}$$

$$\alpha_{1k}^{\max}(N_1 s_{1k}^{\max} d_1 - D_k) + \beta_{1k}^{\max}(N_1 (s_{1k}^{\max})^2 d_1^2 - D_k^2/N_{\text{conv}}) + N_2\alpha_{2k}^{\max} s_{2k}^{\max} d_2 + N_2\beta_{2k}^{\max}(s_{2k}^{\max})^2 d_2^2 \le 0,$$

$$k \in \{\{L_1, L_2, \ldots, L_\ell\} \cap \{Q_1, Q_2, \ldots, Q_q\}\}, \tag{5}$$

$$\alpha_{1k}^{\max}(N_1 s_{1k}^{\max} d_1 - D_k) + \beta_{1k}^{min}(N_1 (s_{2k}^{\max})^2 d_1^2 - D_k^2/N_{\text{conv}}) + N_2\alpha_{2k}^{\max} s_{2k}^{\max} d_2 + N_2\beta_{2k}^{\max}(s_{2k}^{\max})^2 d_2^2 \le 0,$$

$$k \in \{\{L_1, L_2, \ldots, L_\ell\} \cap \{Q_{q+1}, Q_{q+2}, \ldots, Q_K\}\}, \tag{6}$$

$$\alpha_{1k}^{min}(N_1 s_{1k}^{\max} d_1 - D_k) + \beta_{1k}^{min}(N_1 (s_{1k}^{\max})^2 d_1^2 - D_k^2/N_{\text{conv}}) + N_2\alpha_{2k}^{\max} s_{2k}^{\max} d_2 + N_2\beta_{2k}^{\max}(s_{2k}^{\max})^2 d_2^2 \le 0,$$

$$k \in \{\{L_{\ell+1}, L_{\ell+2}, \ldots, L_K\} \cap \{Q_{q+1}, Q_{q+2}, \ldots, Q_K\}\}, \tag{7}$$

$$\alpha_{1k}^{\min}(N_1 s_{1k}^{\max} d_1 - D_k) + \beta_{1k}^{\max}(N_1 (s_{1k}^{\max})^2 d_1^2 - D_k^2/N_{\text{conv}}) + N_2\alpha_{2k} s_{2k}^{\max} d_2 + N_2\beta_{2k}(s_{2k}^{\max})^2 d_2^2 \le 0,$$

$$k \in \{\{L_{\ell+1}, L_{\ell+2}, \ldots, L_K\} \cap \{Q_1, Q_2, \ldots, Q_q\}\}, \tag{8}$$

$$d_1 \ge \mu_k,\; k \in \{L_1, L_2, \ldots, L_\ell\},\; d_1 \le \mu_k,\; k \in \{L_{\ell+1}, L_{\ell+2}, \ldots, L_K\}, \tag{9}$$

$$d_1^2 \ge \nu_k,\; k \in \{Q_1, Q_2, \ldots, Q_q\},\; d_1^2 \le \nu_k,\; k \in \{Q_{q+1}, Q_{q+2}, \ldots, Q_K\}, \tag{10}$$

$$d_1, d_2 \ge 0. \tag{11}$$

Since the subproblem includes two variables, we investigate two cases: whether only one or at least two constraints are active at an optimality. We categorize the constraints into three groups: (5–8); (9–10); and (11), and investigate the two cases for each.

1. Only one of the constraints (5–8) is active at optimal solution. There are two subcases:

   (a) Only one of d_1 and d_2 is positive. Suppose $d_1 > 0$ and $d_2 = 0$. The (l, q) subproblem becomes a single modality problem with d_1 as the only variable. Then, by making constraints (5–8) active one-by-one, we are able to obtain a positive value of d_1 by solving a quadratic equation. Among all such candidate values of d_1, we only keep those that are feasible to the rest of the constraints in (5–8), as well as in (9–10). The same approach is repeated when $d_1 = 0$ and $d_2 > 0$.

   (b) Both $d_1 > 0$ and $d_2 > 0$. We assume each one of the constraints (5–8) to be active one-by-one and the rest of them to be strict inequalities. There are K subcases to consider. In each of the subcases, the Lagrange multipliers for the $K - 1$ inactive constraints among (5–8), for the $2K$ inactive constraints (9–10), and also for the two non-negativity constraints (11) become zero owing to complementary slackness. Thus, the KKT conditions reduce to those identical to the two-modality and single constraint problem with $d_1 > 0$ and $d_2 > 0$. These can be written as a quartic equation that can be solved analytically. After doing this for each of the subcases, we only keep solutions that are feasible to the original subproblem.

2. At least two of the constraints (5–8) are active at optimal solution. As the problem includes two variables, the active constraints will provide a system of two quadratic equations that can be solved for d_1 and d_2. In particular, the intersection of the two active constraints creates a quartic equation in terms of either d_1 or d_2,

which can be solved in closed form. Then by substituting the obtained d_1 or d_2 in either of the constraints, we get a quadratic equation with one unknown variable, which also can be easily solved in closed from. There will be $\binom{M}{2}$ such systems of two quadratic equations. We keep candidate solutions that are feasible for the entire subproblem.

3. At least one of the constraints (9–10) is active. We make each of the constraints (9–10) active one-at-a-time and solve it for d_1. Then we substitute the resulting value of d_1 into constraints (5–8). Note that at least one of the constraints (5–8) must then be active at optimal solution, because the objective function is increasing in d_2. Thus we solve K quadratic equations one-by-one to obtain candidate solutions for d_2. We keep solutions that are feasible for the entire subproblem.

2.3 Experiment Design and Procedure

For brevity, M_1 refers to modality 1 and M_2 is modality 2. For M_1, we used

$$s_1 = 1,$$
$$\alpha_1^\tau/\beta_1^\tau = 10\text{ Gy},\ \alpha_1^\phi/\beta_1^\phi = 2\text{ Gy},$$
$$\alpha_1^\tau = 0.35\text{ Gy}^{-1},\ \beta_1^\tau = 0.035\text{ Gy}^{-2},\ \alpha_1^\phi = 0.35\text{ Gy}^{-1},\ \beta_1^\phi = 0.175\text{ Gy}^{-2}.$$

For M_2, we set

$$\beta_2^\phi = 0.175\text{ Gy}^{-2},\ \beta_2^\tau = 0.035\text{ Gy}^{-2}.$$

N_{conv} was fixed at 25 fractions with $d_{\text{conv}} = 2$ Gy. These values are common in the literature, and yield $B = 35$ as the RHS of constraint (2). We used

$$\gamma(N_1 + N_2) = \frac{[(N_1 + N_2) - 1 - T_{\text{lag}}]^+ \ln 2}{T_d},$$

where T_d and T_{lag} are tumor doubling time and lag time, and $[\cdot]^+ = \max(\cdot, 0)$. This functional form is common, and assumes that repopulation does not start until T_{lag} days after treatment begins [2]. We employed $T_d = 3$ days and $T_{\text{lag}} = 0$ days as representative values since qualitative trends were invariant with these numbers. $N_{\max}$ was fixed at 50 days.

Experiments were conducted for values $\{0.8, 1, 1.2\}$ of a biological parameter $r = \alpha_2^\phi/\alpha_2^\tau$. A biologically superior modality inflicts a higher damage on both the tumor and normal tissue. The ratio r captures the differential in the damage to the two. As r increases, the damage to normal tissue relative to the damage to tumor using M_2 increases and M_2 becomes less desirable.

Section 3.1 studies the effect of uncertainty in s_2 (a physical characteristic of M_2). For example, for protons, this could model the uncertainty in the Bragg peak's location. The uncertainty interval $[s_2^{min}, s_2^{max}]$ is modeled by setting $s_2^{min} = (1-\Delta)s_2$ and $s_2^{max} = (1+\Delta)s_2$, for $\Delta \in \{0, 0.1, 0.2, \ldots, 0.9\}$. Here, s_2 is a nominal value chosen from $\{1, 0.9, \ldots, 0.5\}$, and $\Delta = 0$ corresponds the nominal case.

Section 3.2 studies the effect of uncertainty in α_1^ϕ (a biological characteristic of M_1). For example, for photons, this could model the uncertainty in its biological power. The uncertainty interval $[\alpha_1^{min}, \alpha_1^{max}]$ is modeled with $\alpha_1^{min} = (1-\Delta)\alpha_1^\phi$ and $\alpha_1^{max} = (1+\Delta)\alpha_1^\phi$, for $\Delta \in \{0, 0.1, 0.2, \ldots, 0.9\}$. Here, the nominal value α_1^ϕ is fixed at 0.35 Gy^{-1}. Recall that $\Delta = 0$ corresponds to no uncertainty.

3 Results

3.1 *Uncertainty in Physical Characteristic s_2 of M_2*

Tables 1, 2 and 3 report results for $r = 1, 0.8, 1.2$. Each row reports a different value of s_2. Columns correspond to different uncertainty levels Δ. The tables report the % price of robustness (PR) for each (s_2, Δ) combination. This equals the percentage increase in the optimal number of surviving cells in the robust formulation relative to the nominal formulation. Tables are colored depending on what is optimal: blue if M_1, green if the pair M_1, M_2, and yellow if M_2 is optimal.

Table 1 shows that for each s_2, the PR is nondecreasing as Δ increases. When Δ increases, s_2^{max} increases and M_2 becomes less desirable as it inflicts more damage on the normal tissue. The optimal modality thus switches to M_1 at a sufficiently high value of Δ. After this switch occurs, the optimal solution does not depend on parameters of M_2, and in particular, does not depend on Δ. Thus, the PR is constant for all values of Δ where M_1 is optimal in each row. In the top-left cell where $s_2 = 1$ and $\Delta = 0$, M_1 and M_2 are equivalent. Therefore, there is a tie between M_1 and M_2 in that cell. Since the PR is 0 in that cell, it remains 0 throughout that row of $s_2 = 1$ (note that the PR is 0 as expected when $\Delta = 0$). M_2 becomes more desirable (because it inflicts less damage on the normal tissue) as s_2 decreases. Thus, the switch from M_2 to M_1 in each row occurs at a larger value of Δ (less desirable) as nominal s_2 decreases (more desirable). Similarly, for each value of Δ, M_2 becomes more desirable as s_2 decreases. The optimal modality thus switches from M_1 to M_2. This switch occurs at smaller values of s_2 as Δ increases. For each value of Δ, the PR increases as s_2 decreases, whenever M_1 is optimal. This is because the number of surviving cells with M_1 as the optimal modality is invariant as a function of s_2 in this situation, but the number of surviving cells with M_2 as the optimal modality in the nominal problem ($\Delta = 0$) is decreasing as s_2 decreases.

Qualitative trends in Table 2 are identical to Table 1. However, M_2 is optimal more often in Table 2 as it uses a smaller value of $r = 0.8$ making M_2 more desirable. For this same reason, in each fixed row of Table 2, the switch from M_2 to M_1 occurs at a higher value of Δ. Similarly, for each fixed column of Table 2, the switch from M_1 to M_2 occurs at a higher value of s_2.

Table 1 % Price of Robustness with $r = 1$ for Sect. 3.1

s_2	Uncertainty Δ in s_2									
	0.0	0.1	0.2	0.3	0.4	0.5	0.6	0.7	0.8	0.9
1.0	0.0	0.0	0.0	0.0	0.0	0.0	0.0	0.0	0.0	0.0
0.9	0.0	13.4	14.7	14.7	14.7	14.7	14.7	14.7	14.7	14.7
0.8	0.0	13.3	24.0	28.5	28.5	28.5	28.5	28.5	28.5	28.5
0.7	0.0	13.2	23.8	32.4	39.6	41.4	41.4	41.4	41.4	41.4
0.6	0.0	13.1	23.6	32.2	39.3	45.3	50.3	53.3	53.3	53.3
0.5	0.0	13.0	23.4	31.9	39.0	44.9	50.0	54.3	58.1	61.4

Table 2 % Price of Robustness with $r = 0.8$ for Sect. 3.1

s_2	Uncertainty Δ in s_2									
	0.0	0.1	0.2	0.3	0.4	0.5	0.6	0.7	0.8	0.9
1.0	0.0	13.1	21.5	21.5	21.5	21.5	21.5	21.5	21.5	21.5
0.9	0.0	13.0	23.5	32.1	32.7	32.7	32.7	32.7	32.7	32.7
0.8	0.0	12.9	23.3	31.8	38.9	43.3	43.3	43.3	43.3	43.3
0.7	0.0	12.8	23.1	31.5	38.6	44.5	49.6	53.2	53.2	53.2
0.6	0.0	12.6	22.9	31.2	38.2	44.1	49.2	53.5	57.3	60.6
0.5	0.0	12.5	22.6	30.9	37.8	43.7	48.7	53.0	56.8	60.0

Table 3 % Price of Robustness with $r = 1.2$ for Sect. 3.1

s_2	Uncertainty Δ in s_2									
	0.0	0.1	0.2	0.3	0.4	0.5	0.6	0.7	0.8	0.9
1.0	0.0	0.0	0.0	0.0	0.0	0.0	0.0	0.0	0.0	0.0
0.9	0.0	0.0	0.0	0.0	0.0	0.0	0.0	0.0	0.0	0.0
0.8	0.0	13.7	15.3	15.3	15.3	15.3	15.3	15.3	15.3	15.3
0.7	0.0	13.6	24.4	31.0	31.0	31.0	31.0	31.0	31.0	31.0
0.6	0.0	13.5	24.3	33.0	40.2	45.5	45.5	45.5	45.5	45.5
0.5	0.0	13.5	24.2	32.9	40.0	46.1	51.2	55.5	58.6	58.6

Qualitative trends in Table 3 are also identical to Table 1. However, M_2 is optimal less often in Table 3 because it uses the higher value of $r = 1.2$ making M_2 less desirable. For this same reason, in each row of Table 3, the switch from M_2 to M_1 occurs at a lower value of Δ. Similarly, for each column of Table 3, the switch from M_1 to M_2 occurs at a lower value of s_2.

3.2 Uncertainty in Biological Characteristic α_1^{ϕ} of M_1

Tables 4, 5 and 6 report results for $r = 1, 0.8, 1.2$. Rows report different values of α_2^{τ}, which model the biological power of M_2. Columns correspond to different levels of uncertainty Δ. The tables report the PR for each $(\alpha_2^{\tau}, \Delta)$. This equals the percentage

Table 4 % Price of Robustness with $r = 1$ for Sect. 3.2

α_2^τ	Uncertainty Δ in α_1^ϕ									
	0.0	0.1	0.2	0.3	0.4	0.5	0.6	0.7	0.8	0.9
0.35	0.0	0.4	0.6	0.8	0.8	0.8	0.8	0.8	0.8	0.8
0.40	0.0	4.9	7.9	8.1	8.1	8.1	8.1	8.1	8.1	8.1
0.45	0.0	4.9	9.9	13.8	13.8	13.8	13.8	13.8	13.8	13.8
0.50	0.0	4.9	9.9	14.8	18.4	18.4	18.4	18.4	18.4	18.4
0.55	0.0	4.9	9.9	14.8	19.8	22.2	22.2	22.2	22.2	22.2
0.60	0.0	4.9	9.9	14.8	19.8	24.7	25.1	25.1	25.1	25.1
0.65	0.0	4.9	9.9	14.8	19.8	24.7	27.6	27.6	27.6	27.6
0.70	0.0	4.9	9.9	14.8	19.8	24.7	29.6	29.7	29.7	29.7

Table 5 % Price of Robustness with $r = 0.8$ for Sect. 3.2

α_2^τ	Uncertainty Δ in α_1^ϕ									
	0.0	0.1	0.2	0.3	0.4	0.5	0.6	0.7	0.8	0.9
0.35	0.0	4.9	9.9	10.0	10.0	10.0	10.0	10.0	10.0	10.0
0.40	0.0	4.9	9.9	14.8	17.6	17.6	17.6	17.6	17.6	17.6
0.45	0.0	4.9	9.9	14.8	19.8	23.4	23.4	23.4	23.4	23.4
0.50	0.0	4.9	9.9	14.8	19.8	24.7	27.8	27.8	27.8	27.8
0.55	0.0	4.9	9.9	14.8	19.8	24.8	29.7	31.3	31.3	31.3
0.60	0.0	5.0	9.9	14.9	19.8	24.8	29.7	34.1	34.1	34.1
0.65	0.0	5.0	9.9	14.9	19.8	24.8	29.7	34.7	36.4	36.4
0.70	0.0	5.0	9.9	14.9	19.8	24.8	29.7	34.7	38.3	38.3

Table 6 % Price of Robustness with $r = 1.2$ for Sect. 3.2

α_2^τ	Uncertainty Δ in α_1^ϕ									
	0.0	0.1	0.2	0.3	0.4	0.5	0.6	0.7	0.8	0.9
0.35	0.0	0.4	0.6	0.8	0.8	0.8	0.8	0.8	0.8	0.8
0.40	0.0	0.4	0.6	0.8	0.8	0.8	0.8	0.8	0.8	0.8
0.45	0.0	4.2	4.4	4.6	4.6	4.6	4.6	4.6	4.6	4.6
0.50	0.0	4.9	8.9	9.1	9.1	9.1	9.1	9.1	9.1	9.1
0.55	0.0	4.9	9.9	12.8	12.8	12.8	12.8	12.8	12.8	12.8
0.60	0.0	4.9	9.9	14.8	15.9	15.9	15.9	15.9	15.9	15.9
0.65	0.0	4.9	9.9	14.8	18.5	18.5	18.5	18.5	18.5	18.5
0.70	0.0	4.9	9.9	14.8	19.8	20.7	20.7	20.7	20.7	20.7

increase in the optimal number of surviving cells in the robust formulation relative to that in the nominal formulation.

Table 4 shows that for each α_2^τ, the PR is nondecreasing as Δ increases. In each row, the PR becomes constant after a sufficiently high value of Δ. A closer investigation of optimal doses revealed that at a sufficiently high value of Δ, dose d_1 reaches the value $D_1/N_1 s_1^{\max}$ and this eliminates the effect of uncertainty in α_1^ϕ from the normal tissue constraints. The PR thus remains constant thereafter. Also, in each row, the optimal modality switches to M_1 (or M_1, M_2) as Δ increases. This fact that M_1 becomes more desirable as the uncertainty in its biological power increases may appear counterintuitive at first. However, this scenario is indeed possible, because the uncertainty in α_1^ϕ affects both the LHS and the RHS of the normal tissue constraint. To test this, we removed the dependence of the RHS $\alpha_1^\phi D_1 + \beta_1^\phi D_1^2/N_{\text{conv}}$ on α_1^ϕ by setting this RHS to a fixed constant. We then re-solved the problem. As expected, the optimal modality then did not switch from M_2 to M_1 in any row as Δ increased. For each Δ, the optimal modality switches from M_1 to M_2 (or a combination of M_1, M_2). This is because M_2 becomes more desirable as its α_2^τ increases. For each Δ, the price of robustness increases with α_2^τ as long as M_1 (or M_1, M_2) remains optimal, because, by using M_1, we are forgoing a better-quality M_2 when α_2^τ increases.

Qualitative trends in Table 5 ($r = 0.8$) are identical to those in Table 4 ($r = 1$). But M_2 is biologically superior to M_1 in Table 5 because M_2 causes less damage to the normal tissue owing to a smaller value of r in the base case ($\alpha_2^\tau = 0.35$ Gy^{-1} and $\Delta = 0$). Thus, M_2 is optimal more frequently in Table 5. Consider any fixed (α_2^τ, Δ) pair where M_1 is optimal in both Tables 4 and 5 (either by itself or with M_2). Then, the PR in Table 5 is higher than that in Table 4, because, by using M_1, we are forgoing a better-quality M_2 in Table 5.

Qualitative trends in Table 6 ($r = 1.2$) are identical to those in Table 4 ($r = 1$). But M_2 is biologically inferior to M_1 in Table 6 because M_2 causes more damage to the normal tissue owing to a larger value of r in the base case ($\alpha_2^\tau = 0.35$ Gy^{-1} and $\Delta = 0$). Thus, M_2 is optimal less frequently in Table 6. Consider any fixed (α_2^τ, Δ) pair where M_1 is optimal in both Tables 4 and 6 (either by itself or with M_2). Then, the PR in Table 6 is lower than that in Table 4. This is because, by utilizing M_1, we are forgoing a lower-quality M_2 in Table 6.

Insights derived from our approach will need to be verified via clinical studies.

Acknowledgements This research was funded in part by the National Science Foundation via grant CMMI #1560476.

References

1. Baumann M, Krause M, Overgaard J, Debus J, Bentzen SM, Daartz J, Richter C, Zips D, Bortfeld T. Radiation oncology in the era of precision medicine. Nat Rev Cancer. 2016;16(4):234–49.
2. Fowler JF. Is there an optimal overall time for head and neck radiotherapy? A review with new modeling. Clin Oncol. 2007;19(1):8–27.
3. Halperin EC. Particle therapy and treatment of cancer. Lancet Oncol. 2006;7(8):676–85.
4. Nourollahi S, Ghate A, Kim M. Optimal modality selection in external beam radiotherapy. In: Forthcoming in mathematical medicine and biology 2018.
5. Saberian F, Ghate A, Kim M. Optimal fractionation in radiotherapy with multiple normal tissues. Math Med Biol. 2016;33(2):211–52.

Incentive-Based Rebalancing of Bike-Sharing Systems

Samarth J. Patel, Robin Qiu and Ashkan Negahban

Abstract This paper proposes an incentive-based approach for rebalancing bike-sharing systems where customers are offered discount to pick up bikes from nearby stations that are expected to become full in the near future. The main contribution of this work is twofold: (1) we develop a customized station object in the Simio simulation software to facilitate modeling of bike-sharing systems and reduce the burden on the modeler by eliminating the need to code the basic functionalities of a bike station; and, (2) we develop a discrete event simulation model of a real-world bike-sharing system (CitiBike) using instances of the customized station object to evaluate the effectiveness of pickup incentives in rebalancing the system. The model is calibrated using historic data and the results confirm the effectiveness of such incentive-based rebalancing scheme. More specifically, the results suggest that while incentives help improve bike availability in general throughout the system (i.e., better balance and service), offering too many incentives can in fact reduce total profit due to decreased marginal profit per ride.

1 Introduction

Despite the societal, environmental, and health benefits of bike-sharing systems, their adoption has been relatively slow. In New York City, for instance, only 0.2% of the city's population use the CitiBike system (the largest bike-sharing system in the United States) on a regular basis, i.e., subscribers, despite the fact that in June 2017 alone, CitiBike users offset more than 2 million pounds of carbon emissions and burned more than 165 million calories [1]. One of the main challenges in operating bike-sharing systems is rebalancing. Time-varying and opposite demand patterns for docks and bikes across different regions and unbalanced flows lead to bike/dock shortages, which in turn lead to reduced profit due to lost demand (balks), customer

S. J. Patel · R. Qiu · A. Negahban (✉)
School of Graduate Professional Studies, The Pennsylvania State University, Malvern, PA 19355, USA
e-mail: anegahban@psu.edu

H. Yang and R. Qiu (eds.), *Advances in Service Science*, Springer Proceedings in Business and Economics, https://doi.org/10.1007/978-3-030-04726-9_3

dissatisfaction, and potential loss of customers in terms of continued use or future adoption. The slow adoption of these systems can also be attributed (at least partially) to this problem. To alleviate this problem, service providers redistribute bikes through rebalancing operations. In 2016, CitiBike rebalanced about one million bikes using box trucks, vans, contracted trikes, and bike trains [1]. This traditional method of rebalancing is costly, requires substantial planning and coordination, and is against the green philosophy of these systems in reducing carbon emissions if performed using motor vehicles.

It is therefore crucial to redistribute bikes among stations in a proactive, economical, and environment-friendly way. In this paper, we investigate whether offering incentives to potential riders to pick up bikes from stations that are expected to run out of docks in the near future can potentially reduce the need for such rebalancing operations by helping the system rebalance itself. This work contributes to the simulation and bike-sharing literature as follows: (1) we develop a customized "Station" object in the Simio simulation software to facilitate development of models of bike-sharing systems. To the best of our knowledge, none of the commercial simulation packages has built-in implementation of a bike station that can be directly used in bike-sharing models, hence, modelers often need to spend a significant amount of time and coding effort to do this. The proposed station object contains the logic for basic operation of a bike station allowing the user to simply drag and drop an instance of this object onto the model without having to "code" the basic functionalities. Therefore, this customized object significantly facilitates modeling of large-scale bike sharing systems for researchers and practitioners using the Simio software package; and, (2) we evaluate an incentive-based rebalancing scheme (involving pickup incentives) in terms of the number of lost customers and total profit by experimenting with a simulation model of a real-world bike-sharing system built using instances of the customized station object. The model corresponds to the CitiBike system in Jersey City and its findings can help improve current strategies for rebalancing operations.

The remainder of the paper is organized as follows. Section 2 provides a critical analysis of the literature on bike-sharing systems. Section 3 describes the general logic and operation of the customized bike station object and the simulation model of a real-world bike-sharing system. Section 4 summarizes the experimental results and Sect. 5 provides the conclusions and potential future extensions.

2 Literature Review

Schuijbroek et al. [2] classify the related literature into the following two categories:

- **System design**: Dell'Olio et al. [3] develop a comprehensive methodology to optimize station locations. Martinez et al. [4] and Prem Kumar and Bierlaire [5] use mixed-integer programming models to maximize station performance by identifying demand patterns and resource allocation to stations. Lin and Yang [6] use a mathematical model to account for setup costs and travel paths and optimize the bike-share network design. While these studies provide insights on the initial design (number of stations and their locations), rebalancing operations are not explicitly considered. This is an important gap as consideration of rebalancing operations could potentially affect the optimal design. The model proposed in this paper addresses this gap and could be incorporated into the models proposed in this stream to support a comprehensive analysis of the system design.

- **Demand analysis and rebalancing operations**: These studies involve demand modeling and identifying the important factors for managerial decision-making, especially those related to rebalancing operations. Kaltenbrunner et al. [7] predict the system's future bike inventory to help improve performance by making the information available to potential riders via a website. Vogel and Mattfeld [8] study rebalancing activities using an aggregate feedback loop model and show that active repositioning of bikes improves service quality. Shu et al. [9] develop a stochastic network flow model to improve utilization of bike stations and bike redistribution activities. They also account for the effect of other local public transportation means on bike demand. Schuijbroek et al. [2] use integer programming to determine the service level requirement at each station and optimize routing for rebalancing operations. Jian et al. [10] use simulation-based optimization to minimize bike/dock unavailability by improving bike and dock allocation during different time intervals (e.g., morning and evening rush hours). O'Mahony and Shmoys [11] use integer programming to optimize routing of bike transporting vehicles used for rebalancing during rush hours and over-night.

The paper by Fricker and Gast [12] is of particular interest to us as it considers incentives, where customers are offered discount to drop off bikes at stations that are running out of bikes. However, in order to make the problem analytically tractable, they make several (strong) assumptions. They consider a homogenous system where the demand rate is the same for all stations. They perform a steady-state analysis and further assume the demand rate is constant. In reality, these systems virtually never reach steady state and demand is nonstationary. They consider only two drop-off options chosen at random, meaning that the two candidate stations recommended by incentives may not necessarily be close to the rider's intended destination. Our work relaxes these limiting assumptions. We consider a heterogeneous system with non-stationary time-varying demand and flow patterns and use the actual geographical location of stations to identify nearby stations as potential incentive options. Moreover, we consider various balking behavior, where customers do not necessarily balk right away if there is no bike available at the station. To the best of our knowledge, the work presented here is the first to consider incentives in such settings.

3 Simulation Model Development

In Sect. 3.1, a "customized station object" is developed in the Simio simulation software package. In Sect. 3.2, we use the customized station object to develop a model of the CitiBike system in Jersey City.

3.1 *A Customized Bike Station Object in Simio*

Figure 1 shows the external view of the customized station object and its general logic can be summarized as follows. Customer arrival process is modeled by a "Rate Table" which allows for modeling a non-stationary arrival process where the arrival rate changes over time. When a customer arrives at the station, she will check out a bike if available. Otherwise, if there is no bike available, there are three possibilities: (1) wait for a bike to be dropped off at the station (determined by the customer's "waiting probability"); (2) balk and leave the system as determined by the customer's "balking probability" (in the real world, this is when the customer decides to use an alternative means of transportation); Or, (3) walk to a nearby station (a threshold is used for how far a customer is willing to walk). A customer also has a "waiting threshold" that determines how long she is willing to wait for a bike at an empty station before she balks. The station object uses a mechanism to periodically check the waiting queue and remove those customers with their waiting threshold exceeded. When a bike arrives at the station, the customer drops off her bike if there is a dock available. Otherwise, the customer will attempt to drop off the bike at a nearby station.

The customized station object also tracks several statistics such as the number of bikes/docks, number of customers that balk without or after waiting, number of customers that attempt to pick up a bike from a nearby station, and the number of customers that received incentives. The detailed structure of the customized station,

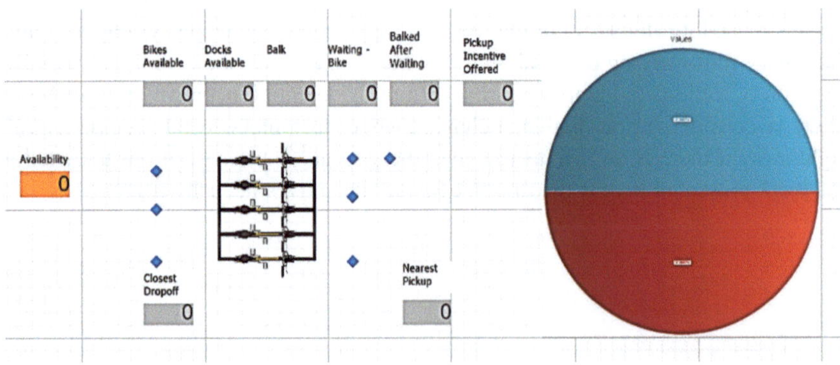

Fig. 1 External view of the station object with a dynamic pie chart that shows bike availability

its properties, state variables, and processes are explained in Patel [13], and the reader is referred to [14] for a general introduction into the Simio software.

3.2 *A Simulation Model of the CitiBike System in Jersey City*

We use instances of the customized station object to develop a simulation model of the CitiBike system in Jersey City to analyze the effectiveness of pickup incentives in reducing balks (i.e., lost customers) and improving profit. As of the date of this study, CitiBike-Jersey City has about 50 stations that are (virtually) isolated from the stations in New York City (i.e., there is almost no bike travel between the two subsystems). This allows us to treat these 50 stations as a separate system. The main reason behind choosing Jersey City was that even with the customized station object, developing a model for the entire CitiBike system with more than 600 stations would still be tedious. We discuss this issue further in Sect. 5.

Historical data on bike pickups and drop-offs are publicly available through the service provider. Here, we use data from December 2016. The dataset includes station ID, station longitude and latitude and address, trip starting and ending time at the corresponding stations, and user type. We use the bike pickup rate as a lower bound for the true demand for bikes. During this month, we observe that weekdays (Monday–Friday) have different patterns than weekends (Saturday and Sunday). Figure 2 shows how the average pickup rate changes during the day for different days of the week. We focus on weekdays and assume Poisson arrival processes for customers. The use of the Poisson distribution is justified in [10] and supported by a set of goodness-of-fit tests performed in [13]. Based on an assessment of pickup data using HistoRIA [15] and ADD-MORE [16] analysis tools for identifying non-stationary stochastic processes, hourly time-varying arrival rates are used. Once a bike is picked up, the destination station is sampled randomly based on the probabilities estimated from the real-world *from-to* trip frequency data, while the trip duration is determined based on the distance and the average bike speed. Station sizes (number of docks) for the fifty stations also correspond to the real system configuration.

Figure 3 illustrates a snapshot of the simulation model of CitiBike-Jersey City with each station represented by a pie chart where the blue color indicates bike availability and red indicates empty docks. Therefore, a red circle means no bike and a blue circle

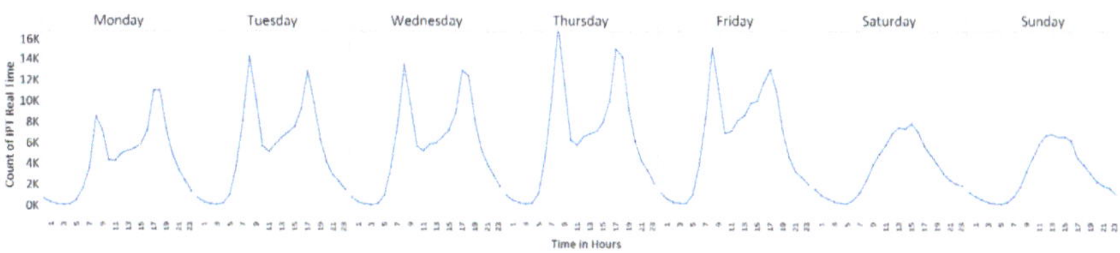

Fig. 2 Hourly patterns of bike pickup rate

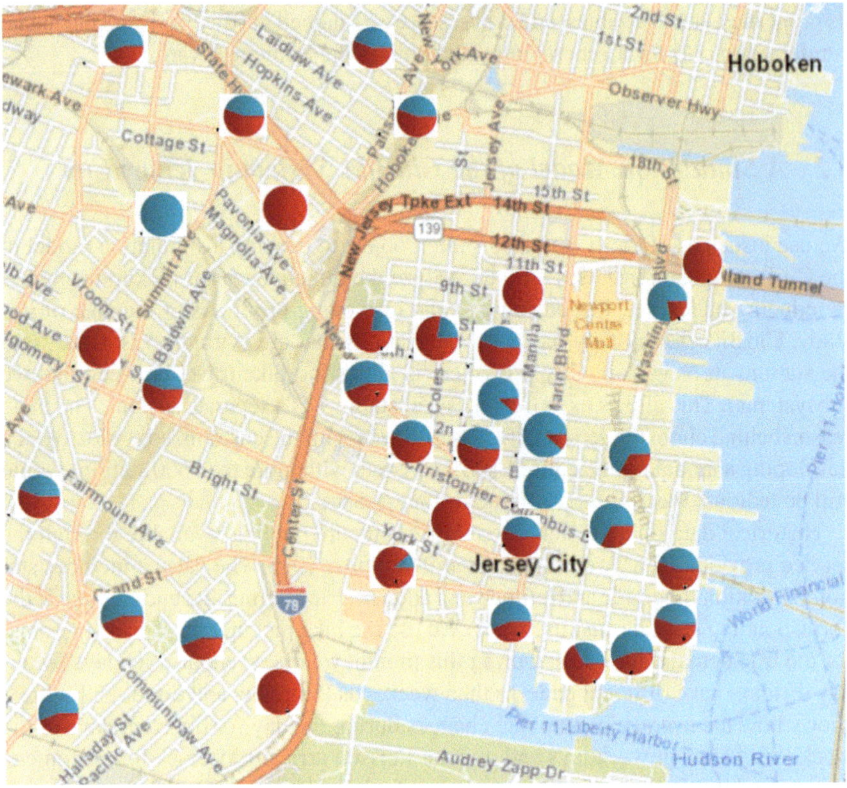

Fig. 3 A snapshot of the simulation model of the CitiBike system in Jersery City

indicates a full station. For each station, the stations' geographic location is used to create a list of nearby stations as potential candidates for pickup incentives. In this model, we use a threshold of 0.5 miles to determine nearby stations.

The logic used for pickup incentives can be described as follows. When a potential customer arrives at a station, an incentive is offered to the customer to pick up a bike from a nearby station that is running out of docks (if any). If there are multiple options, the closest station among the candidate incentive stations is selected. The customer accepts the incentive probabilistically based on her "probability of accepting incentives". In the following section, we perform sensitivity analysis on this probability value as well as the discount level.

4 Experiments and Results

Table 1 summarizes the experimental design. The simulation run is initialized with a perfectly balanced system, with all stations having the same percentage of bike availability. Station sizes and total number of bikes in the Jersey City subsystem are obtained from Citi Bike. The model is run for 100 replications of a 24-hour interval starting at midnight. We use total balks (lost customers) and total profit to evaluate the effectiveness of pickup incentives in rebalancing the system. Total balks is calculated by the sum of lost demand over all stations. Total profit is computed as follows:

$$\begin{aligned} \mathit{Total\ Profit} = \sum_{\mathit{all\ station\ i}} & [(\mathit{All\ bike\ pickups\ from\ station\ i}) \\ & * (\mathit{Profit\ per\ ride\ before\ discount}) \\ & - (\mathit{Bike\ pickups\ from\ station\ i\ with\ incentives}) * (\mathit{Discount\ rate})]. \end{aligned}$$

The results are summarized in Figs. 4 and 5. In both figures, we use the 25th and 75th percentiles to generate the box plots and the 95% confidence intervals for the mean (beige box) as well as the upper and lower percentiles (blue boxes). As shown in Fig. 4, the total balk decreases as the probability of accepting incentives increases. The total balk can be considered as a measure of "how well-balanced the system is". We expect the system to be more balanced as more customers accept pickup incentives.

Figure 5 shows that the total profit follows a concave function of the probability of accepting incentives. We also observe the trade-off between serving more customers versus reduced average profit margin per ride due to incentives. While pickup incentives improve the balance of the system and number of customers served, offering too many incentives decreases total profit as these rides have a smaller profit margin. Moreover, as the discount rate increases, the total profit is maximized at lower levels of the probability of accepting incentives (indicated by the red ovals). For instance, under a 5% discount rate, total profit peaks at probabilities between 0.5 and 0.9. Under a 20% discount rate, however, this range is 0.2–0.4.

Table 1 The parameters of the CitiBike-Jersey City simulation model

Parameter	Description
Initial bike availability	70% bike availability at the beginning of the run for all stations
Station size (number of docks)	Varies per station (determined based on real-world data)
Walking speed	3 mph
Biking speed	8 mph
Waiting threshold	Triangular (0, 5, 10) minutes
Balking probability	Uniform (0, 1)
Waiting probability	1.0 – Balking probability
Discount rate	5, 10, 15, 20%
Profit per ride	$1 (before discount)
List of nearby stations	Within 0.5 miles

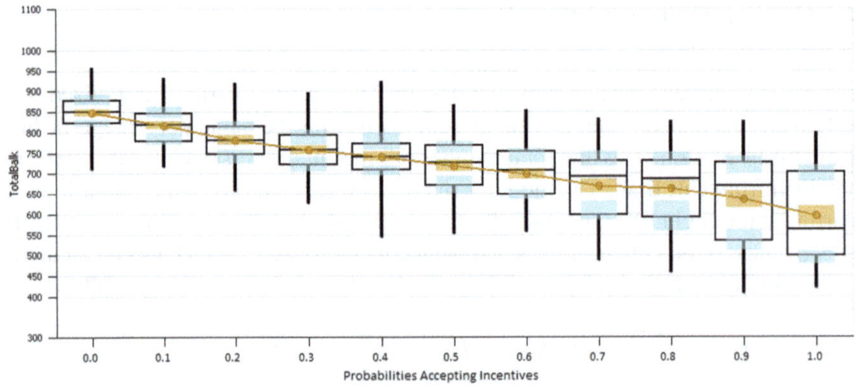

Fig. 4 Total balk based on the probability of accepting incentives

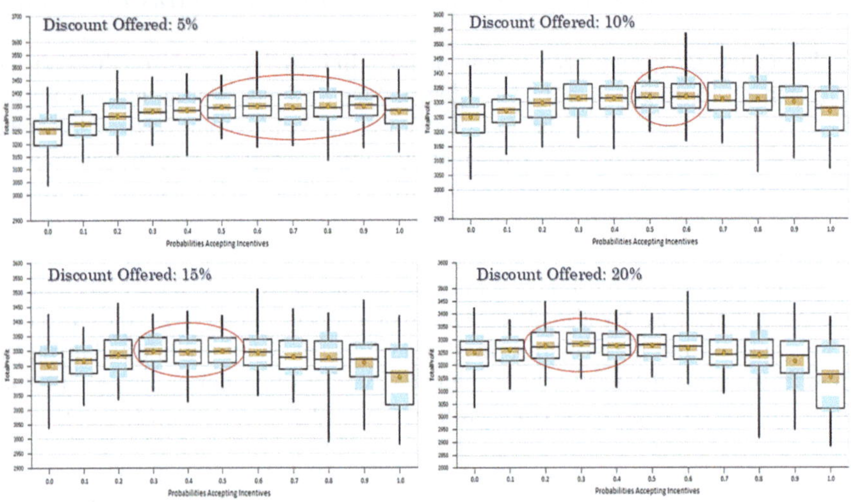

Fig. 5 Total profit for different levels of discount and probability of accepting incentives

5 Conclusions and Future Work

We propose an incentive-based approach for rebalancing bike-sharing systems where customers are offered discount to pick up bikes from nearby stations that are expected to become full in the near future. We develop a customized station object in the Simio simulation software and use instances of it to model the CitiBike-Jersey City system to evaluate the effectiveness of pickup incentives. The results show that while incentives can help improve general bike availability throughout the system (i.e., better balance and fewer balks), offering too many incentives can reduce total profit due to reduced average marginal profit per ride.

While the customized station object significantly reduces modeling effort, the modeler still needs to add input data and additional processes. In our model of CitiBike-Jersey City, each of the fifty stations requires an arrival table to model the non-stationary bike demand for that station, a list of its nearby stations, and additional code for the incentive logic. Therefore, automatic model generation is an important extension that would further facilitate modeling large-scale systems with hundreds of stations. In our analysis, the probability of accepting incentives is independent of the discount rate. Future research on customer behavior is needed to understand the relationship between the two. While not addressed in this paper, explicit modeling of the rebalancing operations and the cost associated with them would be necessary to evaluate how much of the reduced profit due to discounts would be offset by saving on rebalancing costs. Another extension involves modeling drop-off incentives to encourage riders to drop off their bike at stations that are running out of bikes in the near future. A joint analysis of pickup and drop-off incentives may lead to interesting findings.

References

1. Citi Bike NYC. Citi Bike monthly operating reports, Citi Bike NYC. 2017. https://www.citibikenyc.com/system-data/operating-reports. Accessed 20 Oct 2017.
2. Schuijbroek J, Hampshire R, Van Hoeve W. Inventory rebalancing and vehicle routing in bike sharing systems. Pittsburgh, PA: Carnegie Mellon University; 2013.
3. Dell'Olio L, Angel I, Moura JL. Implementing bike-sharing systems. Proc Inst Civ Eng Munic Eng Lond. 2011;164(2):89–101.
4. Martinez LM, Caetano L, Eiró T, Cruz F. An optimisation algorithm to establish the location of stations of a mixed fleet biking system: an application to the city of Lisbon. Procedia Soc Behav Sci. 2012;54:513–24.
5. Prem Kumar V, Bierlaire M. Optimizing locations for a vehicle sharing system. In: Proceedings of the Swiss transport research conference. 2012. p. 1–30.
6. Lin JR, Yang TH. Strategic design of public bicycle sharing systems with service level constraints. Transp Res Part E Logist Transp Rev. 2011;47(2):284–94.
7. Kaltenbrunner A, Meza R, Grivolla J, Codina J, Banchs R. Urban cycles and mobility patterns: exploring and predicting trends in a bicycle-based public transport system. Pervasive Mob Comput. 2010;6(4):455–66.
8. Vogel P, Mattfeld DC. Modeling of repositioning activities in bike-sharing systems. In: Proceedings of the world conference on transport research. 2010. p. 1–13.
9. Shu J, Chou MC, Liu Q, Teo C-P, Wang I-L. Models for effective deployment and redistribution of bicycles within public bicycle-sharing systems. Oper Res. 2013;61(6):1346–59.
10. Jian N, Freund D, Wiberg H, Henderson S. Simulation optimization for a large-scale bike-sharing system. In: Proceedings of the 2016 Winter simulation conference. IEEE; 2016. p. 602–13.
11. O'Mahony E, Shmoys DB. Data analysis and optimization for (citi)bike sharing. In: Proceedings of the twenty-ninth AAAI conference on artificial intelligence. 2015. p. 687–94.
12. Fricker C, Gast N. Incentives and redistribution in homogeneous bike-sharing systems with stations of finite capacity. Euro J Transp Logist. 2014;5(3):261–91.
13. Patel SJ. An incentive-based rebalancing scheme for large bike-sharing systems. Master's paper. The Pennsylvania State University; 2017.

14. Smith J, Sturrock D, Kelton D. Simio and simulation: modeling, analysis, applications. Sewickley, PA: Simio LLC; 2017.
15. Ansari M, Negahban A, Megahed FM, Smith JS. HistoRIA: a new tool for simulation input analysis. In: Proceedings of the 2014 Winter simulation conference. IEEE; 2014. p. 2702–13.
16. Negahban A, Ansari M, Smith JS. ADD-MORE: automated dynamic display of measures of risk and error. In: Proceedings of the 2016 Winter simulation conference, IEEE; 2016. p. 977–88.

A T-shaped Measure of Multidisciplinarity in Academic Research Networks: The GRAND Case Study

David Turner, Diego Serrano, Eleni Stroulia and Kelly Lyons

Abstract Service-science research has long been studying T-shapedness, arguing that service scientists should be T-shaped individuals, deeply knowledgeable in one field and able to collaborate and communicate across disciplines. The value of multidisciplinarity has also been recognized in academic environments, as funding agencies are committing substantial support to large-scale research initiatives that span across disciplines, organizations, academia and industry, even across national borders, and aim to address the major challenges of our time, from climate change, to energy shortage, to pandemics. New incentives and performance indicators are needed to encourage and reward multidisciplinary collaborative work. In this paper, we introduce a metric for multidisciplinarity, based on the notion of T-shapedness and we report on the application of this measure on data collected over four years from the GRAND Network of Centres of Excellence, a large-scale, Canadian, multidisciplinary research network conducting research on digital media with numerous academic and industrial partners. We describe our findings on how the community evolved over time in terms of its T-shaped multidisciplinarity and compare the multidisciplinarity of GRAND researchers to their non-GRAND peers.

1 Introduction

The GRAND Network of Centres of Excellence (NCE) is a Canadian multidisciplinary research network, conducting research on digital media, the technologies that produce them, and their applications in our everyday lives. GRAND was funded from the Canadian government through the NCE (Networks of Centres of Excellence) program and, in its first four years (2010–2014), supported 41 research projects, across 26 Universities, involving over 200 researchers and their trainees. GRAND was

D. Turner · D. Serrano · E. Stroulia (✉)
Department of Computing Science, University of Alberta, Edmonton, AB, Canada
e-mail: stroulia@ualberta.ca

K. Lyons
Faculty of Information, University of Toronto, Toronto, ON, Canada

H. Yang and R. Qiu (eds.), *Advances in Service Science*, Springer Proceedings in Business and Economics, https://doi.org/10.1007/978-3-030-04726-9_4

a highly multidisciplinary network, with researchers from science and engineering, social sciences, health sciences, and arts and humanities. The GRAND digital-media research agenda was also very broad, related to new algorithms and tools to support the production of digital content; constructing platforms to host and enable the efficient access to this content; developing and empirically evaluating applications using digital media in a variety of settings (e.g., entertainment, training, work, and healthcare); and, formulating policies around the use, sharing, and dissemination of digital content.

GRAND is an example of a new breed of large, geographically distributed, multidisciplinary research programs. The international research community and funding agencies are recognizing the need to support large-scale initiatives to address the grand challenges of our time. Driving this trend is the belief that these problems cannot be effectively addressed solely by researchers in a single discipline or a single organization, and that their study has to involve a broad spectrum of expertise across multiple "centres of excellence". However, even as this belief is generally, and increasingly, shared, the questions of when, why, and precisely how these research networks are made effective are still very much open and the subject of considerable debate.

The investigation of these general questions and, more specifically, the study of how digital collaboration tools can contribute to the effectiveness of research networks has been part of the GRAND agenda since its inception. To that end, we developed a software platform, the Forum, to support communication and collaboration across the network members and projects and to streamline the administrative workflows and data collection required by the NCE program that is funding GRAND. The Forum has become a rich repository of data about the activities and research production of the GRAND research community. The availability of this dataset presents a unique opportunity to study some of the core questions around the effectiveness of large-scale research networks in fostering interdisciplinarity.

We base our study on a few key methodological assumptions. Multidisciplinary research integrates understanding, knowledge, techniques, tools, data, etc. from more than one body of knowledge to produce solutions that are beyond the scope of any one field [11]. Multidisciplinarity also emerges as another dimension of research excellence, beyond the more traditional metrics, i.e., publication and citation counts [14]. Even though these statements may intuitively make sense, there is no generally accepted measure of a researcher's multidisciplinarity. The main contribution of this paper is the **formulation of a quantitative measure of multidisciplinarity**, or, more specifically, a measure of fitness with a generally accepted notion of ideal multidisciplinarity, namely T-shapedness [5, 7]. Our measure reflects the multidisciplinarity of the output of a researcher. The second contribution of this paper is our **use of our multidisciplinary metric to analyze the multidisciplinarity of the GRAND community before and after involvement in GRAND against (a sample of) their Canadian peers**. In order to examine the multidisciplinarity of Canadian researchers within and outside GRAND, we use data from Scopus.

The rest of this paper is organized as follows. We first review background research related to our work (Sect. 2). We then describe the GRAND research network as the subject of our study, and present the T-shaped multidisciplinary metric, followed by the methodology, data collection, and analysis of our study (Sect. 3). Next, we present our findings and discuss their implications (Sect. 4). Finally, we conclude by summarizing the lessons we learned through our study (Sect. 5).

2 Related Work

There are many quantitative measures of research and scholarly information broadly called informetrics [1]. More recently, measures of interdisciplinary research have received attention [14] as have alternative measures of scholarly influence in social media and the web [2]. Given our quantitative study of the multidisciplinary GRAND research network, we summarize here the most relevant work on measures of multidisciplinary research.

As a general background to this work, we adopt the general definitions of the various related X-disciplinarity terms, as led out by Jensenius in http://www.arj.no/2012/03/12/disciplinarities-2/. In the context of multidisciplinary work, people from different disciplines work together, each drawing on their disciplinary knowledge; interdisciplinary work integrates knowledge and methods from different disciplines; and transdisciplinary work creates a unity of intellectual frameworks beyond the disciplinary perspectives. A large number of categorizations of interdisciplinarity has been reviewed by [8]; this conceptual framework for interdisciplinary research considers three criteria: the scope of interdisciplinarity, multidisciplinary and interdisciplinary research interactions, and the objectives of the research activity. However, this qualitative framework requires a domain expert and adopts, as the unit of analysis, the research-proposal document.

In contrast, our work aims at a quantitative measure of multidisciplinarity that can be applied to a researcher or a research network. To that end, we have been inspired by the notion of a T-shaped individual—or a person with T-shaped skills—who displays depth in a particular field of study (the stem of the T) and a breadth of abilities and skills across disciplines (the bar or top of the T) [7, 9, 10, 15]. The concept was first attributed to Guest in 1991 [7] but many others have described T-shaped skills as necessary components to building multidisciplinary teams [3, 6]. Intuitively, an academic researcher with a perfect T-shapedness score should have a substantial percentage of their publications in one discipline, while the remainder of their publications should be fairly smoothly distributed over a number of other disciplines [13].

T-shapedness has been studied in the context of service science where it is argued that Service scientists should be T-shaped individuals. Recently, it has been suggested that education systems should develop T-shaped professionals in part to encourage and reward academics to collaborate with colleagues in other disciplines and to research agendas that are transdisciplinary [5].

Two additional studies are very closely related to our own in that they investigate the multidisciplinarity in research projects. First, [4] studied a number of projects receiving two particular National Science Foundation's grants. They analyzed and compared methods of collaboration, in order to highlight gaps in existing collaboration methods and practices. However, this study does not provide a clear metric for multidisciplinarity. The second [16] studied interdisciplinarity by classifying publications into disciplines according to the ISI Subject Categories. The degree of multidisciplinarity was measured through indicators of disciplinary diversity as suggested by [12], which argues that interdisciplinarity requires the consideration of diversity (defined by the variety, balance of the distribution, and disparity of production), and coherence (the degree to which the process of integration is taking place). Our approach follows the diversity aspect of interdisciplinarity, defining the T-shapedness metric to capture variety and balance.

3 The GRAND Network and Its T-shaped Multidisciplinarity

The GRAND NCE is an example of today's large-scale, geographically distributed, multidisciplinary research programs, with numerous academic and industrial partners. The first objective of this study is to gain a deeper understanding of how the GRAND network of researchers worked during the first four years of the network's life and how their collaborative practices changed over time. These insights can potentially be extremely relevant not just to GRAND, but also to other large multidisciplinary networks that may want to encourage similar practices. Our second objective is to compare the GRAND community against a sample of the Canadian research community in terms of the multidisciplinarity of their research outcomes, in order to examine whether the GRAND network led to a higher degree than what is typical of other Canadian researchers.

In order to measure the multidisciplinarity of researchers within and outside GRAND, we enhanced the information collected in the Forum with information from Scopus about the disciplinary range of each researcher's productivity. This information can be retrieved for GRAND researchers and a sample of Canadian researchers from outside GRAND, which enables us to comparatively examine the relative multidisciplinarity of these two groups. Scopus associates each publication with a subset of 26 different subject areas; therefore, for a given researcher, we can identify the union of their publications' subject-area sets for a given year, and the number of publications associated with each of these subject areas. Consider for example, a *Researcher* with nine publications, $pub_1 \dots pub_9$, each one associated with at least one (and possibly more) of four subject areas, labelled S_1, S_2, S_3 and S_4. These publications give rise to the following subject-area sets: pub_1:{S_1, S_2}, pub_2:{S_1, S_2}, pub_3:{S_1}, pub_4:{S_1}, pub_5:{S_1, S_4}, pub_6:{S_1}, pub_7:{S_1}, pub_8:{S_1} and pub_9:{S_1} such that the counts of publications in each subject area are $|S_1| = 9$, $|S_2| = 2$, $|S_3| = 0$, and $|S_4| = 1$. We will use this *Researcher* as an example to illustrate the definition of our T-shapedness multidisciplinarity measure, in a simplified context of a smaller number of subject areas (4 instead of 26).

As we have already discussed, a researcher with a perfect T-shapedness score should have a substantial percentage of their publications in one of the Scopus subject-area, indicating depth of expertise in this primary area (the T stem), while the remainder of their publications should be fairly smoothly distributed over a number of other subject-areas, indicating multidisciplinary breadth of knowledge (the T horizontal bar) [6], balance, and disparity. The breadth of disciplines outside the primary discipline, represented as the horizontal bar of the T in our measure, increases as (a) the number of breadth disciplines increases, and, (b) the amount of work is evenly balanced across these breadth disciplines. The third diversity principle, disparity, refers to the way in which the disciplines are different from, or similar to, each other.

Our analysis relies on the Scopus discipline categories, and we assume that all categories are equally distinct from each other. Therefore, we define our T-shaped metric of multidisciplinarity (referred to as MD henceforth) based on: (a) the ratio of a researcher's productivity in their primary/core subject area to their overall research output, and (b) the degree to which the rest of their production is smoothly distributed over all areas other than their primary subject area. Intuitively, a "perfectly T-shaped" researcher would have an ideal ratio (r_{ideal}) of publications in their primary subject area and the rest of their publications should be smoothly distributed over the other (non-core) subject areas. Intuitively, a value of 0.5 or lower might suggest less expertise in a core area while $r_{ideal} = 0.75$ or higher might signify a higher depth with relatively little productivity outside a core area of expertise. The value of r_{ideal} for the purposes of this study was chosen to be 0.618, which we call the golden ratio. We experimented with a number of different values for r_{ideal} between 0.6 and 0.70 (0.6, 0.618, 0.65, 0.7) and found that the results presented below are consistent using all these alternative values.

We have implemented the MD measure in terms of two vectors: v_{stem} and v_{breath}. The stem discipline vector (v_{stem}) captures the divergence of the researcher's productivity in their primary discipline relative to their overall productivity from the ideal ratio. The breadth vector ($v_{breadth}$) captures the degree to which the researcher's productivity is balanced across the other (non-core) subject areas. Considering n as the number of subject areas, $|s_i|$ as the number of publications associated with discipline S_i, S_{stem} is the subject area associated with the highest number of publications by the *Researcher*, and R as the ratio of productivity in s_{stem} over their overall productivity, we can define the vectors v_{stem} and $v_{breadth}$ as follows:

$r_{ideal} = 0.618$	The "ideal" T ratio
$R = \frac{\lvert s_{stem}\rvert}{\max\{1, \sum_{i=1}^{n} \lvert s_i\rvert\}}$	The ratio of the researcher's productivity in their core area over their overall productivity
$v_{stem} = \left(1, 1 - \frac{\lvert r_{ideal} - R\rvert}{r_{ideal}}\right)$	The vector defined by the *Researcher*'s productivity in their core subject area, in effect the ratio of the difference between the researcher's ratio R over the ideal ratio r_{ideal}
$\vec{v}_{breadth} = (\lvert s_1\rvert, \lvert s_2\rvert, \lvert s_3\rvert, \ldots, \lvert s_n\rvert),\ i = 1..n\ except\ stem$	The breadth vector defined by the *Researcher*'s productivity in all other non-core subject areas

We can identify the theoretical best case, $v_{stemBest}$ and $v_{breadthBest}$ for the two vectors, respectively. The best case is when a researcher's ratio of publications in the stem area to the rest of the publications equals r_{ideal}, in which case $v_{stemBest} = (1, 1)$. Likewise, when a researcher's non-stem publications are evenly distributed among the rest of the subject areas, $v_{breadthBest} = (k, k, \ldots, k)$, where k is $\frac{\sum_{i=1}^{n} |s_i| - |s_{stem}|}{n-1}$. Note that there are many researchers similar to this best-case example, namely all those who have all the rest of their pubs evenly distributed in $n=25$ non-core subject areas. We can also define worst-case vectors when a researcher has no publications at all, in which case, $v_{stemWorst} = (1, 0)$ and $v_{breadthWorst} = (0, 0, \ldots, 0)$.

Given the vectors v_{stem} and $v_{breadth}$, we calculate two angles: (a) the angle between v_{stem} and the vector $v_{stemBest}$ and (b) the angle between $v_{breadth}$ and the $(n - 1)$-dimensional vector $v_{breadthBest}$. Both these angles capture some aspect of the "divergence" of the researcher's productivity profile from the ideally T-shaped profile, whether in the ratio of their productivity in their stem subject area to their overall production, or in the smoothness of the distribution of the rest of their work in all other non-stem subject areas.

Figure 1a, b illustrate the above two angle calculations for our example *Researcher*, with nine publications associated with four subject areas as follows: pub_1:$\{S_1, S_2\}$, pub_2:$\{S_1, S_2\}$, pub_3:$\{S_1\}$, pub_4:$\{S_1\}$, pub_5:$\{S_1, S_4\}$, pub_6:$\{S_1\}$, pub_7:$\{S_1\}$, pub_8:$\{S_1\}$ and pub_9:$\{S_1\}$ such that $|S_1|=9$, $|S_2|=2$, $|S_3|=0$, and $|S_4|=1$. Note that pub_1 contributes to the counts of two subject areas, S_1 and S_2. Our rationale behind this choice is that this publication may be potentially "discovered" by a larger audience, namely all readers interested in either of the two areas; thus, assuming relative independence of these areas and similar sizes of their corresponding communities, the association of publication pub_1 would imply potential readership of twice the size of the readership of pub_6 for example. We then order the researcher's subject areas in descending order according to the counts of publications associated with each area and we identify s_1 as the core area of expertise for this researcher, s_2 as the subject area with the second highest number of publications, and so on until s_n. Following the definition of R, we calculate the ratio of the *Researcher*'s

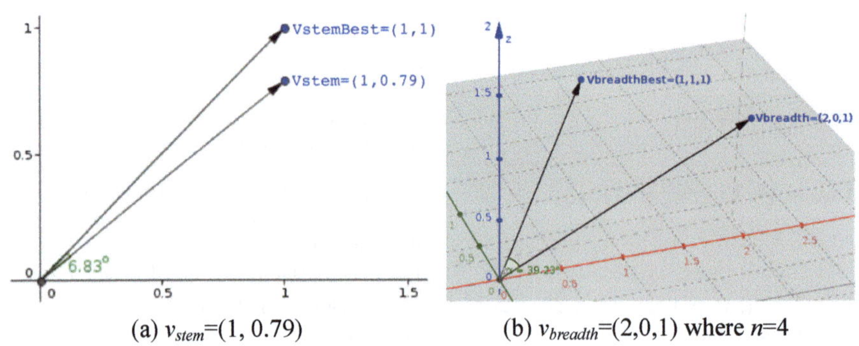

(a) v_{stem}=(1, 0.79) (b) $v_{breadth}$=(2,0,1) where n=4

Fig. 1 T-shapedness vectors of the example *Researcher₁*

publications in the stem area to their total number of publications in all disciplines, which for *Researcher* is $9/12 = 0.75$. Assuming $r_{ideal} = 0.618$ as the ideal T-ratio, the T-shapedness vectors of *Researcher* are $v_{stem} = (1, 0.79)$ and $v_{breadth} = (2, 0, 1)$, and the corresponding T-shapedness angles are $6.83°$ and $39.23°$, respectively.

To further illustrate our T-shapedness MD metric let us consider another researcher, *Researcher*$_2$, who has the following publications: $pub_1{:}\{S_4\}$, $pub_2{:}\{S_1, S_4\}$, $pub_3{:}\{S_4\}$, $pub_4{:}\{S_4\}$, $pub_5{:}\{S_2\}$ and $pub_6{:}\{S_3\}$ such that $|S_1| = 1$, $|S_2| = 1$, $|S_3| = 1$, and $|S_4| = 4$. In this case, S_4 is the stem area of *Researcher*$_2$, and using the same ideal ratio $r = 0.618$, the T-shapedness vectors for *Researcher*$_2$ are $v_{stem} = (1, 0.92)$ and $v_{breadth} = (1, 1, 1)$ and the corresponding angles are $2.262°$ and $0°$.

Next, we calculate the overall multidisciplinary score of a researcher as the normalized weighted average of the above two angles. To normalize the $\angle v_{stem} v_{stemBest}$ and $\angle v_{breadth} v_{breadthBest}$ angles such that they are values between 0 and 1, we divide them by the theoretical worst-case angles, $\angle v_{stemWorst} v_{stemBest}$ and $\angle v_{breadthWorst} v_{breadthBest}$ respectively, which essentially produces a measure of the distance between the researcher and the worst case. Each normalized angle is then weighted based on the value of r_{ideal} and the result is subtracted from 1 to give a positive value between 0 and 1 for the final value of MD, calculated as follows.

$$\angle v_{stemWorst} v_{stemBest} = \cos^{-1}\left(\frac{\sqrt{2}}{2}\right)$$

$$\angle v_{breadthWorst} v_{breadthBest} = \cos^{-1}\left(\frac{\sqrt{n-1}}{n-1}\right)$$

$$\angle v_{stem} v_{stemBest} = \cos^{-1}\left(\frac{v_{stemBest} \cdot v_{stem}}{|v_{stemBest}||v_{stem}|}\right)$$

$$\angle v_{breadth} v_{breadthBest} = \cos^{-1}\left(\frac{v_{breadthBest} \cdot v_{breadth}}{|v_{breadthBest}||v_{breadth}|}\right)$$

$$MD = 1 - \left[r\left(\frac{\angle v_{stem} v_{stemBest}}{\angle v_{stemWorst} v_{stemBest}}\right) + (1-r)\left(\frac{\angle v_{breadth} v_{breadthBest}}{\angle v_{breadthWorst} v_{breadthBest}}\right)\right]$$

Revisiting the researchers in our example, we calculate their two *MD* scores as 0.633 and 0.969; the less productive (in absolute numbers) *Researcher*$_2$ is more multidisciplinary than the original example *Researcher*$_1$ because they have a closer-to-the-ideal distribution of publications over subject areas. Note that *MD* is calculated for a period of time, based on the researcher's publications in this period. When comparing the *MD* scores of a researcher at times t_1 and t_2, where t_2 is after t_1, and both are after an original timestamp t_0, a positive difference $MD(t_2 - t_0) - MD(t_1 - t_0)$ indicates an increase in multidisciplinarity.

To investigate the potential impact of the GRAND NCE on the multidisciplinarity of its members, we had to establish a comparison data set for the community of GRAND researchers. This motivated us to only compare GRAND researchers to researchers with Canadian federal research funding: our sample of Canadian researchers would be a subset of those researchers in Canada who are not part of GRAND but receive funding in the same areas as GRAND researchers and are at the same universities as GRAND researchers. This resulted in a total of 186 GRAND researchers and 534 researchers in the Canadian researcher sample in our dataset. We refer to the Canadian researcher sample as the control group.

For each researcher in GRAND, we searched the Tri-Council funding agency databases to determine if that researcher had received funding during the years of GRAND (2010–2013) and, if so, we identified the evaluation committee within that agency from which the researcher had received funds. For each GRAND researcher, R who received funding between 2010 and 2013, we manually identified all researchers R', who: (a) are not part of GRAND; (b) are at the same university as R; and, (c) received funding between 2010 and 2013 through the same evaluation committee as R. This resulted in 1337 researchers from which we randomly selected 668 (approximately, half). (Note: there were 3 GRAND researchers funded by CIHR between 2010 and 2013 but we did not find any CIHR-funded non-GRAND researchers to include in our non-GRAND sample; therefore, our non-GRAND sample contains only NSERC and SSHRC funded researchers.)

We needed to be able to collect Scopus data for each researcher in our sample and each researcher in GRAND. We searched the Scopus database to retrieve a Scopus ID for each of the 211 researchers in GRAND and the 668 in our sample. We then manually verified their Scopus IDs and found that 25 GRAND researchers and 134 researchers in the non-GRAND sample did not have a Scopus ID. This resulted in a total of 186 GRAND researchers and 534 researchers in the Canadian researcher sample in our dataset. We refer to the Canadian researcher sample as the control group. Table 1 summarizes the dataset of our study.

Table 1 The study data set

GRAND network investigators	211
GRAND network investigators with no Scopus ID	25
GRAND network investigators studied	**186**
Sample Canadian researchers with NSERC and SSHRC funding (from the same committees as the GRAND investigators)	668
Sample Canadian researchers with no Scopus ID	134
Sample Canadian researchers studied	**534**

4 Research Findings and Discussion

We calculated the T-shapedness (MD) of the productivity of the GRAND researchers and the control group for two periods: from 2006 to 2009 and from 2010 to 2013. These two timeframes were chosen to provide us with two data points for each researcher: one for their multidisciplinarity during the four-year period just before GRAND started, referred to as MD@2009, and a second one for the four-year period during GRAND, referred to as MD@2013.

We then considered three questions. *Q1: Has the multidisciplinarity of researchers improved over time (from 2009 to 2013)?* To answer this question, we computed the paired-difference t-test between the MD@2009 and MD@2013 values of every researcher in GRAND and the control dataset. We found that both GRAND and control researchers improved in terms of their multidisciplinarity. For GRAND researchers the p-value was 0.0006256 and for the researchers in the control group the p-value was 0.03286—see Table 2. It appears that both groups of researchers had more multidisciplinary research output in the period from 2010 to 2013, as compared to the period from 2006 to 2009. The two p-values indicate that the phenomenon is slightly stronger for the GRAND community.

We then proceeded to investigate this phenomenon more precisely, asking whether *Q2: the multidisciplinarity increase in GRAND was stronger than the corresponding increase in the control group*. To answer this question, we computed the increase of the MD measure, i.e., MD@2013–MD@2009, for each researcher in the GRAND community and in the control group. An independent-samples t-test revealed support for the hypothesis that "participation in GRAND led to a more pronounced increase in the researcher's multidisciplinarity" (p-value = 0.03474).

Finally, we examined *Q3: whether there were any significant differences in the multidisciplinarity of the two groups in 2009 or in 2013*, effectively asking whether the GRAND community was more (or less) multidisciplinary than the control community in 2009 (or in 2013). An independent-samples t-test between the MD@2009 and the MD@2013 values of GRAND and control researchers revealed that in 2009 the control group was marginally more multidisciplinary than the GRAND Researchers (p-value = 0.08522), but in 2013 this difference was practically eliminated (p-value = 0.7982); the researchers in GRAND had a slightly more pronounced

Table 2 Comparing average MD of GRAND researchers and the control group

	Min.	Max.	Mean	Std. dev.
MD@2009 GRAND	0.00	0.622	0.366	0.171
MD@2009 Control	0.00	0.614	0.390	0.168
MD@2013 GRAND	0.00	0.629	0.398	0.151
MD@2013 Control	0.00	0.640	0.401	0.164
MD@2013–MD@2009 GRAND	−0.25	0.449	0.032	0.134
MD@2013–MD@2009 Control	−0.53	0.499	0.011	0.138

increase in multidisciplinarity during the four years of GRAND participation than the control group, which led to the elimination of this difference.

Our findings indicate that researchers in GRAND benefited from their participation in GRAND in that they became more multidisciplinary than their Canadian peers who were not participating in GRAND. As a community, GRAND researchers started slightly less multidisciplinary than their peers, but at the end of the four years they became slightly more multidisciplinary. This finding implies that, to some degree, GRAND met its objective of pulling expertise from different areas together to produce research that can potentially have impact across areas. At the very least, this finding provides some evidence that the NCE program fulfils its mandate since a researcher's participation in a NCE encourages increasingly multidisciplinary productivity.

5 Conclusions and Future Work

In this paper, we introduced a T-shapedness measure of multidisciplinarity, defined as the relative ratio of one's research production in one's core area of expertise over one's total production. We used this measure in a study of the GRAND research community, a multidisciplinary pan-Canadian NCE, conducting research on all aspects of digital-media technologies.

We found that the GRAND community became increasingly multidisciplinary over time, according to this measure, more so than the control community of their Canadian peers who had obtained research grants from the same NSERC/SSHRC area committees. This result suggests that the GRAND NCE, or, at the very least, the NCE program, has effectively cultivated multidisciplinary research production.

We believe that this work, beyond offering insights in the evolution of the GRAND researcher community, its activities and its research output, puts forward a general methodology for analyzing large research communities and comparing them against each other. In the future, we plan to examine in depth the record of individual researchers who best exemplify (or constitute exceptions to) the trends we discovered in order to gain insights on specific activities and best practices for researchers to take advantage of belonging in such a research network. We feel the T-shaped measure of multidisciplinary should be further applied to other such networks and communities of researchers.

References

1. Bar-Ilan J. Informetrics at the beginning of the 21st century: a review. J Inform. 2008;2(1):1–52.
2. Bar-Ilan J, Sugimoto C, Gunn W, Haustein S, Konkiel S, Lariviere V, Lin J. Altmetrics: present and future. In: Proceedings of the 76th ASIS&T annual meeting: beyond the cloud: rethinking information boundaries. 2013.

3. Brown T. T-shaped stars: the backbone of IDEO's collaborative culture. Morten Hansen, chiefexecutive.net; 2010.
4. Cummings J, Kiesler S. Collaborative research across disciplinary and organizational boundaries. Soc Stud Sci. (Sage Publications) 2005;35(5):703–22.
5. Demirkan H, Spohrer J. Commentary—cultivating T-shaped professionals in the era of digital transformation. J Serv Sci. (INFORMS) 2018;10(1):98–109.
6. Donofrio N, Spohrer J, Zadeh H. Research-driven medical education and practice: a case for T-shaped professional. MJA Viewpoint. Collegiate Employment Research Institute; 2009.
7. Guest D. The hunt is on for the Renaissance Man of computing. Independent. (London) 1991;17.
8. Huutoniemi K, Klein J, Brunn H, Huukinen J. Analyzing interdisciplinarity: typology and indicators. J Res Policy. (Elsevier) 2010;39(1):79–88.
9. Iansiti M. Real-world R&D: jumping the product generation gap. Harvard business School Press; 1999.
10. Oskam I. T-shaped engineers for interdisciplinary innovation: an attractive perspective for young people as well as a must for innovative organisations. In: 37th annual conference—attracting students in engineering, vol 14. 2009.
11. Porter A, Cohen A, Roessner D, Perreault M. Measuring researcher interdisciplinarity. Scientometrics. (Springer Science+Business Media BV) 2007;72(1):117–47.
12. Rafols I, Meyer M. Diversity and network coherence as indicators of interdisciplinarity: case studies in bionanoscience. Scientometrics. Springer(Springer) 2010;82(2):263–87.
13. Stirling A. A general framework for analysing diversity in science, technology and society. J R Soc Interface. (The Royal Society) 2007;4(15):707–19.
14. Wagner C, Roessner J, Bobb K, Klein J, Boyack K, Keyton J, Rafols I, Borner K. Approaches to understanding and measuring interdisciplinary scientific research (IDR): a review of the literature. J Inform. (Elsevier) 2011;5(1):14–26.
15. Wu J, Zou X, Kong H. Cultivating T-shaped engineers for 21st century: experiences in China. In: 2012 ASEE annual conference & exposition. 2012. p. 25–372.
16. Yegros A, Amat C, D'Este P, Porter A, Rafols I. Does interdisciplinary research lead to higher scientific impact. In: STI indicators conference, Leiden. 2010.

A Framework for Delivering Service Differentiation Through Operating Segments: Research Opportunities and Implementation Challenges

Morris A. Cohen and Jose A. Guajardo

Abstract This paper presents a framework for developing a differentiated strategy for the delivery of services. It summarizes the methodology introduced in Guajardo and Cohen in (Manuf Serv Oper Manag 30(3):440–454, 2018 [6]) and discusses modeling and implementation implications associated with application of the framework for the management of value-added services that are bundled with manufactured products. The framework utilizes the concept of operating segments (introduced by Frei and Morriss in (Uncommon service: how to win by putting customers at the core of your business, 2012 [5])) and considers issues associated with the definition of market segments appropriate for differentiated services as well as for the design of such services. The paper also discusses operational processes and the tradeoffs associated with producing and delivering differentiated service products to market segments. This discussion includes a review of a set of representative analytical models for service delivery that illustrate OM Service research opportunities.

Keywords Service differentiation · Operating segments · After-sales services Service operations strategy

1 Introduction

Product design and post-sales services can be used to support a differentiation strategy based on customer perceptions of value created through product use. This approach is more effective when compared to reliance on global standards and outsourced suppliers of manufacturing and initial product fulfillment functions for determining where and how to adopt service differentiation. Since any product can be viewed

M. A. Cohen (✉)
The Wharton School, University of Pennsylvania, Philadelphia, USA
e-mail: cohen@wharton.upenn.edu

J. A. Guajardo
Haas School of Business, University of California, Berkeley, Berkeley, USA
e-mail: jguajardo@berkeley.edu

H. Yang and R. Qiu (eds.), *Advances in Service Science*, Springer Proceedings in Business and Economics, https://doi.org/10.1007/978-3-030-04726-9_5

as a bundle of tangible goods and intangible services, differentiation specific to services must be considered. We note that while the relative contribution of "pure manufacturing vs. pure service" in a bundle, will vary, the emphasis on services has been increasing as firms adopt a customer focus to gain competitive advantage.

Quality thus has two dimensions, i.e. for the tangible product and for the bundled services. Differentiation therefore will be achieved with respect to customers' perception of the quality they derive from multiple product attributes, associated with both the tangible and service dimensions of a firm's products. Our focus in this paper will be on the attributes associated with the value-added services that are associated with customer support. The goal of this paper is to review the framework for developing a strategy based on service differentiation that was introduced by Guajardo and Cohen [6], which also provides a discussion of a case study illustrating implementation of the framework. We will focus on the role of analytical models that can be used to support implementation of a differentiation strategy and use the framework to note research opportunities and challenges for improving the state-of-the-art.

The framework uses the concept of operating segments, (introduced by Frei and Morriss [5]). This concept is based on the observation that companies with successful service strategies have concluded that "you cannot be good at everything". An operating segment is defined as a *list of service priorities shared by a meaningful group of customers*. Customers with similar service priorities are part of the same operating segment. Importantly, not only are the service priorities different across operating segments, but so are the operational capabilities needed to deliver them. Figure 1 illustrates the variation in the relative performance of a firm for its service product when compared to its competition, as a function of the various attributes embodied in the product, from the perspective of a given group of customers. Consider, for example, the case of retailers such as Walmart, Sears and Mom & Pop stores. The operating segments for each can be compared (see Fig. 2 from Frei and Morriss [5]). If the curve is increasing, then we can say that there is consistency between the relative performance and the importance of service attributes, as is the case for Walmart, i.e. they are good at what their customers care most about and less so for those attributes that their customers find to be less important. Sears and Mom & Pop are less consistent.

Our framework for service differentiation, (illustrated in Fig. 3), uses the mechanism of operating segments to consider the following questions: (a) who are members of the segments? (b) what service products are offered? (c) how are these products produced and delivered to the segments? Answers to these questions describe a firm's current approach to delivering differentiated services. Firms also must consider what the answers should be in order to maximize competitive performance. The framework moreover raises the strategic question of, to what extent should a firm adopt differentiation. The answer to this question could range from "not at all", i.e. one size fits all, to "mass customization", where every customer receives a unique product, designed and delivered to them, and thus each segment consists of a single customer. In general market segments are based on demographic, market and product factors.

Each product-service bundle that a firm offers can be defined by a vector of service attributes and, as noted, preferences for these attributes are shared among the

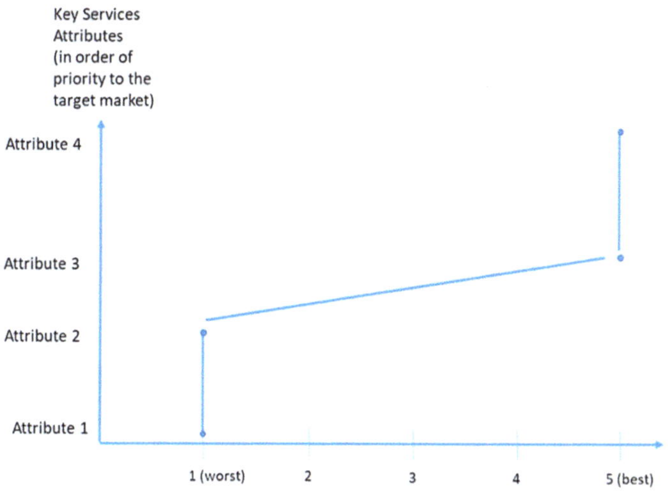

Fig. 1 The attribute map

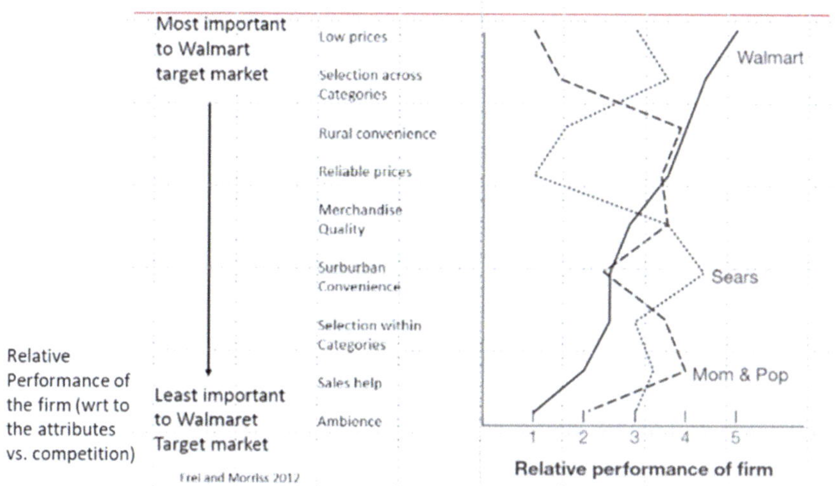

Fig. 2 Retailer attribute maps

members of the relevant customer segment. In particular, we can define the operating segment in terms of the target values for each service attribute metric as well as membership rules for belonging to the segment (based on customer characteristics). The operations challenge is to develop and implement effective policies for the design and control of the processes required to produce and deliver these products to these customers, leading to resource management decisions (capacity, allocations, and prioritization) and performance evaluation (in terms of strategic indicators).

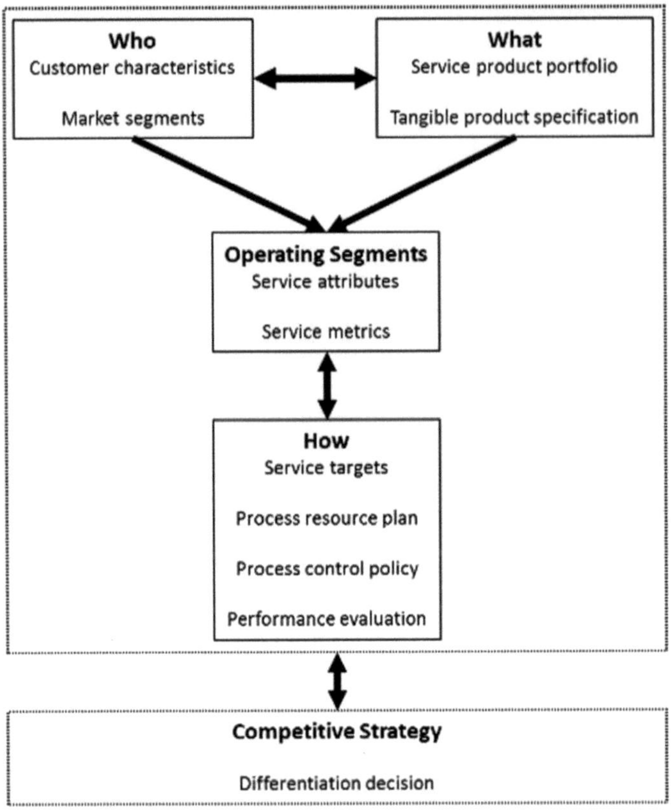

Fig. 3 Overview of the service differentiation framework

A key challenge for managing service differentiation is based on answering the questions introduced above in terms of tradeoffs associated with benefits, costs and risks in order to optimize competitive performance. Note that the questions noted above arise at all stages of a product's life cycle, (from design, through sourcing and production, sales, distribution, after-sales and end-of-life).

The next section introduces and describes the key elements of our framework and discusses methodologies for defining both market and operating segments. Section 3 explores issues associated with modeling and optimizing the operational processes to plan, control and evaluate differentiated service delivery. It also considers the strategic question concerning the decision to adopt differentiation. The final section discusses research and implementation challenges.

2 Defining Market and Operating Segments

The basic elements of a service differentiation strategy can be defined as follows:

Let $\boldsymbol{C} = (c_1, \ldots, c_m)$ be the set of relevant observable customer characteristics; (i.e. based on demographic characteristics such as gender, age, and income) and other customer characteristics such as price sensitivity. We note that other factors also can be used to group customers based on product, market and technology factors.

Let $\boldsymbol{P} = (p_1, \ldots, p_n)$ be the set of relevant product characteristics, based on service priorities. This will include attributes of the services to be delivered that matter to the customers and that are operationally relevant (e.g., response time, ability to resolve problems, technical knowledge). Note that these attributes entail differences in the operational aspects of service delivery (e.g., "fast service" is an attribute that clearly has operational implications such as larger capacity, better-trained service force). In general, there can be considerable heterogeneity in service preferences across market segments and as a result the operational requirements for (optimally) serving different these segments will vary.

Let $\boldsymbol{S} = (s_1, \ldots, s_o)$ be the set of service process policies used to produce and deliver a differentiated set of service products. Each policy, s_i, denotes a process defined by a hierarchy of decisions associated with the design and management of the service delivery process, (e.g., setting capacity levels for different classes of resources, and scheduling and control policies that govern the utilization of these resources).

Step 1: Identification of Meaningful Groups of Customers

The first step in defining segments is to identify the set of "meaningful" groups of customers, i.e. that accounts for commonality of attribute preferences across groups.

Let, $\boldsymbol{CL} = (CL_1, \ldots, CL_j)$ be a vector group identifiers, where each distinct group denotes a collection of characteristics, and membership in the group is based on that group's sensitivity to the service attributes in $\boldsymbol{P}$. Since members of each group share common priorities for different aspects of the service, we can define groups by a membership function that maps $\boldsymbol{C}$ and $\boldsymbol{P}$ into $\boldsymbol{CL}$. Determination of group membership can be accomplished either through application of business rules (e.g. based on demographics or market factors), or through the application of statistical methods, (e.g. cluster analysis).

Step 2: Definition of Operating Segments

We can define the set of rank ordered attributes $P_j = (p_{[1]}, \ldots, p_{[j]})$ characterizing group CL_j's service preferences. Thus (CL_j, P_j) defines an operating segment and $\{(CL_1, P_1), (CL_2, P_2), \ldots, (CL_j, P_j)\}$ defines the set of all potential operating segments. We therefore reduce this set to a subset of "meaningful" segments, if a given operational policy can deliver to more than one segment, i.e., multiple segments are served by a common process or policy. Thus, $\{(\mathbf{CL_k}, \mathbf{P_k})\}$ is the final collection of operating segments for $k = 1, 2, \ldots, k'$ where $k' \leq j$, which represents the firm's service product portfolio. The triple, $(\boldsymbol{CL_k}, \boldsymbol{P_k}, \boldsymbol{s_k})$ defines the k'th group of customers, their rank ordered service performance attributes, and the operational policy required to deliver the service product.

It is necessary to identify which attributes that can be associated with service that will be most relevant to the firm for the management of service differentiation. One approach is to explore the relationship between the performance of the firm, (e.g. likelihood to recommend the brand, customer satisfaction, market share, profit, etc.), customer characteristics and customer perception of service quality. The customer characteristics and quality attributes used to manage differentiation can be based on business rules, or on cluster analysis. Alternatively they can be based on regression analysis (see [6]).

3 Producing and Delivering Differentiated Service

We now consider the interactions between various decisions associated with our framework. These decisions form a hierarchy and are highly inter-dependent as illustrated in Fig. 4.

The definition of an operating segment requires selection of targets for service performance metrics for its associated market segment. These targets can be based on historical or competitive considerations or could be tied to customer preferences for different attributes of the service experience. This would require solicitation of inputs from customers; i.e. asking them to prioritize the value of various attribute metrics. Guajardo and Cohen [6] uses stepwise regression of customer survey responses.

The decision to set targets for selected attributes, should consider how different groups of customers will react to different levels of service performance in terms of their demand for the product-service bundle. Selection of target values, will of course, impact requirements for resources, as well as firm profit and competitive position.

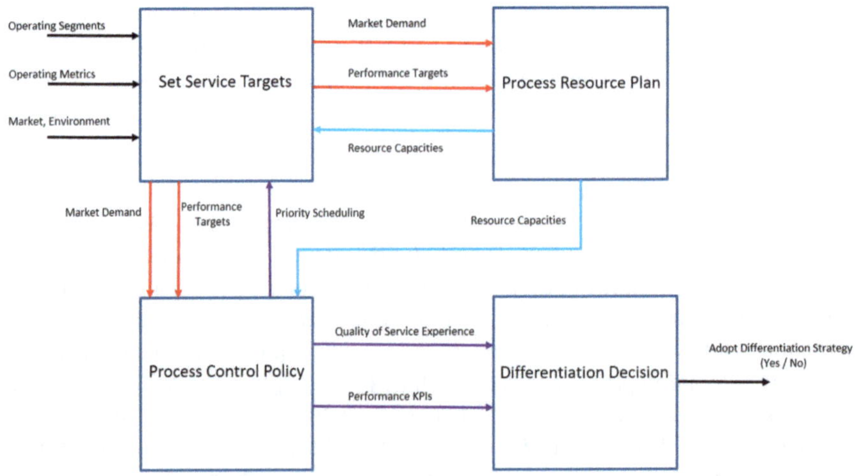

Fig. 4 Management hierarchy for service differentiation

Thus selection of targets could be based on service standards or can be derived from the solution to an optimization problem, which incorporates a relationship between demand (market share) and the relative level of service.

Step 3: Define Process Resource Plan

Management of the processes required to produce and deliver services requires a plan for deployment of resources which determine the capacity and capability of the service providers and a policy for control, i.e. how these resources are allocated to customers demanding service. The process resource plan typically involves setting service capacity (e.g. number of servers). In general customers can have different requirements for service and servers can have different capabilities for fulfilling customer needs. A resource plan thus seeks to match supply with demand. A solution to this problem must take into consideration the fact that some servers are flexible and can server multiple segments while others can only serve one segment. The process resource plan also requires that service capacity be scheduled to meet service targets for each segment over an appropriate planning horizon.

The process control policy provides mechanisms for matching arriving customers to a queue in front of a server capable of meeting that customer's requirements. Real time routing of customers through the service delivery process ultimately is required and a control policy for such routing could be based on priority rules or rationing conditional on the realized state of the system.

Analytical Models—Representative Examples

There is an extensive literature in operations that has looked at both resource planning and control for service processes. We illustrate how such models could be used to solve these problems by considering a number of examples. The examples illustrate specific requirements to support the delivery of differentiated service, i.e. (1) endogeneity of demand with respect to service quality, (2) competition based on different service quality attributes, (3) allocation of resources to multiple segments with different entitlements for service quality, (4) capacity and capability planning for multiple classes of service providers and (5) real time control and prioritization of service delivery.

Our first two examples model the case where there is competition on the basis of the quality of service. The first model, of Ho and Zheng [10] introduced a fixed capacity service queueing model where share in a service market is determined by the relative quality of service delivery. The objective is to maximize the demand rate, based on an equilibrium condition derived from total market demand, where market share is endogenously determined by consumer utility, which is derived from multiple dimensions of service performance. The solution is based on an equilibrium condition where "tomorrow's demand rate" equals today's demand rate and defines performance targets for each service attribute.

The second formulation by Cohen and Whang [3], develops a product life-cycle model where customers purchase a product from a manufacturer who provides after-sales service support in competition with an independent service provider. The manufacturer sets product price and both the manufacturer and the service provider set quality and the price of service. Customer utility here is influenced by both service price and service quality, (which is based on availability of the product, which in turn is determined by the resources and policies used to deliver support services). Both of these model formulations illustrate different ways to formulate a model that captures the impact of competition on the basis of the quality of service.

The next model illustrates customer prioritization in the context of service delivery. It deals with the situation where there are multiple classes of customers with different service entitlements, which clearly will occur when differentiation of service is being considered. Deshpande et al. [4] analyze a (Q, R) inventory stocking problem where there are multiple classes of customers associated with differentiated service targets, as measured by stock fill rate. Stock is issued to customers on a FIFO basis until on hand stock falls below a threshold and from then on stock is only issued to high priority customers while low priority demand is backlogged.

The final modeling situation we consider is captured in the papers by Gurvich et al. [7, 9], Gurvich and Whitt [8] and Mehrotra et al. [11]. This strand of the literature considers joint optimization of resource decisions (number of each type of server) and control decisions (dispatch rule for assigning customers to servers), where both the demand for service and the capabilities of the supply (service agents) are differentiated in terms of speed and capability, i.e. some servers have the capability to serve multiple customer groups (based on their training, experience, and/or incentives).

Step 4: Service Differentiation Decision
Many firms offer one level of after-sales service to their customers. When firms deviate from this strategy they typically deliver a differentiation strategy in the following ways: (1) Price discrimination (aka revenue management) which is based on the willingness of customers to pay for different levels of service quality, (2) Product based, i.e. customers who purchased an expensive product receive a higher level of support service, and (3) Deliver differentiated service based on customer and product attributes, e.g. high rollers in a casino who are given perks and discounts to incentivize them to gamble. It is interesting to note the consumer electronics firm we have worked with recently introduced "Concierge" service for those customers who purchased their most expensive high HD TV ($40 K).

Firms must consider the tradeoff between the benefits of differentiating (better match between supply and demand) and the cost of the increased complexity of the service process. Finally, we note that firms must consider customer perceptions of fairness (when some get a higher level of quality than others).

4 Research Challenges and Opportunities

This paper has described a framework for implementing a strategy based on the delivery of differentiated levels of customer service. Its use raises a number of managerial questions, including assessing the tradeoffs and risks associated with offering differentiated levels of service. Managers also need to determine where/when differentiated service should be offered and how this strategy should be implemented in terms of service product design and delivery.

We note that the framework suggests opportunities for conducting both analytical and empirical research. We observed, in particular, that modeling to support the framework introduced here requires consideration of endogeneity of demand with respect to service quality and competition based on the level delivered service quality based on specific performance attributes. We also noted that decisions for the allocation of resources to the multiple segments introduced by differentiation must consider the different entitlements of each customer segment for service quality. This gives rise to the need to plan the capacity and capability of multiple classes of service providers and to manage real time control and prioritization of service delivery. Needless to say inclusion of all of these factors in a single model is not feasible since the analytical challenges of dealing with the issues noted above will lead to significant modeling and solution algorithm challenges. Thus heuristics and simulation approaches will need to be considered.

While much of the data that is needed to implement our methodology is readily available from customer surveys and CRM software systems, there are considerable gaps in the state-of-the-art of models to optimize decisions in the overall hierarchy of decisions associated with differentiated service delivery. Competition based on relative performance for service attributes delivered to different market segments should also be considered. Another factor that should be considered is the implicit decision hierarchy, ranging from long term strategic and structural decisions to shorter term tactical decisions, all the way to real-time control. Most of the models of service delivery focus on one or two stages of the overall process. An additional factor that should be included is the behavioral response to services. Recent examples along these lines refer to consumer response to operational transparency in services [1] and more generally to the notion of customer compatibility [2]. As our case study illustrated, as reported in Guajardo and Cohen [6], response to service involves complex reactions to a wide range of factors.

There is considerable room to enhance the methodology available for jointly optimizing operations processes design and control throughout the hierarchy. As noted, the framework discussed here can be implemented through readily available data and estimation tools. We also note that the framework suggests a variety of hypotheses concerning the drivers of alternative strategies, which is fertile ground for empirical research. Finally, the underlying strategic question—to differentiate or not to differentiate, and if yes, to what extent has not been adequately addressed in the literature.

References

1. Buell R, Kim T, Tsay C. Creating reciprocal value through operational transparency. Manag Sci. 2017;63(6):1673–95.
2. Buell R, Campbell D, Frei F. The customer may not always be right: customer compatibility and service performance. Harvard Business School Working Paper No. 16-091. 2018.

3. Cohen MA, Whang S. Competing in product and service: a product life-cycle model. Manag Sci. 1997;43(4):535–45.
4. Deshpande V, Cohen MA, Donohue K. A threshold inventory rationing policy for service differentiated demand classes. Manag Sci. 2003;49(6):683–703.
5. Frei F, Morriss A. Uncommon service: how to win by putting customers at the core of your business. Cambridge: Harvard Business Review Press; 2012.
6. Guajardo JA, Cohen M. Service differentiation and operating segments: framework and an application to after-sales services. Manuf Serv Oper Manag. 2018;30(3):440–54.
7. Gurvich I, Armony M, Mandelbaum A. Service level differentiation in call centers with fully flexible servers. Manag Sci. 2008;54(2):279–94.
8. Gurvich I, Whitt W. Service-level differentiation in many-server service systems via queue-ratio routing. Oper Res. 2010;58(2):316–28.
9. Gurvich I, Luedtke J, Tezcan T. Staffing call centers with uncertain demand forecasts: a chance constrained optimization approach. Manag Sci. 2010;56(7):1093–115.
10. Ho TH, Zheng YS. Setting customer expectation in service delivery: an integrated marketing-operations perspective. Manag Sci. 2004;50(4):479–88.
11. Mehrotra V, Ross K, Ryder G, Zhou Y-P. Routing to manage resolution and waiting time in call centers with heterogeneous servers. Manuf Serv Oper Manag. 2012;14:66–81.
12. Wang Y, Cohen MA, Zheng Y-S. Differentiating customer service on the basis of delivery lead-times. IIE Trans. 2002;34(11):979–89.

Higher Education as a Service: The Science of Running a Lean Program in International Business

Joan Lofgren, Oleg V. Pavlov and Frank Hoy

Abstract This chapter contributes to understanding higher education as a service by applying the service science framework to an undergraduate business program in northern Europe. We utilize the Service Science Canvas, which is a new tool for service science analysis. This innovative academic program relies exclusively on visiting faculty from around the world to teach intensive three-week courses. While access to resources and governance structures were found to be similar to business programs elsewhere, several elements were found to be highly unusual, if not unique. First, the intensive curriculum structure promotes value co-creation among faculty and students. Second, access rights to faculty are negotiated annually, leading to agility but also risk. Third, stakeholder networks are broadly dispersed, but on balance serving as a rich resource for the program. Governance has evolved to ensure quality standards in a constantly changing academic community.

1 Introduction

Higher education is undergoing transformation on many levels, one of which is its integration into the marketplace where it is viewed as a public or private service. The variety of administrative arrangements of academic programs and institutions that exist around the world offer opportunities for academic planners. Yet the comparison of alternative arrangements is challenging due to too many choices, too many program attributes to compare. This paper is a step in a continued effort to impose structure on the comparative analysis of higher education.

Service science offers a useful framework, which casts higher education as an ecology of service systems that deliver value by generating and transferring knowl-

J. Lofgren (✉)
Aalto University School of Business, Lönnrotinkatu 5, 50100 Mikkeli, Finland
e-mail: joan.lofgren@aalto.fi

O. V. Pavlov (✉) · F. Hoy
Worcester Polytechnic Institute, 100 Institute Rd, Worcester, MA 01609, USA
e-mail: opavlov@wpi.edu

H. Yang and R. Qiu (eds.), *Advances in Service Science*, Springer Proceedings in Business and Economics, https://doi.org/10.1007/978-3-030-04726-9_6

edge [1–5]. University programs, like other service organizations, must focus on the quality of services provided in order to maintain competitive advantage and attract and retain new customers, i.e., students, plus the faculty, who are the producers of educational services [6]. Higher education has also undergone significant internationalization in recent decades [7–9]. Internationalization in higher education can be seen in the recruitment of faculty, students and staff, the export of education models as well as multinational operations.

We apply the service science framework to evaluate an international undergraduate business program in Finland in Aalto University. This lean program has been very successful, while it is organized quite differently than traditional business degree programs. We describe the program by using the Service Science Canvas, which is a new tool for service science analysis [5]. The Service Science Canvas organizes the 10 elements of the service science theory into a one-page matrix. The Service Science Canvas was inspired by the Business Model Canvas introduced by Osterwalder and Pigneur [10]. The Business Model Canvas has become popular as a planning and visualization tool in the entrepreneurship community due to its convenience and the simplicity of use (see www.strategyzer.com for examples). This paper is a preparatory step that will allow us to compare the Aalto program to more traditional business programs.

2 The Mikkeli Business Program

The Bachelor's Program in International Business located at the Mikkeli Campus of the Aalto University School of Business (formerly Helsinki School of Economics, HSE) is a unique program taught in English almost entirely by visiting faculty from around the world. Courses are run year-round in intensive three-week modules with students taking one course at a time. The student body comprises about 2/3 Finns and 1/3 foreign degree students, mainly from Asia. All students are expected to spend two years doing coursework in Mikkeli and a semester abroad in their third year, usually in the fall. Thus they are set to complete the European three-year Bachelor's degree in about two and half years. Previously tuition-free, the program now charges tuition of students coming from outside the European Union (EU) and European Economic Area (EEA), but it also offers scholarships. A small staff of about 10 persons runs a lean organization that welcomes over 60 faculty to teach 80 courses each year. The Mikkeli program is a prime example of a university program seen as service provision. The staff run a "well-oiled machine" designed to support faculty in their delivery of courses and thesis supervision.

Table 1 Elements of the service science framework

Resources: anything that can be used in service production: human, technological, financial or social
Access rights: control access to resources; may be leased, owned, shared, or privileged
Entities: resource configurations capable of value creation in a service system
Stakeholders: parties affected by or affecting service interactions, commonly a customer, provider (of resources), authority (ensuring compliance) or competitor
Value co-creation: occurs through collective efforts of stakeholders
Networks: patterns of interaction among service systems and between entities in service systems
Ecology: service systems and service entities constitute the service system ecology
Governance: formal and informal mechanisms directing service systems towards certain objectives
Outcomes: activities of service systems lead to outcomes, i.e. value for customers
Measures: stakeholders evaluate the performance of a service system against key benchmarks, such as quality, productivity, compliance and sustainable innovation

3 The Service Science Framework

Service science studies the design and management of service systems [2]. The ten elements of service systems are outlined in Table 1. The Service Science Canvas is a generic service science template developed by Pavlov and Hoy [5]. Its utility is similar to the Business Model Canvas introduced by Osterwalder and Pigneur [10] in the entrepreneurship field. The Service Science Canvas can be applied to any service system.

4 Service Science of the Mikkeli Program

In the discussion below, we apply each of the ten elements of the Service Science framework to the Mikkeli Program. They are also summarized in the Service Science Canvas in Fig. 1.

4.1 Resources

Successful academic programs rely on a supply of adequate resources [11, 12]. As an undergraduate service system, the Mikkeli program has stable budget funding from public and private sources. Faculty are hired for one or several courses at a time and must perform well in order to be invited back. The program has extreme flexibility in human resources and more leverage for maintaining teaching quality

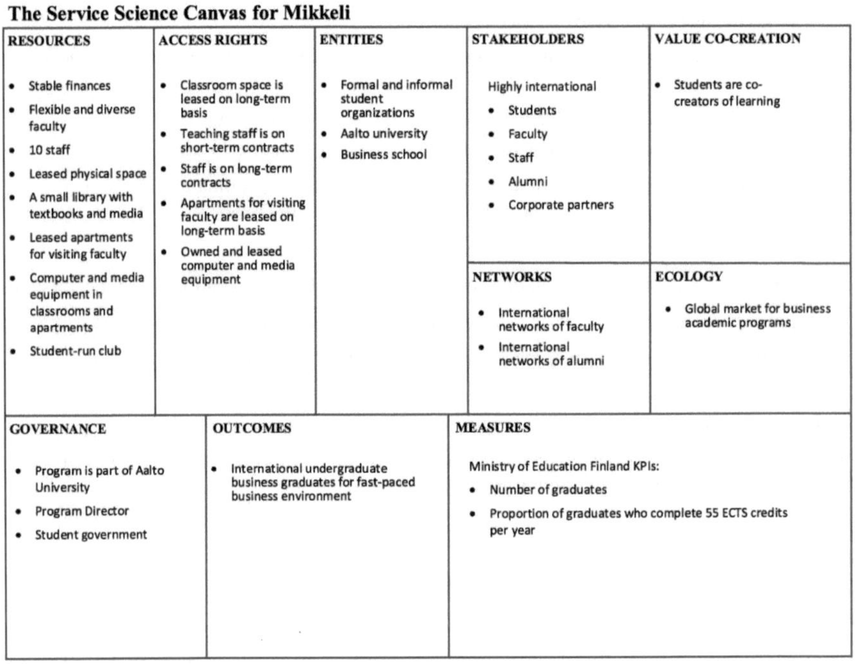

Fig. 1 The service science canvas for the Mikkeli program

than traditional academic departments. Not having permanent faculty (except for the Program Director, the only full-time academic) also involves risk, when instructors cancel last-minute due to illness or other emergencies.

This program leases on a long-term basis two floors in a building on the campus of the Mikkeli University Consortium in the city center. The program manages to be very lean by neither owning nor leasing recreational facilities, student housing or cafeteria facilities. Unlike in the US, there are no university-owned student dormitories. Students typically rent highly subsidized apartments from the nationwide organization that owns and runs student housing in Finland. Students, staff and faculty can choose among many relatively inexpensive gyms around the city.

4.2 Access Rights

Control over resources is determined by access rights. The Mikkeli program is unique in organizing its curriculum entirely around short-term faculty, mainly from outside Finland. Access rights are established through a recruitment process carried out well in advance of the courses delivered. Instructors are offered a short-term contract with Aalto University as private individuals, but access to their time often involves stakeholder consent at their home universities, which can be difficult to obtain in times of financial crisis.

4.3 *Entities*

Academic entities include academic institutions, programs, departments, centers and schools. The Mikkeli program is a fairly autonomous undergraduate program because it is run on a satellite campus. However, it is fully integrated into the undergraduate education of the Aalto School of Business, and several quality processes ensure that it meets university, school, and international accreditation standards.

The student organization Probba is a key entity in the Aalto Mikkeli service system. It is part of the formal student organization of the Aalto School of Business, known as KY. It represents interests of students to the university. Elected Probba representatives give input on a range of issues, from teaching quality to upgrading the learning environment.

4.4 *Stakeholders*

The Mikkeli program has four types of stakeholders: customers, providers, authority, and competitors. First, demand for educational services comes from students, and therefore students are primary customers. The customer base for the Mikkeli program is very broad, with over 700 applications for 80 study places received each year from about 70 countries around the world. Second, faculty, staff, students, alumni and other stakeholders are all providers of educational services, since they are co-creators of learning. Third, academic programs are subject to bodies exercising authority such as ministries of education, boards of trustees, academic affairs committees, etc. that enforce standards of higher education. The Mikkeli program is well integrated into the School of Business and Aalto University, which includes following standards set for accreditation (The School has so-called triple crown accreditation). The Program Director implements various quality initiatives and checks as the primary manager of the visiting faculty. Fourth, the Mikkeli program competes for high quality faculty and students with a range of other programs in Finland and abroad. Its unique model remains a competitive advantage, offsetting its relatively remote location in northeast Europe.

4.5 *Value Co-creation*

Value co-creation dictates that academic programs must provide value for all stakeholders, and all stakeholders participate in the value creation process. The value co-creation of the Mikkeli program includes preparing students for working in international business or further study by providing knowledge and skill development in the field of international business.

4.6 Networks

The Mikkeli program relies on a network of over 100 visiting faculty, many of whom return to Mikkeli year after year. A strong network of alumni also contribute to the program, for example, serving as guest speakers in courses. The international network of visiting faculty is based on relationships developed over years of negotiating access rights. While the flexibility to not invite an instructor back is always in the background, there is a core of veteran faculty teaching the same course each year and they are a source of expertise for development projects.

4.7 Ecology

Undergraduate programs are integral parts of a broader educational ecology, the global landscape of educational services and their stakeholders. The Mikkeli program is part of an ecology of international business programs competing for faculty and students globally.

4.8 Governance

The Governance element shows organizational structures that are used by university programs. Quality processes in the School of Business include data on faculty qualifications, assurance of learning and feedback loops. The program and school are part of a well-established higher education governance system in Finland that some years ago transitioned from completely state-funded to including some hybrid, or foundation models of public-private funding (Aalto is one of them).

4.9 Outcomes

The main outcome of an academic program is the production of qualified graduates. The fast pace of the Mikkeli program shapes graduates who fare well in a rapidly changing business environment.

4.10 *Measures*

Key performance indicators (KPIs) are now commonplace in university programs and Aalto Mikkeli is no exception. Ministry of Education KPIs currently include the number/proportion of students completing over 55 ECTS credits per year as well as degrees granted.

5 Conclusion

Our service science analysis highlights some common issues of educational service systems. For example, flexibility in talent management is important in responding to the needs of the market, and effective governance is key to ensuring quality. Moreover, value co-creation increasingly involves customers. The analysis also indicates the strengths of the Mikkeli program such as its flexible, lean structure and international network. Its weaknesses arise from the risks of negotiating teaching resources on an annual basis. Future work may include an examination of the cultural dimensions and internationality of the program to add to the Service Science Canvas for Mikkeli. More generally, the follow up work will compare the Aalto program to more traditional business programs within the framework imposed by the Service Science Canvas.

References

1. Maglio PP, et al. Service systems, service scientists, SSME, and innovation. Commun ACM. 2006;49(7):81–5.
2. Spohrer J, et al. Steps toward a science of service systems. Computer. 2007;40(1):71–7.
3. Lella, G., et al. Universities as complex service systems: External and Internal perspectives. In: IEEE international conference on service operations and logistics, and informatics (SOLI). 2012. Suzhou, China.
4. Spohrer J, et al. Service science: reframing progress with universities. Syst Res Behav Sci. 2013;30(5):561–9.
5. Pavlov O, Hoy F, Toward the service science of education. In: Maglio PP, et al., editors. Handbook of service science, vol. 2. Springer: New York; 2018.
6. Ali F, et al. Does higher education service quality effect student satisfaction, image and loyalty? A study of international students in Malaysian public universities. Qual Assur Educ. 2016;24(1):70–94.
7. Knight J, Altbach P. The internationalization of higher education: motivations and realities. J Stud Int Educ. 2007;11:290–307.
8. Knight J. Internationalisation: key concepts and elements. In: Gaebel M, et al., editors. Internationalisation of European higher education. An EUA/ACA handbook. Raabe: Stuttgart; 2009.
9. Lofgren J, Leigh E. The Mikkeli programme: international education and flagship response to globalisation. In: Nygaard C, Branch J, editors. Globalisation of higher education, the learning in higher education series, Institute for Learning in Higher Education. Libri: Faringdon; 2017.

10. Osterwalder A, Pigneur Y. Business model generation: a handbook for visionaries, game changers, and challengers. New York: Wiley; 2010.
11. Massy W. Reengineering the university: how to be mission centered, market smart, and margin conscious. Baltimore, MD: Johns Hopkins University Press; 2016.
12. Zaini R, et al. Let's talk change in a university: a simple model for addressing a complex agenda. Syst Res Behav Sci. 2017;34(3):250–66.

A Hypergraph-Based Modeling Approach for Service Systems

Mahei Manhai Li, Christoph Peters and Jan Marco Leimeister

Abstract Currently, research on service science has emerged as its own discipline, where service systems are its basic unit of analysis. However, without a clearly defined modeling approach for service systems, analyzing a service system is challenging. We therefore propose a conceptual hypergraph-based modeling approach, which can be used to model services for both traditional goods-dominant businesses, as well as service-businesses. We define key elements of a service system, while drawing upon hypergraph theory and present three modeling properties which are required to model a service systems graph (SSG). The focus of SSGs is to describe the relationships between the various resources, actors and activities, thus configuring a service system. It provides the foundation for computer graphic simulations and database applications of service business structure for future research.

Keywords Service systems · Service system graphs · Modeling
Service modeling · Service system modeling · Service systems engineering
Hypergraph

1 Introduction

Throughout the emergence of Service Science, service systems have always been a key concept of the discipline [27]. Since the seminal paper on service dominant logic [30] both service-centric businesses and traditionally goods-dominant businesses have begun to apply a service perspective on their organization in order to remain competitive and innovate [20]. Especially research on the transition to services has gained traction by terms, such as servitization [16].

M. M. Li (✉) · C. Peters · J. M. Leimeister
Information Systems, Research Centre for IS Design (ITeG),
University of Kassel, 34121 Kassel, Germany
e-mail: mahei.li@uni-kassel.de

C. Peters · J. M. Leimeister
Institute for Information Management, University of St.Gallen, 9000 St. Gallen, Switzerland

H. Yang and R. Qiu (eds.), *Advances in Service Science*, Springer Proceedings
in Business and Economics, https://doi.org/10.1007/978-3-030-04726-9_7

Service as a way of thinking has gradually evolved and is used by both manufacturing and service-businesses, since production can be also be seen as internal services for providing an end-customer a value proposition [15]. Thinking in service systems can help identify service innovation potentials [5, 8]. However, we model service systems using a multitude of modeling approaches focusing on actors and a processual perspective [6, 19, 26, 28] or more technical perspectives. Yet, it can be challenging for practitioners to utilize the concept of service systems from a business perspective [1]. Since service scientists "study, manage, and engineer service systems, solving problems and exploiting opportunities to create service innovations" [24], our research goal is to provide a tool to model and analyze service systems. Our research question is therefore as follows: How can we model basic service systems both correctly and graphically?

2 A Service Systems Perspective

The service system's inherent focus lies in finding the right configurations of resources for actors in order to create value in the right context (value-in-context, formerly referred to as value-in-use) through the use of services to customers [7, 16, 30]. Vargo and Lusch [30] have addressed the configuration in applying their concept of service dominant logic and called it resourcing. A service system is guided by a value proposition, which in turn has a corresponding configuration of actors and resources [7, 30].

We define a service system as a value co-creation configuration of resources [22]. This perspective is rooted in service dominant logic [23]. Resources include both operand and operant resources [21]. Different configurations of resources are connected by respective value propositions, sometimes also seen as service exchanges [29].

Recent research revisits the importance of value propositions and engagement of service systems [11], in which organizations seek to find the right constellation of actors ("who") within a service system that enables actors to find the correct resources ("who" and "with whom") for a specific context ("when") in order to co-create value [11, p. 1], whereas the creation of value ("value-in-use") happens through activities between actors, also referred as interactions [30]. This coincides with an input-output perspective, in which the realization of value happens through a transformation process of resources by actors [11].

The actors are essential to realize the initially proposed value. They act upon the resource configurations to achieve the value proposition. Since a service system includes different types of resources and actors, who create value to a customer, we define the term "service objects" that pairs corresponding resources and actors, for a value proposition. Realizing the value proposition for the customer is imperative. From a service-provider perspective, it is decisive to know the constellation of resources that actors require. An actor can be individuals, teams, organizations or even software systems, if they mobilize the required resources. The service system therefore needs to be orchestrated to bring all resources and actors together. Hence, our service system graph is focused on service orchestrators as key stakeholder.

In conclusion, the constituent elements of a service systems are: resources, actors, service objects and activities. These elements should be configured to realize value. These elements serve as form of lightweight ontology for our service system modelling approach. Key contribution of this paper is the definition of their relationships using hypergraph theory.

3 Developing Service Systems Graphs (SSG)

First, we define the key concepts of our service system using hypergraph theory, which has its origins in graph theory and generalizes upon the concept of graphs [3]. A hypergraph $G = (V, E)$ exists as a pair of edges E and set of vertices V, where the edges $e \in E$ does not only connect two, but any number of vertices $v \in V$, thus calling E a set of hyperedges. A hyperedge $e \in E$ is therefore a subset of all vertices V, which are connected by it, $e \subseteq V$. Additionally, E is a subset of $P(V)\backslash\emptyset$, where P(V) is the power set of V.

Since service systems require resources as input factors, we define a set R with $r \in R$ as all required forms of resources. Service systems also require actors [9]. We define actors as a set A with $a \in A$ representing an actor. Since we define A as the set of required actors, it would be better to consider A as a team or organizational unit that is required for providing the service. An actor a can thus be an individual, a group, an organizational unit or even software systems.

A service system for a specific value proposition requires both actors and related resources. We called a pair of actor and required resources, "service objects" and define all service objects as a set O with $o \in O$ being a single service object. Formalized, a service object is a tuple of the required resources and the required actors specific to a value proposition. Hence, service objects are the subject matters of service systems, which are defined in a specific context as input sets of respective outputs. Let $O \neq \emptyset$ be the set of required service objects of any service-driven organization, with $o \in O$ defined as a service object. Thus, a service object is a tuple consisting of resources and actors. Formalized, service objects are defined as follows:

Definition 1 A finite non-empty set O with tuple of (R, A) is called service object where

i. R is a finite set of resources with $R = \{r_1, r_2, \ldots, r_n\}$;
ii. A is a family of subset actors of R with $A = (a_i)$ in which $a_i \subset R$ and $R = \bigcup_{i=1}^{n} a_i$ for $i \in \{1, 2, \ldots, n\}$.

Definition 1 shows that service object O is essentially a hypergraph [3]. Therefore, the service object O, the tuple (R, A) is a hypergraph of service objects, which inherently represent all possible value propositions of said service system. In other words, the potential of a service system can be unlocked by reconfiguring its resources and assigning it a suitable actor.

Hypergraph theory has extensively focused on its sets of vertices [10], whereas we put equal importance to its hyperedges. Due to the roles of actors in service science, we inscribe the semantic meaning of actors into hyperedges. A service object includes both actors and resources, both paramount for the realization of the service.

The service object is the static part of a service system. It constitutes the necessary input resources R, which actors A require, before an actor can provide value to a service consumer. In other words, it represents the potential value an actor can provide to a potential consumer.

The vertices (sometimes known as nodes) of a hypergraph G_i represent resources and hyperedges of G_i represent actors, whereas we define required actors and resources as service objects. Following hypergraph theory, hyperedges can intersect with each other, illustrating shared resources.

The service object can be the end result, as well as intermediate results of any service, each representing value propositions in terms of service exchanges. If the above mentioned elementary object is put together with other service objects, another service system can be configured. This is an analogous characteristic to "traditional" manufacturing cases (e.g., [15]), in which outputs are used as inputs for other processes, thus creating a path. We will revisit the path characteristic shortly, when introducing service activities.

An element graph is a graph of order = 1, that is, $|G_i| = 1$ for $i \in \{1, 2, \ldots, n\}$ [3]. It represents a service object $o_0 \in O$ with tuple (a_0, R_0) where $|R_0| = 1$. It is apparent that the elementary graph itself has edges. We changed the representation from a solid dot by adding a circle around it to indicate that it also has an hyperedge and hence constitutes a service object, as depicted in Fig. 1. We argue that single resources can always be considered as element graphs. However, we recommend only drawing the hyperedge if it is either an explicit output of a service object or if the element graph itself is a single input.

Fig. 1 Hypergraph G

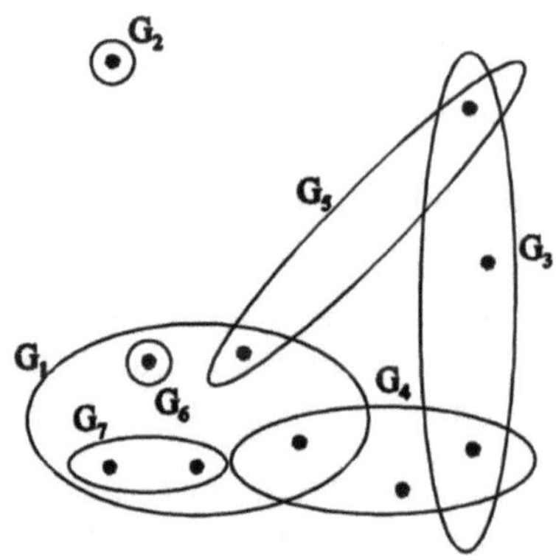

Although we have mapped service objects, which consist of actors and resources, we have yet to map a sequence of activities into our modeling approach. Up until now, a hypergraph can be used to model non-directional set of elements that contains the information of relationships among elements with the help of hyperedges. To map the relationship of elements of different hypergraphs, directional hypergraphs can be used to map the relationship of elements towards other elements of different graphs [13]. However, it is not possible to map entire hypergraphs towards other hypergraphs or toward elements of other hypergraphs. We need to expand upon the existing definitions of hypergraphs. We do so by introducing an approach to map a service object to other service objects with ψ. In the following section we will define how to map hypergraphs to other hypergraphs. This enables us to model service systems with hypergraphs.

Definition 2 O is a finite non-empty set of service object and O is a hypergraph of service objects. A mapping $\psi(\psi^+, \psi^-)$

$$\text{with} \quad \psi : \mathrm{O} : \mathrm{O} \times \mathrm{O} \rightarrow \text{Boolean} \quad \text{where } \mathrm{O} \times \mathrm{O} \subset 2^{\mathrm{O}}$$

is called a service activity of service objects.

Service activities for service objects are represented by the binary mapping between different service objects. One service object is seen as input, whereas the other is seen as output, while the value realization is enabled by an activity that makes the transition from one service object to another possible. The mapping ψ is a tuple of (ψ^+, ψ^-), which is a directed or counter-directed mapping of hypergraphs. In this paper, ψ and ψ^+ are used synonymously for directed mapping (Fig. 2), accompanied by the drawing of an arrow line. This is not to be confused with directed hypergraphs, which only allows relationships between elements of different hyperedges [13].

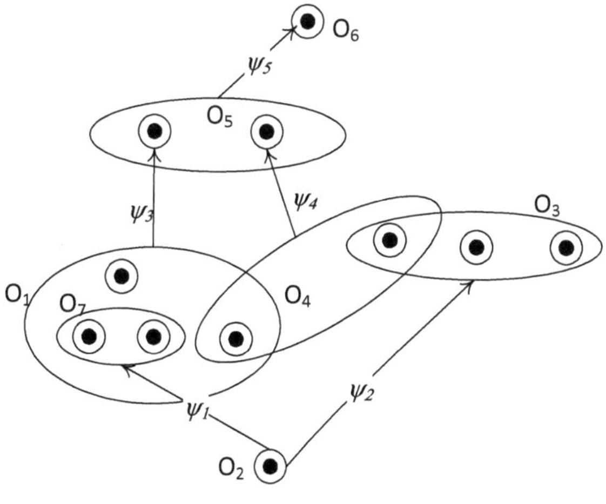

Fig. 2 Directed mapping of hypergraphs

Fig. 3 Example service system graph with $R_i \Leftrightarrow O_i$

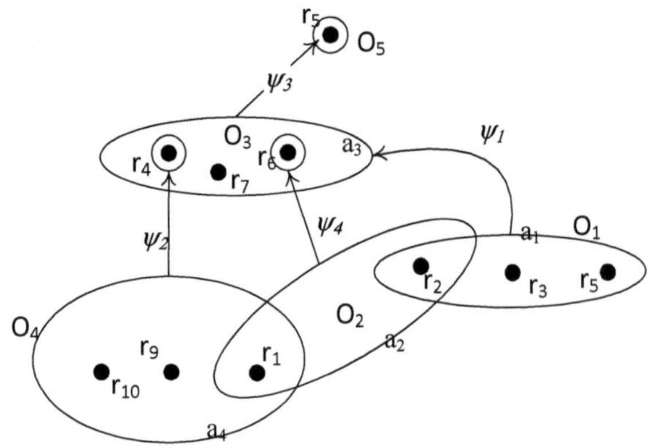

Definition 3 Let R be a finite nonempty set of resources, A a finite nonempty set of actors and O a set defined as tuple (R, A) be a hypergraph of a service object and Ψ be a set of functions as service activities. Then the tuple (R, A, Ψ) is called the SSG or service system graph,

$$\text{where} \quad \Psi : \Psi(O) \rightarrow \Psi^{+}(O) \quad \text{with } \exists\, o \in O \mid \Psi^{-}(o) \cap \Psi^{+}(o) = \emptyset.$$

Function $\Psi(O)$ defines which service objects are required as input factors and function $\Psi^{+}(O)$ defines the output service objects.

A service system graph is a directed graph, which models the value creation and value propositions of a chain of services. The service system is a family of subset service objects. Thus, strictly speaking, a single service object can include a configuration of service object and corresponding activities. This means that service systems can consist of service systems.

To illustrate the relationships of a service system, we present the detailed example of a SSG (R, A, Ψ): Fig. 3 shows the SSG with a set of resource $R = \{r_1, r_2, \ldots, r_{10}\}$, a family of the subset $A = (a_1, a_2, a_3, a_4, a_5)$; $a_1 = R_1 = \{r_2, r_3, r_5\}$; $a_2 = R_2 = \{r_1, r_2\}$; $a_3 = R_3 = \{r_4, r_6, r_7\}$; $a_4 = R_4 = \{r_1, r_9, r_{10}\}$; $a_5 = R_5 = \{r_5\}$; and the function $\Psi = (\psi_1, \psi_2, \psi_3, \psi_4)$ where $\psi_1 = ((a_1, \{r_2, r_3, r_5\}), (a_5, \{r_4, r_6, r_7\}))$; $\psi_2 = ((a_4, \{r_1, r_9, r_{10}\}), (a_0, \{r_4\}))$; $\psi_3 = ((a_5, \{r_4, r_6, r_7\}), (a_0, \{r_8\}))$; $\psi_4 = ((a_3, \{r_4, r_6, r_7\}), (a_0, \{r_5\}))$.

4 Properties of Modeling Service Systems

Compared to the definition of hypergraphs, SSGs allow the existence of a predicate between two hypergraphs, which is represented by ψ_t. In order to model, we will present a selection of three modeling properties in the following section:

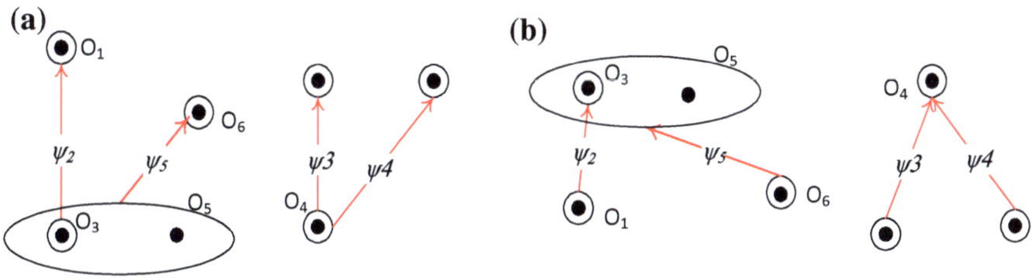

Fig. 4 **a** Multiple required service object and **b** multi-delivered service object

Multi-required (MR) service object: In an application environment, one service object is required by multiple activities, that is, it can be the input of more than one activities. According to the definition of SSG ⟨R, A, Ψ⟩ and $\Psi(\Psi^-, \Psi^+)$ we said a service object is a multi-required service object if $\bigcap_{i=1}^{n} \psi_i(o) \neq \emptyset$ where $\psi \in \Psi^-$ and $\exists$ o $\in$ O, n ≥ 2. In Fig. 4a the service object O_3, O_4 are multi-required service object for activities ψ_2, ψ_5 and ψ_3, ψ_4.

Multi-delivered (MD) service object: One service object can be delivered by more than one activity. It is similar to or operators. According to the definition of SSG(R, A, Ψ) and tuple $\Psi(\Psi^-, \Psi^+)$ a service object is called a multi-delivered service object when $\exists \bigcap_{i=1}^{n} \psi_i(o) \neq \emptyset$, where $\psi \in \Psi^+$ and $\exists$ o $\in$ O, n ≥ 2. In Fig. 4b the service object O_3, O_4 are multi-required service object for activities ψ_2, ψ_5 and ψ_3, ψ_4.

Sequence of service process: To model the service system, we still require to define the sequence of activities: Based on service system graph SSG(R, A, Ψ) and the service object O, subset of activities $\Psi_{after} \subset \Psi$ and $\Psi_{before} \subset \Psi$ with

$$\Psi_{before} = \left\{ \bigcup_{i=1}^{n} \psi_i \mid \bigcap_{i=1}^{n} \psi_i(o) \neq 0 \text{ where } \psi \in \Psi^+ \text{ and } \exists\, o \in O \right\} \text{ and}$$

$$\Psi_{after} = \left\{ \bigcup_{i=1}^{n} \psi_i \mid \bigcap_{i=1}^{n} \psi_i(o) \neq 0 \text{ where } \psi \in \Psi^- \text{ and } \exists\, o \in O \right\},$$

then $a \in \Psi_{after}$ follows each $b \in \Psi_{before}$.

We employ service system graph as a modeling approach to both formalize the relationships of configurations and visualize them using the inherent graphical notation. The next section discusses possible applications and areas of future research.

5 Application Scenario

Our modelling approach SSG has several application scenarios. The most evident one lies in its role as a tool to analyze both the organization's status quo and to structure possible alternative service system configurations. This chapter includes a detailed modeling example to show how this tool can be utilized. We focus on presenting both the graphical representation and the formal realization of a real-world service system scenario. The graphical illustration helps service systems engineers and business decision makers to structure their current business using a service systems perspective. It can also help authors make different system configurations of the same service apparent, thus giving decision makers the option to choose "paths" to reach their desired goal.

To illustrate our SSG, we modeled a possible service system based on our research project, an implementation of a CRM system at a mid-size German company "PowerCorp". PowerCorp faces the challenge of implementing a complex CRM system, for which they have tasked a team of IT consultants, service support providers, experts from the software provider and researchers. For the success of the IT-enabled organizational change project [25], four core services have been commonly understood as crucial: First, the technical task of installing and configuring the CRM system based on the PowerCorp's existing IT-infrastructure. Second, the users require sufficient training using workshops or online courses that are specifically tailored to the needs of both the user's and the system's technical configurations. Third, the implementation of the CRM system requires extensive organizational analyses, which are usually provided by IT-consultants (including system tests). Fourth, the organization requires extensive after sales service support in case new requirements or questions arise. Part of the support requirements is also realized by a crowd support approach, which utilizes the potential of peer-advice hidden among PowerCorp's business units [18].

Figure 5 shows a service system with its service objects and activities. O_1–O_4 are the key service objects that are required for a successes project O_{14}. They cover the above-described core services. Each service object consists of resources and an actor. See below for a complete and detailed list of all elements.

For a successful CRM system implementation, an organization also requires involvement from business units not just an external project team consisting of consultants. By relying on key users, valuable contextual domain-knowledge can

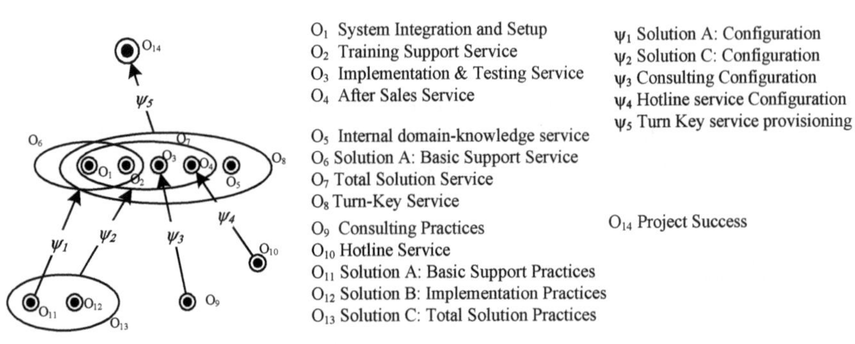

Fig. 5 Service system graph of the CRM implementation project

be integrated to overcome potential organizational pitfalls. We therefore define the service object O_5 as an internal domain-knowledge service.

All five service objects are necessary, while using a SSG helps in clearly denoting where they come from. We view O_1 and O_2 as basic support services, whereas a total solution would include all implementation and testing services, as well as ongoing after sales services (O_7). If these services are guaranteed and the necessary adaptations were made by incorporating rich domain-specific knowledge provided by O_5, the reins can finally be handed over in terms of a turn-key service (O_8) to the board of directors. The provisioning of the turn-key service (O_8) to project success is represented by ψ_5.

To understand how we reach O_7, we see them as MD service objects. This helps us realize that there are two possibilities in providing a total solution to PowerCorp: (A) By configuring total solution practices (O_{13}) to the contextual conditions ψ_2. (B) Configuring the basic support practices O_{11} accordingly ψ_1 and configuring both the hotline services O_9, ψ_3 and consulting portfolio O_{10}, ψ_4. It lies with the decision maker (e.g. service systems engineer) to choose which path to order to reach project success. As Fig. 6 shows, the different service system configurations are very similar to "paths", with two alternative paths highlighted in the graphic.

The resulting service system is linked to a detailed list of resources and actors. This information is important for implementing a SSG as a software system. Relying on our formal definitions, we use can derive machine-readable data formats to integrate other systems. The following paragraphs give a detailed information on the service systems structure and thus according to the application scenario, SSG(R, A, Ψ) is described as follows:

R = {r_1: CRM Software, r_2: Hardware, r_3: Crowd Support System, r_4: Netware; r_5: Customer Specialist, r_6: Project Specialist; r_7: Software Trainer, r_8: Hardware Engineer; r_9: Software Developer, r_{10}: System Analyst, r_{11}: Project Leader; r_{12}: Telephone Service Support, r_{13}: Server with OS; r_{14}: Documents, r_{15}: Tester; r_{16}: System Plat-

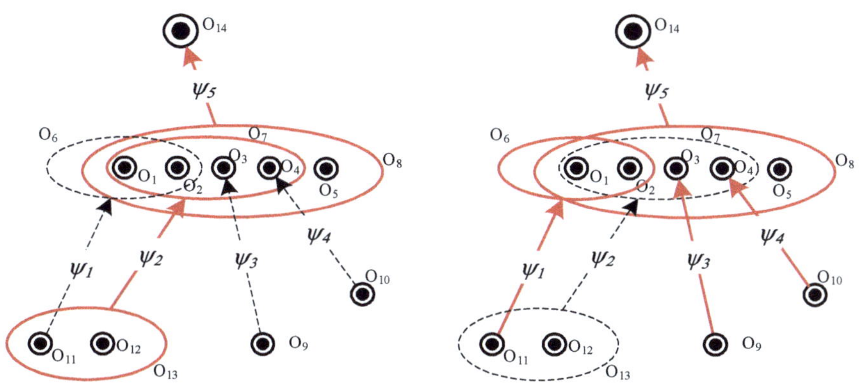

Fig. 6 Service activities with optional path

form r_{17}: Key Users; r_{18}: CRM Platform, r_{19}: Crosse Support/Online Platform, r_{20}: Running Application Hotline}.

A = (a_1 = {r_1, r_2, r_3, r_4, r_7}: Package A definition, a_3 = {r_6, r_8, r_9, r_{10}, r_{11}, r_{12}, r_{13}, r_{14}, r_{15}}: Implementation Team, a_4 = {r_6, r_8, r_9, r_{10}, r_{11}, r_{12}, r_{13}, r_{14}, r_{15}, O_{11}}: Total Solution Concept with Team, a_5 = {O_1, O_2}: Solution A Team with Training, a_6 = {O_1, O_2, O_3, O_4}: Project Solution Concept with Team, a_7 = {O_7, O_5}: Service Contract and Project Team, a_8 = {r_6, r_9, r_{10}, r_{11}, r_{14}, r_{15}}: Business Analyze and Implementation Team, a_9 = {r_{12}, r_{13}}: Project Team, a_{10} = {r_5} Customer Project Team).

Ψ = {ψ_1 Solution A Configuration, ψ_2 Solution C Configuration, ψ_3 Consulting Configuration, ψ_4 Hotline service Configuration, ψ_5 Turn Key service provisioning} with $\psi_1(O_{11}) = O_6$; $\psi_2(O_{13}) = O_7$; $\psi_3(O_9) = O_3$; $\psi_4(O_{10}) = O_4$; $\psi_5(O_8) = O_{14}$}.

O = {$O_1(a_0, \{r_{16}\})$, $O_2(a_0, \{r_{17}\})$, $O_3(a_0, \{r_{18}\})$, $O_4(a_0, \{r_{19}\})$, $O_5(a_{10}, \{r_5\})$, $O_6(a_5, r_{16}, r_{17})$, $O_7(a_6, \{r_{16}, r_{17}, r_{18}, r_{19}\})$, $O_8(a_7, \{r_{16}, r_{17}, r_{18}, r_{19}, r_5\})$, $O_9(a_8, \{r_6, r_9, r_{10}, r_{11}, r_{14}, r_{15}\})$, $O_{10}(a_9, \{r_{12}, r_{13}\})$, $O_{11}(a_1, \{r_1, r_2, r_3, r_4, r_7\})$, $O_{12}(a_3, \{r_6, r_8, r_9, r_{10}, r_{11}, r_{12}, r_{13}, r_{14}, r_{15}\})$, $O_{13}(a_4, \{r_1, r_2, r_3, r_4, r_6, r_7, r_8, r_9, r_{10}, r_{11}, r_{12}, r_{13}, r_{14}, r_{15}\})$, $O_{14}(a_0, r_{20})$}.

6 Discussion and Future Work

In conclusion, a SSG can be utilized to model complex service systems. Consider Fig. 3, where O_5 is the service object for the end-customer a5. We see the equivalents of "OR" and "AND" operators. To create service object O3, one can either choose (ψ_2 AND ψ_4) or one can choose activity ψ_1. For the first, both actor a_2 and a_4 are required, whereas a_2 utilizes shared resources from two different actors. As an alternative path to O_3, ψ_1 is also an option. This enables us to model different configurations of one service system as sub-systems, leveraging the systems of a systems principle [4].

This enables service engineers to model their service systems both from a process perspective and from a structural perspective. Theoretically, we employed the input-output perspective for our modelling approach thus strengthening a service approach that does not differentiate between traditionally goose-dominant logic [16].

Furthermore, the formal description can be transferred into a corresponding database. This would enable an integration of our service system model with existing Enterprise Systems and applications. Similarly, the SSG approach would also greatly benefit from a set of computer-aided tools to model a service system based on our concept. Such a tool would greatly benefit from an interface to other databases.

Future SSG research should also consider focusing on manufacturing [13] and operations applications. SSG enables us to cross the divide of process and structural models, with the latter often including bill of materials [14]. Since both approaches are based on simple graphs, a hypergraph based SSG enables the combination of both, whereas future research could focus on projections from SSG, which could be used to map the relation of traditional process or structural graphs to a SSG [2].

To sum up, the paper presents the foundation of a hypergraph-based service system graph, which can be used to model service systems both formally, as well as graphically. Our concept grounds the concepts of a service systems using hypergraph theory and helps to demarcate the distinction between service systems and service ecosystems [12]. Finally, future research could also include applying SSG into real-world scenarios and the limitation of our service system conceptualization through additional specifications.

References

1. Alter S. Service system axioms that accept positive and negative outcomes and impacts of service systems. Int Conf Inf Syst. 2017;38:1–21.
2. Baget J-F. Simple conceptual graphs revisited: hypergraphs and conjunctive types for efficient projection algorithms. Int Conf on Concept Struct. 2003;2746:229–42.
3. Berge C (ed). Hypergraph: combinatorics of finite sets. Elsevier; 1989.
4. von Bertalanffy L. The history and status of general systems theory. Acad Manag J. 1972;15(4):407–26.
5. Beverungen D, Lüttenberg H, Wolf V. Recombinant service systems engineering. Bus Inf Syst Eng. 2018;21(1):50.
6. Bitner MJ, Ostrom AL, Morgan FN. Service blueprinting: a practical technique for service innovation. Calif Manag Rev. 2007.
7. Böhmann T, Leimeister JM, Möslein K. Service systems engineering. Bus Inf Syst Eng. 2014;6(2):73–9.
8. Breidbach CF, Maglio PP. A service science perspective on the role of ICT in service innovation. In: European conference on information systems (ECIS), AIS Electronic Library; 2015. http://aisel.aisnet.org/ecis2015_rip/33.
9. Breidbach CF, Maglio PP. Technology-enabled value co-creation: an empirical analysis of actors, resources, and practices. Ind Mark Manag. 2016;56:73–85.
10. Bretto A. Hypergraph theory. Heidelberg: Springer International Publishing; 2013.
11. Chandler JD, Lusch RF. Service systems: a broadened framework and research agenda on value propositions, engagement, and service experience. J Serv Res. 2015;18(1):6–22.
12. Frost R, Lyons K. Service systems analysis methods and components: a systematic literature review. Serv Sci. 2017;9(3):219–34.
13. Gallo G, Scutellà MG. Directed hypergraphs as a modelling paradigm. Decis Econ Finan. 1998;21(1–2):97–123.
14. Hegge HMH, Wortmann JC. Generic bill-of-material: a new product model. Int J Prod Econ. 1991;23(1–3):117–28.
15. Hill TP. On goods and services. Rev Income Wealth. 1977;23(4):315–38.
16. Lightfoot H, Baines T, Smart P. The servitization of manufacturing: a systematic literature review of interdependent trends. Int J Oper Prod Manag. 2013;33(11/12):1408–34.
17. Leimeister JM. Dienstleistungsengineering und -management. Berlin, Heidelberg: Springer; 2012.
18. Li MM, Peters C, Leimeister JM. Designing a peerbased support system to support shakedown. In: International conference on information systems (ICIS). South Korea: Seoul; 2017.
19. Lim C-H, Kim K-J. Information service blueprint: a service blueprinting framework for information-intensive services. Serv Sci. 2014;6(4):296–312.
20. Lusch RF, Nambisan S. Service innovation: a service-dominant logic perspective. Manag Inf Syst Q. 2015;39:155–75.
21. Maglio PP, Kieliszewski CA, Spohrer JC, editors. Handbook of service science. Boston, MA, US: Springer; 2010.

22. Maglio PP, Spohrer J. Fundamentals of service science. J Acad Mark Sci. 2008;36(1):18–20.
23. Maglio PP, Spohrer J. A service science perspective on business model innovation. Ind Mark Manag. 2013;42(5):665–70.
24. Maglio PP, Srinivasan S, Kreulen JT, Spohrer J. Service systems, service scientists, SSME, and innovation. Commun ACM. 2006;49(7):81.
25. Markus ML. Technochange management: using IT to drive organizational change. J. Inf. Technol. 2004;19(1):4–20.
26. Patricio L, Fisk RP, Falcão e Cunha J, Constantine L. Multilevel service design: from customer value constellation to service experience blueprinting. J Serv Res. 2011;14(2):180–200.
27. Spohrer J, Maglio PP, Bailey J, Gruhl D. Steps toward a science of service systems. Computer. 2007;40(1):71–7.
28. van Eck P, Gordijn J, Wieringa R, editors. Value-based service modeling and design: toward a unified view of services. 21st ed. Berlin: Springer; 2009.
29. Vargo SL, Akaka MA. Value cocreation and service systems (re)formation: a service ecosystems view. Serv Sci. 2012;4(3):207–17.
30. Vargo SL, Lusch RF. Service-dominant logic: continuing the evolution. J Acad Mark Sci. 2008;36(1):1–10.

Zone of Optimal Distinctiveness: Provider Asset Personalization and the Psychological Ownership of Shared Accommodation

Anita D. Bhappu and Sabrina Helm

Abstract In this paper, we conceptually explore how value co-creation in peer-to-peer lodging services is constrained by a fundamental tension between a provider's personalization and psychological ownership of a shared accommodation and customers' psychological ownership of it. Our conceptual framework contributes to theory development about the unique nature of service in the sharing economy. Drawing on literature about service-dominant logic, service interactions, the theory of extended self, the experience economy, psychological ownership, and strategic management, we contend that there is a zone of optimal distinctiveness wherein both a provider's and a customer's psychological ownership of shared accommodation, as well as the provider's personalization of this asset, are optimized. It is beneficial for providers of peer-to-peer lodging services to stay within this zone of optimal distinctiveness when personalizing their leveraged assets in the sharing economy. Optimizing the psychological ownership of shared accommodation is desirable because it helps to satisfy the basic human needs of both a customer and a provider involved in a peer-to-peer lodging service, thereby increasing value co-creation in their service relationship. Also, a customer with high psychological ownership of shared accommodation will take better care of this asset, thereby reducing a provider's risk for property damage.

1 Introduction

Whereas ownership has traditionally been the normative consumption ideal [1], customers in the sharing economy access goods through services rather than actually acquiring and continually owning material assets. Sharing platforms have proliferated

A. D. Bhappu (✉)
Ernest and Julio Gallo Management Program, University of California, Merced, Merced, CA 95343, USA
e-mail: abhappu@ucmerced.edu

S. Helm
Retailing and Consumer Sciences, University of Arizona, Tucson, AZ 85721, USA

H. Yang and R. Qiu (eds.), *Advances in Service Science*, Springer Proceedings in Business and Economics, https://doi.org/10.1007/978-3-030-04726-9_8

access-based consumption of private property by promoting the shared possession or short-term rental of material assets during peer-to-peer service delivery [2, 1]. In particular, the sharing economy has had a significant impact on the travel and hospitality industry [3] with successful platform businesses—such as Airbnb and HomeAway—offering peer-to-peer rented accommodations that compete with traditional business-to-consumer hotel chains. Airbnb has four million rental lodging options across more than 191 countries, which represent more accommodations than offered by the top five hotel chains combined [4]. HomeAway has two million peer-to-peer vacation rentals in 190 countries listed on its platform [5]. This volume of available shared lodging is a commanding example of the market disruption posed today by peer-to-peer services that comprise the sharing economy.

In this paper, we conceptually explore how value co-creation in peer-to-peer lodging services is constrained by a fundamental tension between a provider's personalization and psychological ownership of a shared accommodation and customers' psychological ownership of it. Our conceptual framework contributes to theory development about the unique nature of service in the sharing economy. Drawing on literature about service-dominant logic, service interactions, the theory of extended self, the experience economy, psychological ownership, and strategic management, we contend that there is a *zone of optimal distinctiveness* wherein both a provider's and a customer's psychological ownership of shared accommodation, as well as the provider's personalization of this asset, are optimized.

We begin this paper by outlining why shared accommodation is a service. We then describe how a provider's asset personalization creates competitive advantage in the sharing economy for their peer-to-peer lodging service. Next, we discuss the effects of provider asset personalization on the psychological ownership of shared accommodation. We then present three propositions that comprise our conceptual framework about the zone of optimal distinctiveness for shared accommodation, and define it. Finally, we end by offering some concluding remarks about expanding and testing our conceptual framework.

2 Shared Accommodation Is a Service

Peer-to-peer lodging can be considered a service because it relies on the sharing of accommodations, and goods sharing is one of the competencies that service is often based on [6]. The act of renting one's house, apartment, or room—a lodging—to a peer is legally a non-ownership service "in which consumers acquire some property rights to an asset and are offered a certain degree of freedom in using this asset for a specified period while the burdens of ownership remain with the owner" ([7], p. 172). Service interactions between peers who share accommodations and exchange some form of compensation resemble a traditional service relationship [8, 9]. In a service relationship, both customers and providers are assumed to have unique requirements, resources, and competencies that are applied and integrated to co-produce the service. Through personal interaction involving feedback that is both direct and informal, a customer and a provider acquire knowledge about each other, as individuals and role occupants, which enhances their service co-production. They may develop trust, goodwill, and a felt sense of obligation towards each other, which creates an expectation of future interaction. As such, reciprocity is both evident and valued in services such as peer-to-peer lodging that are based on relationships [10].

Service-dominant logic [11] is a paradigm for understanding value co-creation during service delivery. It posits that providers create superior value propositions about goods and services, and that customers determine this value when they use or consume these goods and services. The co-created value that they ultimately realize is dependent not only on their co-production but also on value-in-use [12], which is determined by the customer during consumption. Therefore, the co-creation of value extends beyond a provider inviting a customer to participate in the processes of production or design [13]. Service-dominant logic asserts that a customer must incorporate a provider's offering into their life for value co-creation to exist [14]. As conceptualized, co-production goes beyond a dyadic exchange between a customer and provider. It can include a complex combination of resources provided by other organizations or persons [12, 11]—a service system—that is comprised of entities that "interact by granting access rights to one another's resources" ([15], p. 666). In the case of peer-to-peer lodging, the co-production of shared accommodation is facilitated by a provider of lodging, a network of personal and organizational resources provided by a commercial sharing platform, and a customer who quite literally incorporates the provider's shared accommodation into their lives for the duration of their vacation or business stay.

3 Asset Personalization Offers Competitive Advantage

Customers in the sharing economy seek one-of-a-kind lodging experiences, which are shaped in large part by staying in places that showcase the personality and social identity of individual providers. Providers (whether strategically or unintentionally) design the service experience that customers can co-create [14]. Reference [16] identifies four dimensions of a service experience—entertainment, education, escapism, and esthetics, which can increase its value to customers. Even though all of these dimensions could be relevant to peer-to-peer lodging services, we focus on esthetics because it specifically relates to providers' personalization of shared accommodation and customers' interpretation of their distinctive design [3]. "Experiences mark the next step in the progression of economic value, requiring businesses to shift from a *delivery-focused* service paradigm to one that recognizes that service is simply the *stage* and goods the *props* to engage individual customers in a personal way" ([3], p. 2379). The architecture, furniture, and decorations of shared accommodation can help providers of peer-to-peer lodging services differentiate their offering in a given market location [16]. In other words, asset personalization can be a source of competitive advantage for providers in the sharing economy because it enables them to signal their shared social identity with specific customer segments. In offering such superior value propositions to these target customers, providers should have increased opportunities for value co-creation with customers and realize both monetary and non-monetary benefits [14]. This is why providers who list rooms, apartments, and homes for rent on Airbnb and HomeAway are encouraged to showcase their individuality by personalizing the space to reflect their individual tastes and preferences. It creates high variance in peer-to-peer lodging services offered on these sharing platforms, which enhances their own perceived uniqueness and competitive advantage relative to business-to-consumer hotel chains.

4 Asset Personalization Influences Psychological Ownership

Although a provider's asset personalization can distinguish their shared accommodation from other peer-to-peer and business-to-consumer lodging services, it can also hinder a customer's psychological ownership of the shared accommodation. Psychological ownership—the state of having possessive feelings about material objects [17]—is instrumental for both providers and customers of peer-to-peer lodging services to feel "at home" in shared accommodation [18]. A customer may have difficulty being comfortable, or even visualizing themselves, in a space that it is highly personalized with a provider's possessions. Some Airbnb and HomeAway accommodations have no room for guests to unpack and organize their belongings; every surface and wall is filled with the provider's personal artifacts and memorabilia, which reinforce that the lodging is not owned by the customer. At the same time, mimicking the generic esthetic of business-to-consumer hotel chains could pose similar constraints for a provider's psychological ownership of the property where they permanently reside and/or that they legally own. A provider's psychological ownership of their residence or vacation home increases as they personalize the place. If a shared accommodation becomes too depersonalized, it may lose its competitive advantage and the distinctiveness that engenders a sense of belonging for customers who temporarily occupy it. Therefore, optimizing the psychological ownership of shared accommodation is arguably very important for both customers and providers of peer-to-peer lodging services.

The motivational forces that feed psychological ownership are grounded in three basic human needs that have been categorized by [17, 18] into efficacy, identity, and place. First, possessing something gives individuals the opportunity to satisfy their need for being in control, including space [19]. This affords individuals a sense of power and security or, in other words, a sense of efficacy. Second, ownership serves an important function in terms of personal and social identity. Possessed objects and spaces can serve as symbolic expressions of the self for the purposes of both self-definition and self-image projection to others [20, 21]. When individuals invest time to know and get familiarized with an object or a space, it becomes part of their extended self [17, 18], which ([20], p. 140) describes as a "metaphor comprising not only that which is seen as 'me' (the self), but also that which is seen as 'mine'." Lastly, possessed objects and spaces can provide a sense of place or belonging. According to ([22], p. 124), possessed objects and spaces provide a sense of home, symbolically speaking, or "a fixed point around which to construct one's daily activities." This sense of home affords individuals with a metaphorical refuge, thereby creating a personal history and sense of location in society [17, 18]. Possessed objects and spaces, thus, serve to symbolically satisfy the human need to be physically anchored.

When a possessed object or space fulfills one of these three basic human needs, it allows an individual to develop feelings of psychological ownership towards it. It is still unclear, however, whether both providers and customers in the sharing economy are able to sufficiently satisfy their needs for efficacy, identity, and place with the objects and spaces that they share during peer-to-peer lodging services. For them to feel psychological ownership, they must be attracted to the shared accommodation and be able to experience and manipulate it. Furthermore, the shared accommodation needs to be open—available, receptive, and hospitable—to enable the them to feel at home in it [18]. Reference [1] argues that having shared or temporary possession of an object or a space engenders an individual's "proprietary feelings" towards it, which increases its perceived value [23]. Reference [1] has even suggested that such shared physical assets can become part of a customer's extended self. Whereas ownership has been the normative consumption ideal [1], customers in the sharing economy primarily appreciate physical assets for their value-in-use [2]. These customers may prefer immaterial or "light" possessions and consumption practices associated with sharing [2] and form more fluid relationships with objects and spaces. Regardless, value co-creation in peer-to-peer lodging services is constrained by the ability of both a customer and a provider to feel sufficiently at home in, and take psychological ownership of, the accommodation that they share.

Therefore, we now offer the following propositions, which are also depicted in Fig. 1:

P1: A provider's psychological ownership of shared accommodation is positively related to their personalization of this asset.

P2: A customer's psychological ownership of shared accommodation is curvilinearly related to a provider's personalization of this asset; it increases initially but then maximizes and starts to decrease thereafter.

P3: There is a *zone of optimal distinctiveness* wherein both a provider's and a customer's psychological ownership of shared accommodation, as well as a provider's personalization of this asset, are optimized.

5 Zone of Optimal Distinctiveness

Optimal distinctiveness refers to the strategic management paradox that requires entrepreneurial ventures to be both different from and like their competitors in a given market location [24, 25]. As discussed earlier, asset personalization can help

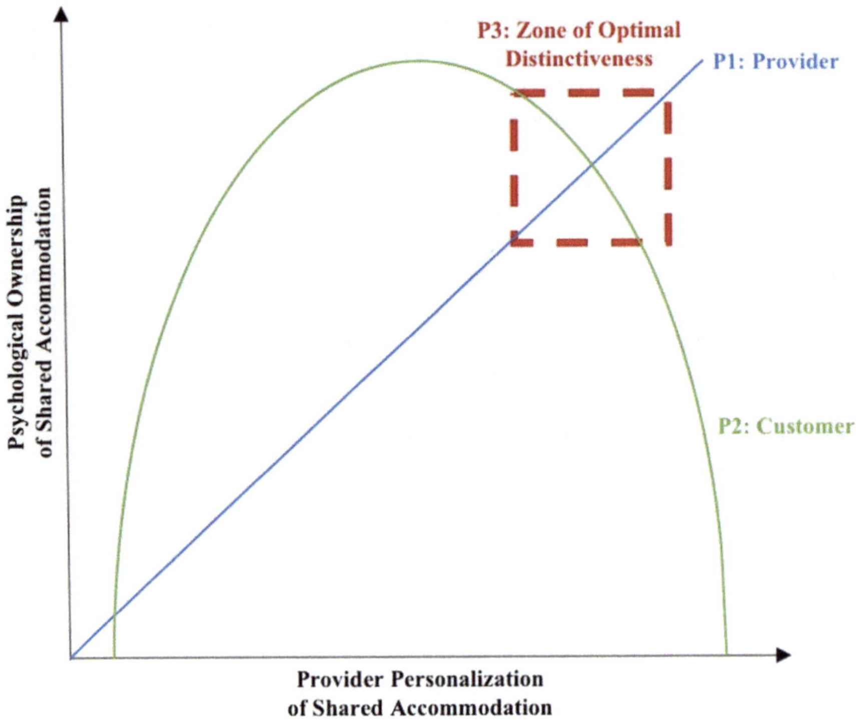

Fig. 1 The zone of optimal distinctiveness for shared accommodation

a peer-to-peer lodging provider to gain competitive advantage. However, providers have to balance their service differentiation with the need for customers to perceive their offering as legitimate. This is especially critical when service providers are new market entrants and have limited reviews and ratings to help customers assess their reputation [24, 25]. In our conceptual framework, we expand the strategic management paradox of optimal distinctiveness by including the need for both a provider and a customer to also have psychological ownership of a shared accommodation. This additional complexity is specific to the dynamic and entrepreneurial environment of peer-to-peer lodging services; business-to-consumer hotel chains are not subject to this constraint when trying to achieve optimal distinctiveness. Furthermore, we do *not* assume "that there is a single, relatively static convergence point in organizational characteristics from which distinctiveness is judged in a market" ([25], p. 98). Rather, we assume that there is a zone of optimal distinctiveness to reflect that peer-to-peer lodging providers "employ a repertoire of actions to construct relations and reinforce boundaries while acquiring reputation and legitimacy" ([24], p. 282) for their service offerings over a period of time.

It is beneficial for providers of peer-to-peer lodging services to stay within this zone of optimal distinctiveness when personalizing their leveraged assets in the sharing economy. Increasing the psychological ownership of shared accommodation is desirable because it helps to satisfy the basic human needs of both a customer and a provider involved in a peer-to-peer lodging service, thereby increasing value co-creation in their service relationship. Also, a customer with high psychological ownership of shared accommodation will take better care of this asset, thereby reducing a provider's risk for property damage. Psychological ownership engenders feelings of responsibility that manifest as protective and nurturing behaviors. When customers feel responsible as the caretakers of a property, even though they are not its legal owners, they act as the "psychological principals" or stewards of this asset [26, 18].

As we have conceptualized it, the zone of optimal distinctiveness should apply to providers who reside or vacation in their shared accommodation irrespective of whether they legally own this asset. Normatively, we assume that providers and customers are engaged in service relationships and exchange some form of compensation related to peer-to-peer lodging. Our conceptual framework, therefore, excludes property management and rental companies that employ functionally-equivalent providers who do not occupy the shared accommodation; they engage in service encounters with customers rather than service relationships [8, 9].

6 Concluding Remarks

By drawing from literature across multiple disciplines, we have laid out a provocative conceptual framework about value co-creation in peer-to-peer lodging services. Our defined zone of optimal distinctiveness expands the strategic management paradox of seeking differentiation and legitimacy in entrepreneurial ventures to also include the need for psychological ownership of shared accommodation by both customers and providers in this service domain. In doing so, it highlights how the co-creation of value in shared accommodation is different than in business-to-consumer lodging services. From a theoretical perspective, it would be interesting to investigate how value co-creation, as conceptualized by service dominant logic, directly impacts psychological ownership. Reference [18] asserts that individuals become invested in the ownership of a product when engaging in the act of creating it; the co-created offering essentially represents an investment of their resources, values, and identity. It would also be interesting to explore how service relationships between customers and providers affect the zone of optimal distinctiveness over time. "Socially embedded exchange relations may leave buyers 'stuck' in suboptimal long-term relationships … The result is a 'dark side' to relationships … where buyers benefit from relationships in the present but at the cost of neglecting to identify or to choose a set of suppliers better suited to future needs" ([27], p. 894). Finally, future research could also extend our conceptual framework to other service domains in the sharing economy, such as carsharing, by exploring how psychological ownership is related to the stewardship of leveraged assets [26].

To empirically test our conceptual framework, future research can employ different methods. First, secondary data collection can provide first insights into perceptions of asset personalization and psychological ownership among providers and customers of peer-to-peer lodging services. For example, qualitative content analysis of publicly-available descriptions and photographs posted by providers on a shared accommodation platform such as Airbnb [28, 29], as well as of online customer reviews posted on the same sites [30], can serve as an unobtrusive approach to access honest customer and provider opinions [31, 32] for assessing whether provider asset personalization, psychological ownership, and optimal distinctiveness emerge as themes in these publicly-available communications. In addition, experimental designs can allow for the testing of causal relationships between provider asset personalization, psychological ownership, perceived value co-creation, and the stewardship of shared assets. These relationships could also be psychometrically assessed by surveying providers and customers of peer-to-peer lodging services. All of these research opportunities would contribute to further theory development about the unique nature of service in the sharing economy.

Acknowledgements We thank anonymous reviewers and session participants at the 2018 Macromarketing Conference for their constructive feedback on an earlier version of this paper.

References

1. Belk R. You are what you can access: sharing and collaborative consumption online. J Bus Res. 2014;67(8):1595–600.
2. Bardhi F, Eckhardt GM, Arnould EJ. Liquid relationship to possessions. J Consumer Res. 2012;39(3):510–29.
3. Mody MA, Suess C, Lehto X. The accommodation experiencescape: a comparative assessment of hotels and Airbnb. Int J Contemp Hospital Manage. 2017;29(9):2377–404.
4. https://press.atairbnb.com/about-us/.
5. https://www.homeaway.com/info/about-us/company-info.
6. Maglio PP, Spohrer J. Fundamentals of service science. J Acad Market Sci. 2008;36:18–20.
7. Moeller S, Wittkowski K. The burdens of ownership: reasons for preferring renting. Manag Serv Qual. 2010;20(2):176–91.
8. Bhappu AD, Schultze U. The role of relational and operational performance in B2B customers' adoption of self-service technology. J Serv Res. 2006;8(4):372–85.
9. Gutek BA, Welsh TM. A brave new service strategy: aligning customer relationships, market strategies, and business structures. Amacom;2000.
10. Proserpio D, Xu W, Zervas G. You get what you give: theory and evidence of reciprocity in the sharing economy. In: Quantitative marketing and economics conference;2016. pp. 1–46.
11. Vargo SL, Lusch RF. Evolving to a new dominant logic for marketing. J Market. 2004;68:1–17.
12. Cabiddu F, Lui TW, Piccoli G. Managing value co-creation in the tourism industry. Annal Tour Res. 2013;42:86–107.
13. Vargo S, Maglio PP, Archpru A. On value and value co-creation: a service systems and service logic perspective. Eur Manag J. 2008;26(3):145–52.
14. Payne AF, Storbacka K, Frow P. Managing the co-creation of value. J Acad Mark Sci. 2008;36(1):83–96.
15. Maglio PP, Spohrer J. A service science perspective on business model innovation. Ind Mark Manage. 2013;42:665–70.
16. Gilmore JH, Pine BJ II. Differentiating hospitality operations via experiences: why selling services is not enough. Cornell Hospitality Quarterly. 2002;43(3):87–96.

17. Pierce JL, Kostova T, Dirks K. Toward a theory of psychological ownership in organizations. Acad Manag Rev. 2001;26(2):298–310.
18. Pierce JL, Kostova T, Dirks KT. The state of psychological ownership: integrating and extending a century of research. Rev General Psychol. 2003;7(1):84–107.
19. Rudmin FW, Berry JW. Semantics of ownership: a free-recall study of property. Psychol Record. 1987;37:257–68.
20. Belk RW. Possessions and the extended self. J Consumer Res. 1988;15(2):139–68.
21. Wattanasuwan K. The self and symbolic consumption. J Amer Acad Bus. 2005;6(1):179–18.
22. Jussila I, Tarkiainen A, Sarstedt M, Hair JF. Individual psychological ownership: concepts, evidence, and implications for research in marketing. J Market Theory Pract. 2015;23(2):121–39.
23. Jiménez FR, Voss K, Frankwick GL. A classification schema of co-production of goods: an open-systems perspective. Eur J Mark. 2013;47(11):1841–58.
24. Snihur Y. Developing optimal distinctiveness: organizational identity processes in new ventures engaged in business model innovation. Entrepreneur Reg Develop. 2016;28(3–4):259–85.
25. Zhao EY, Fisher G, Lounsbury M, Miller D. Optimal distinctiveness: broadening the interface between institutional theory and strategic management. Strateg Manag J. 2017;38:93–113.
26. Davis JH, Schoorman FD, Donaldson L. Toward a stewardship theory of management. Acad Manag Rev. 1997;22:20–47.
27. Elfenbein DW, Zenger T. Creating and capturing value in repeated exchange relationships: the second paradox of embeddedness. Organ Sci. 2017;28(5):894–914.
28. Ert E, Fleischer A, Magen N. Trust and reputation in the sharing economy: the role of personal photos in Airbnb. Tour Manag. 2016;55:62–73.
29. Hum NJ, Chamberlin PE, Hambright BL, Portwood AC, Schat AC, Bevan JL. A picture is worth a thousand words: a content analysis of Facebook profile photographs. Comput Hum Behav. 2011;27(5):1828–33.
30. Yang Z, Fang X. Online service quality dimensions and their relationships with satisfaction: a content analysis of customer reviews of securities brokerage services. Int J Serv Ind Manag. 2004;15(3):302–26.
31. Gbrich C. Qualitative data analysis: an introduction. 1st ed. London: Sage Publications; 2007.
32. Kolbe RH, Burnett MS. Content-analysis research: an examination of applications with directives for improving research reliability and objectivity. J Consumer Res. 1991;18(2):243–50.

Data Mining Methods for Describing Federal Government Career Trajectories and Predicting Employee Separation

Kimberly Healy, Dan Lucas and Cherilyn Miller

Abstract Data mining methods can be applied to human resources datasets to discover insights into how employees manage their careers. We examine two elements of career trajectories in federal government HR data. First, we apply association rule mining and sequential pattern mining to understand the prevalence and direction of interdepartmental transfers. Then we apply logistic regression and decision tree induction to understand and predict employee separation. In this specific application, we find that interdepartmental transfers are uncommon, except between branches of the armed services and out of these branches to the Department of Defence. We also find that demographics, compensation, and political transitions are significant factors for retention, but they account for only a small portion of the probability of a federal employee leaving service. We expect these methods would perform better in industry with a small amount of additional data gathered upon hiring and exit interviews.

1 Introduction

Prior to the emergence of service science as a distinct discipline [1], service and operations research were conducted without consideration of human resource management. However, many problems afflicting a service have human issues as their root causes [2]. Some work has been done since to model this interaction, including policy models and mathematical/statistical models such as the Markov model [3]. This history is summarized well in [4]. These models require significant business understanding to set parameters. For this reason, students of service management may be interested in data mining approaches requiring less initial configuration to help them discover under what circumstances employees will leave a department, either for other departments within the same employer, for employment elsewhere, or for retirement. These insights will allow service managers to consider turnover risks as they develop their service models. A recent release of data by the Office of Per-

K. Healy · D. Lucas (✉) · C. Miller
Engineering Division (Great Valley), Pennsylvania State University, Malvern, PA 19355, USA
e-mail: dj1252@psu.edu

H. Yang and R. Qiu (eds.), *Advances in Service Science*, Springer Proceedings in Business and Economics, https://doi.org/10.1007/978-3-030-04726-9_9

sonnel Management within the United States federal government under the Freedom of Information Act provides an excellent dataset for demonstrating these techniques, including association rule and sequential pattern mining, logistic regression, and decision tree induction.

1.1 Problem Definition

Over 40 years of United States federal government employment data was released to the public for the first time in downloadable format in May 2017. The data was made available through three Freedom of Information Act inquiries by BuzzFeed News. This data can now be retrieved from the Internet Archive at http://www.archive.org [5]. The records span from 1973 to March 2017. They include federal employees' employment details, such as occupation title, salary, and supervisory status, along with employee demographic details such as age, education level, and location.

1.2 Objective of Report

- Objective 1—Describe career trajectories of federal employees
- Objective 2—Predict employee separation over the course of a calendar year using data available in the third quarter of the previous year.

2 Data Understanding

The Federal Employment Dataset comes from the U.S. Office of Personnel Management via the Freedom of Information Act (FOIA). The data files contain four decades of the United States federal payroll spanning the years 1973–2017. The data files within the three dated chunks are partitioned by Department of Defense (DoD) data and Non-Department of Defense (Non-DoD). The data includes quarterly snapshots.

Status files give static data about employees during the quarter of the dated file. Attributes of the Status data files include ID, Employee Name, Date of Filing, Agency and Sub Agency, Station, Age, Education Level, Pay Plan, Pay Grade, Length of Service (LOS), Occupation, Occupation Category, Adjusted Basic Pay, Supervisory Status, Type of Appointment, Work Schedule, and Non-Seasonal or Full-Time Permanent Indicator.

Dynamic files give activity data about employee turnover. The files indicate whether an employee moved into (Accession) or out of (Separation) a position during the quarter of the dated file. Attributes of the Dynamic data files include ID, Employee Name, Agency and Sub Agency, Accession or Separation Indicator, Effective Date (of Accession or Separation), Age, Pay Plan, Pay Grade, Length of Service, Station,

Occupation, Occupation Category, Adjusted Basic Pay, Type of Appointment, and Work Schedule.

There are several limitations to this dataset, including:

- The dataset does not include detailed salary data, including bonuses or additional compensation.
- Thousands of employees' data are withheld by the U.S. Office of Personnel Management, including:
  - Name and duty stations of employees from the Department of Defense agencies, FBI, Secret Service, DEA, IRS, U.S. Mint, Bureau of Alcohol, Tobacco, Firearms, and Explosives, law enforcement officers, nuclear engineers, and some investigators.
  - No data is provided for employees from the White House, Congress, Judicial branch, CIA, NSA, the Department of State's Foreign Service, the Postal Service, Congressional Budget Office, Library of Congress, Panama Canal Commission, and among others.
- The data obtained from the last two FOIA inquires does not include the primary identification attribute, the pseudo-employee ID. This precludes tracking individual careers from September 2014 to March 2017.

3 Analysis I—Career Trajectories

The Career Trajectory analysis used stratified samples of the data to track the movement of employees from 1973–2012. We used sequential pattern mining and association rule mining techniques. We focused on tracking the movement of employees from one department to another. We relied on the employee ID number attribute to track an employee's movement from quarter to quarter.

3.1 Sample Description

The sampling of employees was done in two parts. In the first phase, strata were created based on government agencies. Then, systematic random stratified sampling was used to select 0.3% of government employees from each stratum. In order to remove the sampling bias for employees retained for longer periods, the sampling is done from a list of distinct employees. In the second phase, the stratified sample of randomly selected employees were joined with complete history (from 1973 to 2012) of every selected employee.

Table 1 Association rules for non-department of defense staff

Con	Ant	Support	Support (%)	Rule conf (%)	Lift (%)
HE	SZ	211	1.0715	42.5	4.0901
HS	TD	335	1.7012	38.7	4.4391
HS	EM	90	0.45704	33.7	3.8637
TD	HS	335	1.7012	19.5	4.4391
AG	IN	277	1.4067	16.7	1.1755
HS	DJ	187	0.94962	13.5	1.5521
DJ	HS	187	0.94962	10.9	1.5521
SZ	HE	211	1.0715	10.3	4.0901

3.2 *Modeling and Analysis*

The first technique we used was association rule mining. This technique can be used to identify frequent patterns among antecedent and consequent observations. Reference [6] provides an introduction to this method. Criteria such as lift, support, and confidence were used to determine the importance and prevalence of each rule. Association rule mining was primarily used in order to determine how likely it was that an employee transferred to a particular agency, given their previous employment at another agency. The support of an association rule is the percentage ratio of the records that contain both the antecedent and consequent to the total number of observations in the data set.

The confidence is the percentage ratio of the number of records that contain both the antecedent and the consequent to the number of records that contain just the Antecedent: $\alpha(A \rightarrow C) = P(C|A) = P(A \cap C)/P(A)$. The lift is the percentage of the records that contain both the Antecedent and the Consequent—to the percentage of records that contain the Consequent and to the percentage of records that contain the Antecedent. The formula for the lift is as follows: $l(A \rightarrow C) = s(A \cup C)/s(A) \cdot s(C)$. The lift tells us how much better the association rule is at predicting the result rather than simply assuming the result on its own.

Table 1 provides a summary of the association rules worth noting from the non-dod data sets and Table 2 provides a summary for the dod data sets. Note that agency codes are provided here for brevity. Meanings of agency codes can be found at the Office of Personnel Management's website [7]. We began with the rules that had the greatest support. Then, we evaluated the confidence of the rules. We were most interested in the rules with the higher confidence percentages. Note that while there are some agency transitions that have a substantial confidence and support, we should also focus on the lift. If the lift score is below 1, this indicates that the antecedent and the consequent appear less often together than expected. We filtered out any rules that had a lift significantly less than 1. If the lift was near 1, then we evaluated the support of the rule. The larger the support, the more actionable the rule.

Table 2 Association rules for department of defense staff

Con	Ant	Support	Support (%)	Rule conf (%)	Lift (%)
AR	AR	4151	42.129	42.1	1
NV	NV	2776	28.174	28.2	1
AF	AF	2434	24.703	24.7	1
DD	DD	2150	21.821	21.8	1
DD	[AF, AR]	74	0.75104	24.2	1.1083
DD	[NV, AR]	70	0.71044	23.3	1.0693

Table 3 Sequential rules for department of defense staff

Pattern	Support
<AR, AR, AR, AR, AR>	726
<AF, AF, AF>	725
<NV, NV, NV, NV, NV>	669
<AR, AR, AR, AR, AR, AR, AR>	639
<AF, AF, AF, AF>	630
<NV, NV, NV, NV, NV, NV>	603
<AR, AR, AR, AR, AR, AR, AR, AR>	564
<NV, NV, NV, NV, NV, NV, NV>	555
<DD, DD, DD>	553
<AF, AF, AF, AF, AF>	549
<NV, NV, NV, NV, NV, NV, NV, NV>	514
<AR, AR, AR, AR, AR, AR, AR, AR, AR>	501

The second technique used in our analysis was sequential pattern mining which focuses on the discovery of rules in sequences. Reference [6] provides an introduction to this method. The sequence of an employee's transition from one agency to the next is very helpful to know for making career trajectory predictions. The rules presented in Tables 1 and 2 were found using a method which did not take time into account; sequential pattern mining considers the sequence of agencies in an employee's career. Once again, we evaluated the results with special attention given to the confidence and the support of the rules. In this part of our analysis, the sequential pattern mining rules were created by utilizing the GSP algorithm in the SPMF software.

First, the data was condensed to a year-to-year basis before importing into the SPMF program which allowed for a summarized view of the patterns in the data. The rules generated did not describe any switching between agencies. Therefore, we were able to conclude that on a year to year basis, employees remained in their current agencies at a rate higher than would be expected by chance. The Army, Navy, and the Veterans Administration were associated with very long rules showing continued employment. These rules can be seen in Tables 3 and 4.

Table 4 Sequential rules for non-department of defense staff

Pattern	Support
<VA>	62
<VA, VA>	45
<VA, VA, VA>	40
<VA, VA, VA, VA>	32
<VA, VA, VA, VA, VA>	27
<AG>	25
<VA, VA, VA, VA, VA, VA>	24
<HS>	23
<IN>	22
<VA, VA, VA, VA, VA, VA, VA>	21
<DJ>	20
<HE>	19
<HS, HS>	19
<IN, IN>	19

Table 5 Sequential rules for inter-departmental transitions among DoD staff

Pattern	Support
<AR, DD>	125
<NV, DD>	87
<AF, DD>	82
<DD, AR>	79
<NV, AR>	73
<AR, NV>	72
<AR, AF>	69
<AF, AR>	57
<DD, AF>	50

Table 6 Sequential rules for inter-departmental transitions among non-DoD staff

Pattern	Support
<DJ, HS>	3
<HE, SZ>	3

Next, we ran another sequential pattern mining analysis, however we only considered the instances when an employee switched agencies. The results in Table 4 show some of the most common transitions and their support. These results provide further evidence that interdepartmental transfers are rare. Most of the rules are associated with transfers between the branches of the armed forces or between one of the branches and the Department of Defense (Tables 5 and 6).

4 Analysis II—Identifying Factors Contributing to Separation and Predicting Separation

We examined data from 1973–2012 to see what factors predict that a non-seasonal full time permanent federal employee will leave employment (separate) within a year. We relied on the ability to use employee Pseudo-IDs to track an employee across multiple quarters. Since the data from 2013 onward did not have Pseudo-IDs, the data was removed from this analysis.

In addition, we sought to understand whether presidential elections influence employees' retention. We added indicators to the data indicating whether that year was an election year, whether control of the White House transitioned from one political party to another, and, if so, which party assumed control.

4.1 Sample Description

We randomly selected employees employed in the third quarter of each year. We then counted the quarters in which each was employed during the next calendar year. A count of 4 indicated that the employee was retained (1). A count of 0–3 indicated that the employee was not retained (0). This is an imprecise method as some agencies were not subject to the Freedom of Information Act request, so transfers into these agencies would be indicated as a non-retention outcome. We then removed from the sample any employees exhibiting rare values (fewer than 30 observations) for the variables agency, appointment type, or pay. We adjusted for inflation by assuming a constant rate of 3% annually since 1980.

4.2 Modelling and Analysis

We trained a logistic regression model to determine which attributes have a statistically significant impact on retention when considered together. This model outputs a probability that a record belongs to the target class. Reference [8] provides an introduction to the algorithm and a tutorial for training a model using R. In order to improve our understanding of the minority class, we undersampled from the majority class at a rate of 15%. We found the variables listed in Table 7 to be statistically significant predictors at the given coefficients.

This model has a McFadden R^2 of only 0.081, suggesting that most factors contributing to separation are not represented in this dataset. We do find a statistically significant result indicating that employees are less likely to be retained in the year following the transition from Democratic control of the White House to Republican control, all other factors held constant. It is important to note that there is also a statistically significant result indicating that transitions in party-control are correlated with

Table 7 Coefficients of logistic regression model

Coefficients	Estimate	Std. error	z value	Pr(>|z|)
(Intercept)	−1.94E+01	2.50E+00	−7.765	8.19E−15
Grade	1.24E−02	4.96E−03	2.493	0.012656
appt_type15	−2.80E−01	4.46E−02	−6.27	3.60E−10
appt_type30	−4.05E−01	9.81E−02	−4.127	3.68E−05
appt_type32	−6.41E−01	1.27E−01	−5.042	4.60E−07
appt_type38	−6.63E−01	4.30E−02	−15.43	<2e−16
appt_type40	−5.63E−01	2.92E−01	−1.927	0.054039
appt_type50	−3.87E−01	2.47E−01	−1.569	0.116739
appt_type55	−3.55E+00	1.05E+00	−3.37	0.000751
Year	8.22E−03	1.27E−03	6.47	9.78E−11
Election year	−1.64E−01	3.67E−02	−4.474	7.68E−06
Transition of white house to other party	2.66E−01	5.69E−02	4.677	2.92E−06
Transition of white house to republican control	−2.05E−01	7.15E−02	−2.87	0.004111
Education level	9.44E−02	2.64E−02	3.582	3.41E−04
Pay	3.54E−05	4.97E−06	7.12	1.08E−12
Age	5.84E−02	5.96E−03	9.806	<2e−16
Length of service	1.94E−01	1.68E−02	11.576	<2e−16
Age:Length of service	−3.96E−03	3.17E−04	−12.49	<2e−16
Pay:Age	−4.87E−07	1.00E−07	−4.862	1.16E−06
Pay:Length of service	7.59E−08	2.07E−07	0.366	0.714046
Education level:Age	−2.03E−03	5.81E−04	−3.489	0.000485
Education level:Length of service	1.69E−03	4.09E−04	4.128	3.66E−05
Education level:Pay	−1.79E−06	3.60E−07	−4.957	7.15E−07
Pay:Age:los.numeric	−7.92E−09	3.89E−09	−2.038	0.041536
Education level:Pay:Age	3.44E−08	7.10E−09	4.845	1.26E−06

an increase in the likelihood that an employee is retained. This transition coefficient is greater in magnitude than the Republican coefficient. As such, the reasonable interpretation would be that transition to a Democratic White House increases employee retention more than transition to a Republican White House. Neither has a negative impact on employee retention.

Education level appears in several interaction terms in the logistic regression. Some of these terms have negative coefficients while others have positive coefficients. Given the expected magnitude of the Education variable (approximately 10^1), the magnitude of its coefficient (approximately 10^{-1}), the magnitude of the Education:Length of Service interaction variable (approximately 10^2) and its coefficient (10^{-3}), a unit increase in the education variable would be expected to increase the log likelihood of retention by a factor of 1.1 before considering the negative coefficients on interaction effects including education. The expected magnitude of Education level:Age is 10^2 with a coefficient of 10^{-3} and the magnitude of Education level:Pay is 10^6 with a coefficient of 10^{-6}. In the case of a unit increase in education with all other factors held constant, these factors would decrease the log likelihood of retention by a factor of approximately 1.1. This means the expected impact of a unit increase in education on retention is approximately zero. We are left with a weak conclusion that the effect of education on retention is idiosyncratic based on the employee's age, length of service, and pay.

We also developed a decision tree model using the C5.0 algorithm in order to develop rules that might help understand these interaction effects. This model is described in [9]. The decision tree developed for all employees is provided in Fig. 2. As with the logistic regression, this model was trained with an undersampled set to improve our understanding of the minority case. It has an overall accuracy on unseen data of 65.8%, and its accuracy given that the actual class is 0 is 55.5%. Its accuracy given the actual class is 1 is 75.0%. This gives an average by class of 65.2%. This compares favorably to the null model in which we always predict that an employee will be retained, yielding an overall accuracy of 90.0% but an average by class of 50.0%. The pruned tree does not reference education level. It describes a complex interaction between age, length of service, and pay. As age and length of service increase, retention generally drops. For employees with non-permanent appointments paid less than \$43,000 per year who are between the ages of 40–54, however, retention increases with length of service greater than 9.5 years. As pay increases, retention generally increases. However, for employees with permanent appointments paid more than \$43,000 per year between the ages of 50 and 54, retention drops significantly above 30 years of service. The retention of employees under age 54 with non-permanent appointments is not affected by length of service. This suggests that employees with a permanent appointment tend to work until they can earn their pension, unless they began their service later in life. A plot of variable importance is shown in Fig. 1.

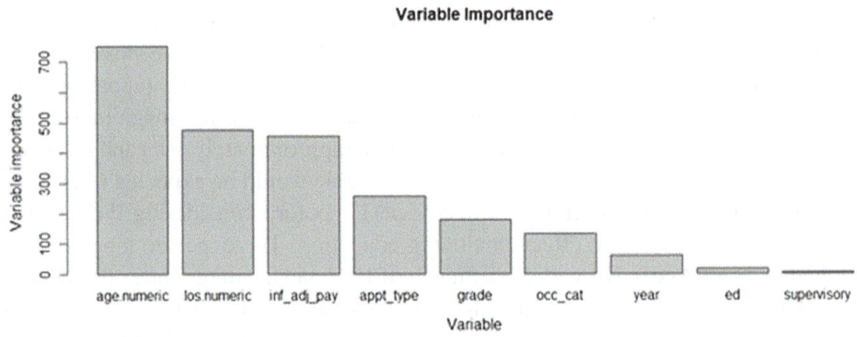

Fig. 1 Variable importance for decision tree

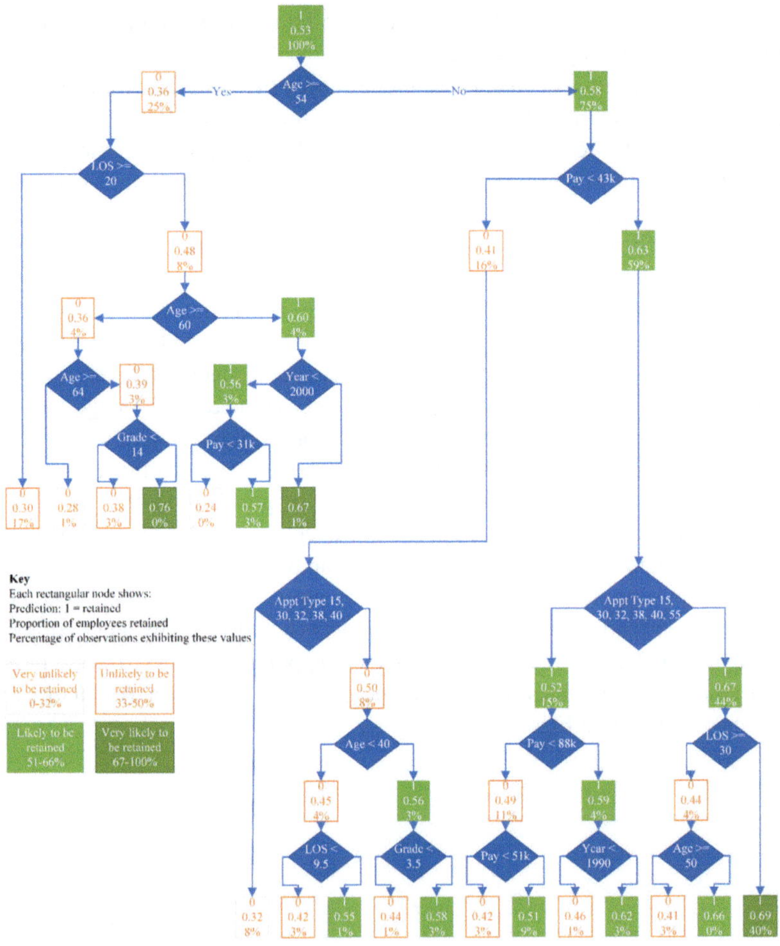

Fig. 2 Decision tree predicting separation

5 Summary

Through pattern mining, we discovered that interdepartmental transfers are rare in the federal government. They tend to be more common in the Department of Defense. Outside of the Department of Defense, we find only two significant sequential rules: employees from the Department of Justice are more likely than most to transition to the Department of Homeland Security, and employees from the Department of Health and Human Services are more likely to transition to the Social Security Administration.

Through logistic regression and a C5.0 decision tree, we determined that the most important factors for predicting departure of employees of the federal government are length of service and age. Among employees who are started their careers later in life, pay becomes an important factor.

These methods could be applied within a company using a richer dataset. For the pattern mining approach, significant insights could be added by including data about the previous employer and the next employer for departing staff. There are some attributes which could enrich the logistic regression and decision tree induction, including previous employer(s), disciplinary actions, employee performance review scores, and supervisor performance review scores.

Acknowledgements This paper is a summary of the results of our winning submission to Penn State's university-wide Data Analytics Challenge, which was chaired by Dr. Robin Qiu with the support of the following committee members from Penn State's Smeal College of Business, College of Engineering, College of Information Sciences and Technology, and the Great Valley Engineering Division. We are grateful for their support. Thank you, Jason Acimovic, Saurabh Bansal, Adrian Barb, Guoray Cai, Terry Harrison, Ashkan Negahban, Robin Qiu, Kathleen Riley, Chris Solo, Satish Srinivasan, Hui Yang, and Tao Yao.

References

1. Moussa S, Touzani M. A literature review of service research since 1993. J Serv Sci. 2010;2(2):173–212.
2. Boudreau J. On the interface between operations and human resources management. Manuf Oper Manag. 2003;5(3):179–202.
3. Lagard M, Cairns J. Modelling human resources policies with Markov models: an illustration with the South African nursing labour market. Health Care Manag Sci. 2012;15(3):270–82.
4. Hafeez K, Aburawi I. Planning human resource requirements to meet target customer service levels. Int J Qual Serv Sci. 2013;5(2):230–52.
5. Internet Archive. Federal employment data from the offices of personnel management. https://archive.org/details/opm-federal-employment-data. Accessed 20 Feb 2018.
6. Penn State. SWENG 545: Data Mining—7.2 Discovering Frequent Sequential Patterns on a Computer, Online Course, Accessed May 2018.

7. Office of Personnel Management. Federal Agencies List. https://www.opm.gov/about-us/open-government/Data/Apps/Agencies/. Accessed 20 Feb 2018.
8. Forte R. Logistic regression. In: Mastering predictive analytics with R. Packt Publishing, Birmingham;2015. p. 93–109.
9. Forte R. Tree-based methods. In: Mastering predictive analytics with R. Packt Publishing, Birmingham;2015. p. 201–8.

Using the Service Science Canvas to Understand Institutional Change in a Public School System

Shari Weaver and Oleg Pavlov

Abstract Reforming STEM education in the United States continues to be a topic of active discussion and research. Why do some school districts succeed while others fail at implementing similar educational interventions? To answer this question, we apply the service science theory to characterize a pK-12 district that is viewed as a complex educational system. Our analysis utilizes the Service Science Canvas, which is a convenient methodological tool that includes common elements of the service science framework.

1 Introduction

In recent years, much attention has been given to improving STEM education in the United States. According to the 2015 TIMSS, the U.S. ranks 10 internationally in math achievement and 11 in science, which is well below countries such as Singapore, Korea and China [11]. While there was some growth in math when comparing the 5 years prior, there has been no improvement in science in that timeframe. This limited progress remains despite the substantial financial contributions of government agencies such as the National Science Foundation (NSF) committed to this effort. In fiscal year 2016, the Department of Education allocated $71.698 billion and the National Science Foundation allocated $7.463 billion to STEM Education Initiatives [8].

In a more specific example, over $575 million was spent in a multiyear initiative to improve low-income minority student outcomes by ensuring access to effective teaching through the development of a system to better measure teacher effectiveness. While the initiative achieved its goals around implementation of a more targeted and rigorous teacher professional development program and evaluation system, student

S. Weaver (✉)
STEM Education Center, Worcester Polytechnic Institute, Worcester, MA 01609, USA
e-mail: sweaver@wpi.edu

O. Pavlov
Social Science & Policy Studies, Worcester Polytechnic Institute, Worcester, MA 01609, USA

H. Yang and R. Qiu (eds.), *Advances in Service Science*, Springer Proceedings in Business and Economics, https://doi.org/10.1007/978-3-030-04726-9_10

outcomes did not improve. The report summarizing these findings concluded that goals around student outcomes may not have been achieved because of the narrow focus of the study on teacher effectiveness alone and not taking into account the broader academic context. The authors suggested that other factors "ranging from early childhood education to students' social and emotional competencies, the school learning environment, and family support" may need to be addressed [12].

In this paper, we propose that in order to increase the effectiveness of STEM initiatives, school districts should be analyzed systemically and holistically. Our method involves using service science principles to design a mechanism to characterize school systems that have implemented STEM initiatives. Once characterized, we can compare school systems that have had ranges of implementation success as determined by sustainability of programmatic change and observable outcomes. The system analysis will then enable us to ascertain key system components that influence initiative implementation success with the purpose of better informing school and community leaders seeking to effect change.

Service science is a field devoted to studying service systems and value co-creation [6]. A service system is defined as a system in which interacting components work together to achieve measurable goals [1, 13]. For instance, pK-12 school systems can be classified as a service system when exchange, application or acquisition of knowledge is considered a benefit. This classification enables us to describe a school system using the ten foundational concepts on which service science is based.

2 Service Science Canvas

Pavlov and Hoy [10] developed a methodological tool, the Service Science Canvas (Fig. 1), that enables the analysis of any system and then applied it to an entrepreneurship program in a higher education institution. We propose applying the Service Science Canvas to examine a pK-12 public school district implementing a STEM initiative as a result of participating in the STEM Integration for Education Leaders program at Worcester Polytechnic Institute (Fig. 2). In this yearlong program, the STEM Education Center works with teams of educational leaders from school districts to support the development of a strategic STEM integration plan. In the past six years that the program has existed, 22 districts in Massachusetts have participated with varying degrees of sustainable success as observed by STEM Education Center staff.

In the section below, we will define each element of the Service Science Canvas as it would apply to a pK-12 public school district. Then, as a case study in the application of this methodological tool, we will consider a pK-3 elementary school in a district that participated in the STEM Integration for education leaders program that has demonstrated success in implementing STEM initiatives.

The Service Science Canvas

<table>
<tr>
<td rowspan="2">Resources

What resources are part of the service system?

Which resources are physical and which ones are not?

Resources can be physical (e.g., technology), non-physical (e.g., intellectual, financial), with-rights (e.g., human), or no-rights (e.g., bits).</td>
<td rowspan="2">Access Rights

Are the resources shared, owned, leased or privileged?</td>
<td rowspan="2">Entitites

What are the entities that are part of the service system?

Entities are dynamic.

Entities can be formal or informal.</td>
<td>Stakeholders

Who is affected by the service system?

The fundamental stakeholders are customers, service providers, authority and competitors.</td>
<td>Value Co-Creation

How does each stakeholder contribute to the value co-creation? What value propositions do they offer others and seek agreement on?</td>
</tr>
<tr>
<td>Networks

What are the patterns of interactions between service systems and between entities? How are they nested?</td>
<td>Ecology

Are there multiple interacting service systems and entities?</td>
</tr>
<tr>
<td>Governance

How are activities coordinated?

How are contracts enforced?

How are disputes resolved?</td>
<td colspan="2">Outcomes

What are the outcomes of the activities by the service system?

Examples of outcomes include value created, contracts agreed on, disputes resolved, or unresolved.</td>
<td colspan="2">Measures

What are the appropriate tangible measures of quality, productivity, compliance and sustainable innovation?</td>
</tr>
</table>

Fig. 1 The service science canvas

STEM Education

<table>
<tr>
<td rowspan="2">Resources
• Faculty
• Staff
• Administrators (school and district)
• Physical space
• Budget
• Curricular materials
• Financial support from funding agencies</td>
<td rowspan="2">Access Rights
• Shared access to classroom, computer lab, and library space
• Shared access to student time
• Access to faculty and staff (full-time, part-time)
• Privileged access to intellectual property</td>
<td rowspan="2">Entitites
• Academic departments
• Department of Elementary and Secondary Education
• School Board</td>
<td>Stakeholders
• Students
• Families
• Faculty
• Staff
• Administrators
• Teacher unions
• Education support professionals
• Department of Elementary and Secondary Education
• Funders
• Local businesses
• Faith-based organizations
• Community members</td>
<td>Value Co-Creation
• Value to students
• Value to parents
• Value to the faculty
• Value to the community
• Value to the school/district</td>
</tr>
<tr>
<td>Networks
• Interaction networks of faculty within and across grade-levels and disciplines
• Interaction networks of school and community-based organizations (businesses, faith-based, and social service)
• Interaction networks of k-12 and higher education faculty and staff</td>
<td>Ecology
• Professional organizations
• Schools within a district
• School consortiums including multiple districts
• Higher education partnerships
• Teachers union</td>
</tr>
<tr>
<td>Governance
• Stakeholder
• Business</td>
<td colspan="2">Outcomes
• Graduating student interested in STEM
• Students taking STEM classes
• College and career ready
• Successful implementation of STEM initiatives</td>
<td colspan="2">Measures
Quality
• Advanced STEM courses
• STEM electives
• Student engagement
Productivity
• Performance on MCAS and other standardized exams
• Graduation rates
Compliance
• Alignment to MA STE and Math Standards
Sustainable Innovation
• Level of STEM integration
• STEM budget allocation</td>
</tr>
</table>

Fig. 2 The service science canvas adapted for STEM education for a pK-12 public school

2.1 Resources

Potential resources in a system have been defined as anything or anyone that is useful in service production [13]. In an educational system, resources include faculty and staff, physical classroom and shared space, funding and curricular materials [10]. There is an ongoing debate about the effect of school resources on student performance/achievement. Greenwald et al. [3] asserts that resources as measured by per pupil expenditure are related to student achievement while Hanushek [5] found no consistent correlation between school resources and student achievement. However, Hanushek [5] concluded that adequate resources are needed to ensure student success but that adding resources without changing the decision-making process in how those funds are used will not result in an increase in achievement. The general consensus is that it is not just the amount of resources but how those resources are allocated or implemented that impacts improvement in student outcomes. The major flaw in this debate is that resources are defined monetarily rather than considering the people resources such as the capacity of teachers, families, and communities. In terms of STEM program implementation, targeted allocation of funding to address STEM initiatives through selected hiring, purchasing of curricular materials and space allocation ensures success.

The pK-3 school analyzed invested in resources by hiring a STEAM teacher who teaches all students in the building on a 6-day rotating schedule. This teacher also collaborates with grade level teams to assist in STEAM integration in each classroom. The library has a dedicated maker space with activities that rotate on a monthly basis. A summer STEAM camp will be developed and implemented in the summer of 2018 for early elementary students.

2.2 Access Rights

Access rights involve the policies and procedures that govern resource access and usage and are categorized as owned outright, leased-contracted, shared, and privileged [9b]. Academic programs share access to buildings and equipment [10]. While competition for this shared access exists in pK-12 public schools, it is often limited to common spaces such as computer labs, the library, etc. Additionally, access to students' time is viewed as shared access. Teachers, particularly in the newly emerging STEAM classes, compete for time to interact with students within the limitation of the school day. The school studied ensures access to students by building the STEAM class into the schedule as special similar to music or library.

2.3 Entities

Entities have been defined as any resource configuration that is able to initiate actions that can, through acquiring, sharing or applying resources, improve its own state [1, 13]. Entities in the pk-12 academic system include academic departments, grade level teams, the Department of Elementary and Secondary Education, and the local school board. Grade level teams in the studied elementary school meet regularly to develop curricular materials that integrate STEAM. The STEAM coordinator meets with each team to provide input and feedback.

2.4 Stakeholders

Spohrer et al. [13] identify customer, provider, authority and competitor-criminal as the four primary types of stakeholders. While not as clear-cut as an economic system, stakeholders in school systems can be identified as students, faculty and staff, and administrators, and charter and private schools respectively. Pavlov and Hoy [9] details three stakeholders in a university educational system as students, faculty, and administrators. Public k-12 schools have another dimension of stakeholders which include students' families and community members [2]. Many mechanisms have been put in place by the building principal in the school studied to ensure that stakeholders are involved in decision-making processes and in STEAM initiative implementation. One example includes involving community members in the development of an outdoor education space in the school's courtyard that includes a garden and a future fishpond.

2.5 Value Co-creation

For any service system initiative to be sustainable, all stakeholders should derive some value from the collective activities of all entities of that system [1, 10, 15]. The value must be perceived by students, faculty, parents, and the community. In pK-12 systems, value is inherent in education as it relates to economic return in the long run. There is also value in the relationship between innovative, highly effective curricular activities and student and teacher motivation and engagement.

In the school studied, students and teachers participate in content-rich, engaging STEAM curriculum that results in value co-creation. Increased engagement with parents and community through direct involvement in school activities and in explicit communication about the value of STEAM education would strengthen confidence and sustain service excellence.

2.6 Networks

Networks develop as entities form patterns of interactions with each other as well as with stakeholders [10, 13]. Networks can be defined as relationships and result from collaborative advantages and the development of cooperative strategies [1]. In pK-12 schools, networks develop between faculty within and across grade levels, schools and community based organizations, and districts and higher education institutions.

Participating in the STEM Integration for Academic Leaders provided opportunities to network on multiple levels both as a direct component of the program and as district initiatives that resulted from program participation. In the STEM Integration program, cohorts are designed to bring together leadership teams from at least three different districts. This allows for networking for faculty and administrators across districts. Within a district, the teams include representatives from various grade bands which enables vertical networking between elementary, middle, and high school faculty. This type of vertical teaming enables the generation and communication of a clear vision for STEM education throughout the district.

Within the school, grade level teams meet regularly to develop STEAM curriculum. Additionally, the STEM coordinator meets with individual teachers as well as the grade level teams to support them in the development, implementation, and evaluation of STEAM curricular modules into their existing curriculum. Networks that are being explored include local faith-based organization and businesses.

2.7 Ecology

Ecology refers to networks of different types of service systems and their entities. In pK-12 education, faculty can be involved in professional education organizations such as National Science Teacher Association (NSTA) or National Council of Teachers of Mathematics (NCTM). Department leaders or school administrators form networks with individual schools in the same district. In some locations, consortiums have formed that include department heads or school administrators from multiple districts. This is particularly valuable in geographic areas that have multiple small districts that can benefit greatly from sharing resources.

2.8 Governance

Governance mechanisms between an authority entity and other governed entities provide a structure to a service system [13] moving the system toward a goal by defining a process to proceed toward that goal and resolve any disputes that might arise [1]. McCrone et al. [7] describe two governance models displayed in schools in the UK. The first is a business model in which a school leader, the principal or superintendent, is responsible for governance making the directives that are followed by faculty, staff, and students. A second model, shared governance, enables stakeholders to be involved in school governance [4]. An example seen in the U.S is that of the school council which generally includes a representative group of administrators, faculty, parents, students, and community members. This council meets with district administrators and the school board to discuss and make recommendations on a variety of educational issues. We assert that the success of this governing system varies widely depending upon resources, entities, access rights and value-co creation interactions.

2.9 Outcomes

Spohrer et al. [13] outlines ten possible outcomes. The main desired outcome is that the value is realized although often when implementing educational initiatives, the value proposition is not understood, agreed to, or able to be realized. The main outcome sought in pK-12 schools is to graduate students who are college or career ready. The Massachusetts Board of Elementary and Secondary Education define students who are college and career ready as those who can "demonstrate the knowledge, skills and abilities that are necessary to successfully complete entry-level, credit-bearing college courses, participate in certificate or workplace training programs, enter economically viable career pathways, and engage as active and responsible citizens in our democracy".

When implementing innovative STEM initiatives, outcomes sought include the successful implementation of those STEM initiatives and an increase in the number of students who take STEM classes throughout their school experience and who, on graduation, express an interest in pursuing a STEM career.

2.10 Measures

Stakeholders evaluate services systems based on four primary types of measures; quality, productivity, compliance, and sustainable innovation [10, 13]. The quality of a pK-12 system can be measured by the number and type of STEM courses offered, teacher effectiveness, and student engagement. School productivity is often rated by graduation rates and achievement on standardized test and performance based assessments. Compliance can be evaluated by adherence by faculty and staff to district and school initiatives. Level of STEM integration and the commitment by districts to financially support STEM initiatives in the annual budget provide evidence of sustainable innovation.

3 Conclusion

This article builds on previous literature which proposed the utilization of service science to characterize education systems [14]. Specifically, we modified the Service Science Canvas, designed by Pavlov and Hoy [10], to identify the 10 general elements and principles as displayed in the particular service science system of a pK-12 school. As a case study, this tool was applied to a STEM initiative in a pK-3 public elementary school following their participation in the STEM Integration for Academic Leaders program administered by Worcester Polytechnic Institute.

While this article contributes to the understanding of the pK-12 public school as a service science system, broader application of the Service Science Canvas requires further study. Future research should include application to multiple participating districts to determine if comparisons can be made between these districts. Future research may include developing computational models to simulate service system dynamics which would enable us to identify points of leverage and potential unintended consequences to inform potential policy implementation.

References

1. Barile S, Polese F. Smart service systems and viable service systems: applying systems theory to service science. Serv Sci. 2010;2:21–40.
2. Epstein J. School/Family/Community partnerships: caring for the children we share. Phi Delta Kappan. 1995;76(9):701–12.
3. Greenwald R, Hedges L, Laine R. The effect of school resources on student achievement. Rev Educ Res. 1996;66(3):361–96.
4. Hanberger H.: Evaluation in local school governance: a framework for analysis. Educ Inq. 2016;7(3).
5. Hanushek E. Assessing the effects of school resources on student performance: an update. Educ Eval Policy Anal. 1997;19(2):141–64.
6. Lyons K, Tracy S. Characterizing organizations as service systems. Hum Factors Ergon Manuf Serv Ind. 2013;23(1):19–27.
7. McCrone T, Southcott C, George N. Governance models in schools. Slough: NFER; 2011.
8. National Research Council: Rising Above the Gathering Storm, Revisited: Rapidly Approaching Category 5: Condensed Version. The National Academies Press;2011.
9. National Center for Education Statistics (2016). Digest of Education Statistics; 2016. https://nces.ed.gov/programs/digest/2016menu_tables.asp..
10. Pavlov O, Hoy F. Toward the service science of education. Handb Serv Sci. 2018;2.
11. Provasnik S, Malley L, Stephens M, Landeros K, Perkins R, Tang JH. Highlights from TIMSS and TIMSS advanced 2015: mathematics and science achievement of U.S. students in grades 4 and 8 and in advanced courses at the end of high school in an international context (NCES 2017-002). Washington, DC: U.S. Department of Education, National Center for Education Statistics;2018. http://nces.ed.gov/pubsearch.
12. Stecher BM et al. Improving teaching effectiveness: final report: the intensive partnerships for effective teaching through 2015–2016. RAND Corporation;2018. https://www.rand.org/pubs/research_reports/RR2242.html.
13. Spohrer J, Anderson L, Pass N, Ager T. Service science and service-dominant language, vol. 2. Otago Forum 2: Academic Papers;2008. pp. 1–18.

14. Spohrer J, Giuisa A, Demirkan H, Ing D. Service science: reframing progress with universities. Syst Res Behav Sci. 2013;30:561–9.
15. Vargo S, Maglio P, Akaka M. On value and value co-creation: a service systems and service logic perspective. Eur Manag J. 2008;26:145–52.

Data-Driven Capacity Management with Machine Learning: A Novel Approach and a Case-Study for a Public Service Office

Fabian Taigel, Jan Meller and Alexander Rothkopf

Abstract In this paper we consider the case of a public service office in Germany that provides services such as handling passports and ID card applications, notifications of change of addresses, etc. Their decision problem is to determine the staffing level for a specific staffing time-slot (e.g., next Monday, 8 am–12.30 pm). Required capacity is driven by features such as the day of the week, whether the day is in school vacations, etc. We present an innovative data-driven approach to prescribe capacities that does not require any assumptions about the underlying arrival process. We show how to integrate specific service goals (e.g., "At most 20% of the customers should have to wait more than 20 min") into a machine learning (ML) algorithm to learn a functional relationship between features and prescribed capacity from historical data. We analyze the performance of our integrated approach on a real-world dataset and compare it to a sequential approach that first uses out-of-the-box ML to predict arrival rates and subsequently determines the according capacity using queuing models. We find that both data-driven approaches can significantly improve the performance compared to a naive benchmark and discuss benefits and drawbacks of our approach.

1 Introduction

In this paper we consider the problem of finding the right level of capacity for service operations. We use the case of a public service office in Germany that provides services such as the application and issuance of passports or ID cards, notifications of change of addresses, etc. Their decision problem is to determine the staffing level for a specific time slot (e.g., next Monday, 8 am–12.30 pm). Practitioners' intuition is that required capacity depends on the day of the week, whether this day falls

F. Taigel (✉) · J. Meller
Universitaet Wuerzburg, 97074 Würzburg, Germany
e-mail: fabian.taigel@uni-wuerzburg.de

A. Rothkopf
Center for Transportation and Logistics, Massachusetts Institute of Technology, Cambridge, USA

H. Yang and R. Qiu (eds.), *Advances in Service Science*, Springer Proceedings in Business and Economics, https://doi.org/10.1007/978-3-030-04726-9_11

on school vacations, etc. Our case is a typical example of over-the-counter service industries: multiple servers/stations that process customer orders in a first-come-first-served manner. Customer/order arrivals and the service time are uncertain and arrival rates are typically time-dependent. For-profit firms and governmental organizations face the same problem of determining the right capacity (i.e., number of servers) for different time intervals. Customers expect good service in terms of short waiting times and decision-makers want to avoid excessive costs for idle capacity.

Many well-established approaches in the literature determine capacity levels based on distributional assumptions for the inter-arrival and inter-departure times of the customers. Such an approach, however, ignores the uncertainty around an estimated distribution parameter and in many practical instances the approach lacks the suitability to be implemented.

We present a novel, data-driven approach to prescribe optimal capacity levels by directly modeling the functional relationship between capacity decision and features that potentially drive the required capacity. Our integrated approach does not require any assumptions about the underlying arrival process. Given a sufficiently large data set of historical observations of features and associated arrival processes, our approach derives a decision rule that directly prescribes the minimal capacity to fulfill given service objectives. In this paper, we consider a single objective (e.g., at most 20% of the customer should have to wait more than 20 min), but we note that our approach can be extended to simultaneously incorporating additional service goals (e.g., at most x% abandonment rate or y minutes average waiting time).

2 Literature

Closely related to our approach is the work by [1] who propose a data-driven approach to determine capacities in a call-center model with multiple customer classes and multiple server pools using historical call-arrival data. In their approach arrival rates of incoming calls are not assumed to be constant or known. Instead of making assumptions about the distribution of the arrivals, they use empirical estimates for the arrival rates which they derive from samples of historic call-arrival-epochs with similar characteristics. Based on these estimated distributions they can determine the expected penalty costs from abandonments with respect to a chosen capacity and hence minimize the sum of the expected penalty costs and the costs for capacity.

They can show that with an increasing amount of available data, their data-driven approach approximately achieves the same costs as one using a simulation-based approach with known arrival rates. However, their results also show that with a decreasing amount of observation, the average costs of their approach increase. We consider this as critical, since [1] requires samples of historic call-arrival-epochs with similar characteristics. Let for example a set of such similar epochs contain all Monday mornings without vacations, in the first week of a month, with no special weather event. This still rather broad specification limits the amount of similar observations to less than 10 given we have one year of data available. Hence, the choice of relevant characteristics and how we determine similar observations will influence the decision. In contrast to [1], we integrate these considerations in our decision model. Our model groups historical demand observations such that they allow for the best decision. Another methodical difference is that our approach does not require to estimate arrival rates, since we directly consider the capacity decision that would have been optimal given past arrivals.

The model presented in [2] also considers a call-center staffing problem and proposes a data-driven approach that determines capacities for each unit period of the planning period (e.g., each hour of a day) by minimizing the mean cost over given historical arrival rates. They do not require explicit assumptions about the distribution of arrivals, however, they implicitly assume, that all historic observations from a specific time slot/unit period are similarly valuable for making the capacity decision for an upcoming period. Hence, they do not consider that external features could potentially explain parts of the variations in the historical data which is the main structural difference to the approach we present in this paper. Furthermore, their approach requires specific costs for waiting and abandonment which are not available in a setting like the public service office where specific service goals related to waiting time are more adequate.

3 Methodology

In this chapter we present a novel, data-driven approach to prescribe optimal capacity levels by directly modeling the functional relationship between capacity decision and features that potentially drive the required capacity. We first formulate the general model and show the flexibility of our approach. In the second subchapter we describe an implementation based on the machine learning technique of decision tree learning.

3.1 Distribution-Free Approach for Feature-Based Capacity Decisions

In this section we introduce a novel approach to prescribe a capacity level $\mu(\boldsymbol{x})$ for a time-slot given a feature vector $\boldsymbol{x}$ that represents information characterizing this particular time-slot, e.g., day of the week, whether the time-slot falls on a school holiday, etc. These prescribed capacities should fulfill certain service objectives $\boldsymbol{G}$ (e.g., ratio of staffing time-slots where at least 80% of the customers are served within a certain time). The actual capacity level $\mu(\boldsymbol{x})$ is then determined by minimizing the capacity level that is required to fulfill the service-level objectives $i = 1, \ldots, O$ for at least a ratio of G_i^{target} of the observations:

$$\min_{\mu(.)} \mu(\boldsymbol{x}) \tag{1}$$

$$s.t.\, G_i(\mu(\boldsymbol{x})) \geq G_i^{target} \tag{2}$$

We can interpret Eq. (2) as second-level service goals that allow to consider the trade-off between capacity and specific service-level objectives that are measured on a time-slot basis, e.g., the maximum waiting time or the average waiting time per customer. We note that Eq. (2) allows to control for multiple service goals independently which is a main difference compared to classical queuing approaches. Traditionally, decision makers have to focus on a single service goal. In the setting of our case study, the decision maker seeks to achieve that at most 20% of the customers within a certain time-slot should have to wait for more than 20 min. This is the only service goal, hence, $O = 1$. Such a constraint can be controlled and relaxed via Eq. (2). E.g., if $G_1^{target} = 0.95$, we allow the service goal to be missed in 5% of the cases. This makes the approach more robust against outliers.

Our data-driven approach learns the functional relationship $\widehat{\mu}(\boldsymbol{x})$ from a set of historical data $T = \left\{\left(\mu_n^{(*)}, \boldsymbol{x}_n\right)\right\}_{n=1,\ldots,N}$ where each observation consists of an ex-post optimal decision $\mu_n^{(*)}$ and a feature vector $\boldsymbol{x}_n$ for each historical time-slot $n = 1, \ldots, N$.

In order to determine the ex-post optimal decisions $\mu_n^{(*)}$, we evaluate the historical arrival processes $\boldsymbol{y}_n$ which consist of the individual arrival times of each customer for each historical time-slot $n = 1, \ldots, N$. Hence, we solve the data-driven counterpart of Eqs. (1)–(2) for a given set of learning data $\boldsymbol{T}$:

$$\min_{\widehat{\hat{\mu}}(.)} \sum_{n=1}^{N} \hat{\mu}(\boldsymbol{x}_n) \tag{3}$$

$$s.t.\, \hat{G}_i\left(\widehat{\hat{\mu}}(.), T\right) \geq G_i^{target} \tag{4}$$

Clearly, solving Eqs. (3)–(4) for a general function $\mu(.)$ is infeasible due to too many degrees of freedom. For this reason, we need to specify a certain form of the functional relationship. For our approach we chose a tree-based model which we find highly suitable due to its high flexibility in modeling complex feature-demand relationships as well as integrated feature selection mechanism. Besides these methodological properties, tree-based models have proven to perform well in various settings (see, e.g., [3, 4]).

3.2 Tree-Based Implementation

The general idea of tree-based machine learning algorithms is to partition the input feature space into disjunct "regions" by recursively finding the feature along with a split value that minimizes an objective function over a given set of historical "training data" $\boldsymbol{T}$. This procedure is recursively repeated until either an additional split would not lead to a substantial improvement or a minimum number of observations is reached. The interested reader is referred to the excellent presentation of tree-based models in [5] for further details.

The intuition behind this approach is that the decision we make for a specific time-slot is based on the decisions that would have been optimal in "similar" segments in the past. Our algorithm determines what is "similar" such that it allows for the best decisions (instead of mean predictions as with the standard tree-learning algorithm). Our solution encompasses the following four steps:

1. *Data preprocessing*: To make the algorithm computationally feasible, we build a $N \times M$-dimensional look-up table W where N is the number of available historical staffing segments and M is the maximum number of servers that is available per time-slot. The entries in W are the ratios of waiting times violating the service target for the arrival process in a particular (historical) staffing time-slot given a specific capacity μ, i.e.:

$$w_{n,\mu} = \sum_j \mathbf{1}\left(z_{nj}\left(\mu, \mathbf{y}_n\right) > t\text{max}\right) / \left|\mathbf{y}_n\right|$$

where $\left|\mathbf{y}_n\right|$ is the number of customers that arrived in time period n and $z_{nj}(.,.)$ is the waiting time of arrival j in time period n and $\delta_{t_{max}}(z) = 1\, if\ z > t_{max}$ and 0 otherwise. The evaluation of the arrival process, i.e., computing $z_{nj}(.,.)$ is the computationally expensive part. With the look-up table, we have to do this only once for each capacity and historical time-slot. We can use this look-up table to obtain the ex-post optimal capacity decisions that we need as a training data set for our algorithm and to evaluate the resulting decisions. We note that additional service goals would require additional look-up tables.
2. *Ex-post optimization*: From W we can obtain the ex-post optimal capacity decision for each time slot $n = 1, \ldots, N$ by:

$$\mu_n^{(*)} = \min_{\mu}\left\{w_{n,\mu} < (1 - \alpha)\right\}$$

where the service level α is the ratio of customers that are supposed to be served on time. We use these capacities in the learning data set $T = \left\{\left(\mu_n^{(*)}, \mathbf{x}_n\right)\right\}_{n=1,\ldots,N}$.
3. *Tree*-learning: We "learn" the structure of the tree by determining the partition of the parameter space that allows for the best capacity decisions. In detail, we recursively apply the following splitting step:

$$(\mathrm{x}_\mathrm{p}^*, \mathrm{s}^*) = \underset{(\mathrm{x}_\mathrm{p},\mathrm{s}):\mathrm{p}\in\{1,\ldots,\mathrm{k}\}\wedge \mathrm{s}\in\mathcal{X}_p}{\operatorname{argmin}} \left(\mathbb{L}\left(\left\{\left(\mu^{(*)}, \mathbf{x}\right) \in \mathrm{S}_\mathrm{T} | \mathrm{x}_\mathrm{p} \le \mathrm{s}\right\}\right) + \mathbb{L}\left(\left\{\left(\mu^{(*)}, \mathbf{x}\right) \in \mathrm{S}_\mathrm{T} \middle| \mathrm{x}_\mathrm{p}\right\} s\right\}\right) \quad (5)$$

where $\mathcal{X}_p$ is the set of all values of x_p, i.e., the p-th feature, in the learning data and the loss function $\mathbb{L}(S_T)$ for a set $S_T \subseteq T$ is the aggregated excessive capacity defined as follows:

$$\mathbb{L}(S_T) = \sum_{n:\left(\mu_n^{(*)},\boldsymbol{x}_n\right)\in S_T} \left(\mu_{S_T} - \mu_n^{(*)}\right)^+ \tag{6}$$

and

$$\mu_{S_T} = \min_{\mu}\left\{\mu \middle| \frac{1}{|T|} \sum_{n:\left(\mu_n^{(*)},\boldsymbol{x}_n\right)\in S_T} \mathbb{I}\left(\mu_n^{(*)} \geq \mu\right) \geq G^{target}\right\} \tag{7}$$

where G^{target} is the ratio of time slots where the service level goal should be reached. Equation (6) is the unutilized capacity if μ_{S_T} is the capacity assigned to all historical observations in a set S_T, which replaces the MSE as the basic loss function. Hence, Eq. (5) determines the split that allows for the best decision by grouping possibly similar situations. If $G^{target} = 100\%$ then Eq. (5) yields the maximum capacity in that subset. Essentially, the algorithm tries all possible splits (i.e., all combinations of x_p and s) and finds the combination that minimizes the sum of the losses from the subsets resulting from the split.

4. *Apply staffing function*: Given the feature vector $\boldsymbol{x}'$ for a new, unseen, staffing time-slot, we now obtain the staffing decision by sorting $\boldsymbol{x}'$ into a region r by comparing the splits in the tree with the associated values of $\boldsymbol{x}'$. More formally,

$$\hat{\mu}\left(\boldsymbol{x}'\right) = \sum_{r=1}^{R} \mu_r \mathbb{I}\left(\boldsymbol{x}' \in r\right)$$

where $r = 1, \ldots, R$ are the partitions of the feature space that were learned in the previous step. Figure 1 shows an example for a decision tree representation of the integrated learning approach. The obvious difference to a regression tree as depicted in Fig. 2 are the leaf labels that are prescribed staffing levels of the integrated tree and predicted quantities for the classical regression tree.

Our main contribution is the integration of the specific optimization problem (minimizing capacity subject to certain service goals) into the estimation of a model that learns the functional relationship between features and output. We expect that this approach is especially useful if (a) arrival rates are not stationary and if (b) the non-stationarity is feature dependent. To clarify these two conjectures, we consider the following simple example. On average, there are 100 customers per shift, but only 25 arrive in the first half of the shift whereas the second half of the shift sees on average 75 customers. Without any features, a separated approach bases the decision on the 100 estimated mean arrivals and typically misses the service goal due to the higher number of arrivals in the second half of the shift. Our integrated approach, would prescribe a capacity that would have achieved the service goal for past realizations of these arrival processes. Hence it would take the non-stationarity into account.

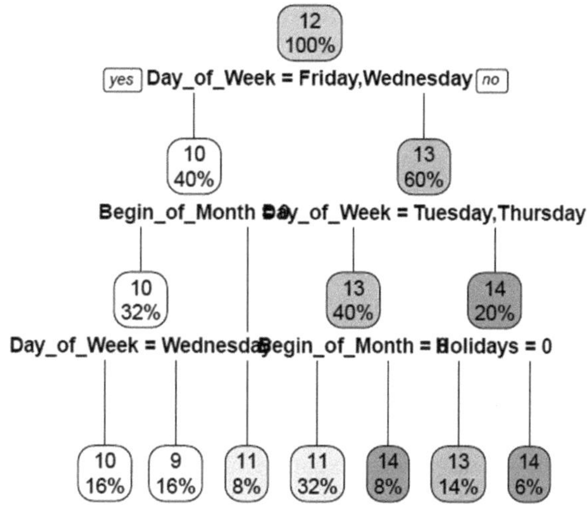

Fig. 1 Exemplary tree representation of an integrated model. The labels in the leaves are prescribed capacities

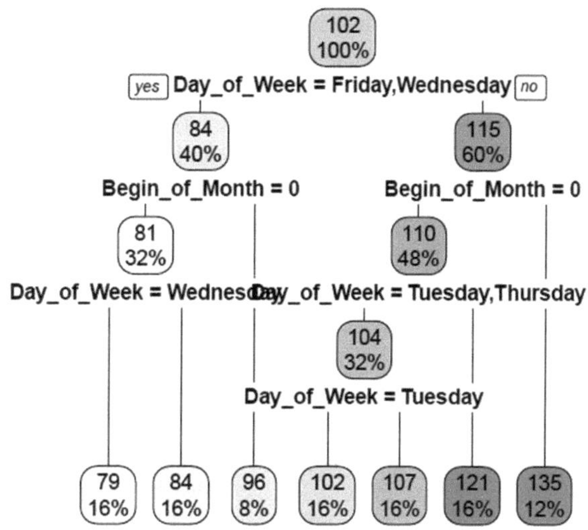

Fig. 2 Exemplary tree representation of a standard prediction tree. The labels of the leaves are predicted numbers of arrivals

In order to clarify conjecture (b), suppose we have a single binary feature, e.g., school holiday: yes/no that affects the arrival rates in the following way: During school holidays, arrival rates are constant throughout the shift with on average 100 arrivals. Without school holidays, we have non-stationarity as described above. In such a setting, a standard estimation model that aims at predicting the mean arrivals would not consider the school holidays feature, since it does not affect mean demand. Whereas our integrated approach would consider the feature if it improves the prescribed decisions, i.e., if it reduces the overall unutilized capacity, if different capacities are assigned to the subsets that are split by the school holiday feature. This is the main effect of the modified splitting function in Eq. (5) in step 3 of our procedure.

4 Case Study: Staffing Service Counters at a Public Services Office

In this section, we validate our approach from the previous section by applying it to the problem of finding optimal capacity levels for the staffing problem at a public services office in Germany. At this office citizens can apply and collect passports and ID cards, change their address, etc. We compare the results of our integrated approach with the more traditional separated approach that uses a standard decision tree model to estimate arrival rates and subsequently applies the Erlang-C formula to optimize capacities. While the separated approach based on Erlang-C may not be the most sophisticated solution available in the literature it is a relevant benchmark due to its prevalence in practice. For more details on the Erlang-C model, see for example [6].

In our case study the current labor agreements force employers to assign employees to fixed shifts which is a time window, for example, from 8 am to 12.30 pm. Hence, we have one 4.5-h staffing time-slot per day. We have one year (251 working days) of historical data including for each individual customer the time-stamp the customer arrived. These time-stamps are generated by an automated ticketing system: customers enter and draw a ticket and are called first-come-first-served once a server is free. Applying the service target to 'serve 80% of the customers in a waiting time below 20 min' to this historical data set, only for 35% of the staffing time-slots the service goal was reached. We denote this ratio by $G^{actual} = 35\%$. That is, for more than 65% of the days, respectively shifts, more than 20% of the customers had to wait more than 20 min.

Labor laws in Germany prohibit employers to track the individual service times at service desks. However, we know that a typical service task takes around 20 min, the minimum service time is 5 min and maximum service time can be 'substantially longer than the typical time'. Hence, for each arrival we draw a service time from a triangular distribution with min = 5, max = 60 and peak = 20 min.

As features we use day of week, whether the day is a school holiday, in the first week of the month or a bridge day (i.e., a working day between weekend and a single holiday). The prediction model achieves an out-of-sample MAPE of 13.5% in predicting the number of arrivals per time-slot. Just using the mean as prediction would result in a MAPE of 20%.

To evaluate both approaches on the given real-world dataset we use leave-one-out cross validation. I.e., one-by-one we take one observation from the data-set which we do not use for training the model, train the model and then evaluate the performance for the left-out observation. As performance measures, we consider the ratio of time-slots, where the service target was achieved, i.e.:

$$\hat{G}\big(\hat{\mu}(.), T\big) = \frac{1}{|T|} \sum_{\left(\mu_n^{(*)}, \boldsymbol{x}_n\right) \in T} \mathbb{I}\big(\mu_n^{(*)} \leq \hat{\mu}(\boldsymbol{x}_n)\big)$$

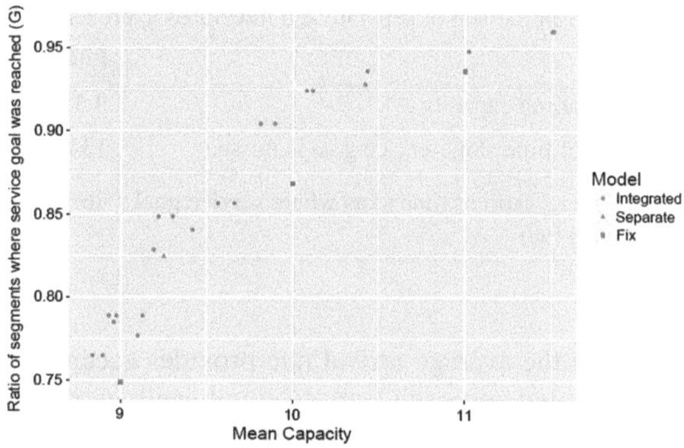

Fig. 3 Shows the achieved service target and required capacity. The integrated approach does not dominate the separate approach, but allows for trade-off between service target and capacity

For $G^{target} = 1$ the integrated approach yields $\hat{G}(\hat{\mu}(.), T) = 96\%$ with a mean assigned capacity of 11.5. The benchmark approach with separate estimation yields $\hat{G}(\hat{\mu}^{separate}, T) = 82.4\%$ with a mean assigned capacity of 9.25. Considering the service target, the integrated approach is clearly better. However, it also requires higher capacity. Using the parameter G^{target} we can trade-off required capacity and achieved service target in a controlled manner.

Figure 3 shows the ratio of achieved service target with respect to the required mean capacity. Compared to the naive approach, where we assign a fix capacity for all days, we can reduce the number of days where the service target is missed from 16 to 13 days (with 11 as fixed capacity respectively mean capacity in the integrated approach), from 33 to 24 days (fixed/mean capacity 10) and from 63 to 53 (fixed/mean capacity 9) using the integrated data-driven approach. The separated approach does not allow to assess different service targets. Hence, we can only compare the single result we obtain with the sequential approach which is $\hat{G}(\hat{\mu}^{separate}, T) = 82.8\%$ with a mean capacity of 9.19. With the integrated approach we achieve similar results with this capacity. We note that comparing both approaches is difficult since the flexibility of the separated approach is limited due to the lack of an adequate model parameter to evaluate different combinations of capacity and $\hat{G}(\hat{\mu}^{separate}, T)$.

Our previous analysis considered that a decision maker needs to assign a single capacity level for a whole shift (from 8 to 12:30 am). In the following, we also evaluate integrated and separate approach for hourly time-slots and find that our integrated approach clearly outperforms the separated benchmark based on Erlang C. The following table shows the detailed results. We see that with the same capacity requirement our approach reduces the number of time slots where more than 20% of customers have to wait more than 20 min by 90 which is a 40% improvement in the performance criterion (Table 1).

We suppose that the better relative performance of the integrated approach for hourly staffing segments compared to full shifts, where the performance is similar, is due to the following reason: For the longer staffing segments the fluctuations in the arrival processes average out. Since the typical pattern is an increasing arrival rate between 8 and 9 am, a peak between 9 and 11 am and a decline until 12.30, planning

Table 1 Comparison of separate and integrated approach for hourly time-slots

	Integrated	Separated
Mean required capacity	9.381	9.383
Number of time slots service goal is missed	134	224
$\widehat{G}(., T)$, i.e., ratio of time slots where service goal is reached (%)	86.7	77.7

based on the average arrival rate provides acceptable results. For hourly planning, the separated approach leads to significantly worse results since it would assign the same capacity to time-slots with increasing and decreasing arrival rates, as long as the average rate is similar. A more detailed examination is part of our future research.

5 Conclusion and Further Research

In this paper we present a novel, data-driven approach to prescribe optimal capacity levels by directly modeling a functional relationship between features that potentially drive the required capacity and the actual capacity decision. Our main contribution is the integration of the specific optimization problem (minimizing capacity subject to certain service goals) into the estimation of a model that learns the functional relationship between features and decision. We expect that this approach is especially useful if (a) arrival rates are not stationary and if (b) the non-stationarity is feature dependent. For the staffing problem at a public services office we find that integrated approach significantly outperforms the commonly used benchmark approach in the case of hourly planning time-slots.

Based on the basic model presented in this paper, our next steps for the case of the public service office will be to analyze the effect of the length of a planning segment on the relative performance of the integrated approach and the separated benchmark. We will also consider more complex service targets since, for example, from a customer's perspective, the mean waiting time is more relevant than the ratio of time-slots where an arbitrary service target is achieved. Our model allows to simultaneously take multiple service targets into account.

We will also extend our approach to other important capacity planning problems such as call-centers, where we can consider abandonments and multiple agent and customer classes. Call center typically track exact time-stamps for incoming, answering and ending calls. Hence, historical service times are given and we can avoid to work with generated service times. Since service times might as well be feature-dependent, we expect additional potential for integrated data-driven approaches like the one we present in this paper.

Furthermore, the comparison with more sophisticated benchmarks such as the data-driven approach described in [1] is a topic of further research. We expect that given a clustering of similar historical observations the involved optimization procedure described in [1] will lead to competitive results. However, in a complex practical setting, finding such a clustering might be challenging. We will investigate whether the clustering that comes as a byproduct of our approach can be used for the data-driven approach in [1].

References

1. Bassamboo A, Zeevi A. On a data-driven method for staffing large call centers. Oper Res. 2009;57(3):714–26.
2. Bertsimas D, Doan XV. Robust and data-driven approaches to call centers. Eur J Oper Res. 2010;207(2):1072–85.
3. Caruana R, Niculescu-Mizil A. An empirical comparison of supervised learning algorithms. In: Proceedings of 23rd international conference on machine learning. New York, NY: ACM; 2006. p. 161–8.
4. Caruana R, Karampatziakis N, Yessenalina A. An empirical evaluation of supervised learning in high dimensions. In: Proceedings of 25th international conference on machine learning, 2008. p. 96–103.
5. Hastie TJ, Tibshirani RJ, Friedman JH. The elements of statistical learning: data mining, inference, and prediction, 2nd. In: Springer Series in statistics. New York, NY: Springer; 2013.
6. Gans N, Koole G, Mandelbaum A. Telephone call centers: tutorial, review, and research prospects. Manuf Serv Oper Manag. 2003;5(2):79–141.

Harnessing Big Data and Analytics Solutions in Support of Smart City Services

Shailesh Kumar Pandey, Mohammad Tariq Khan and Robin G. Qiu

Abstract Connecting and leveraging different types of electronic data sources (e.g., mobile and networked sensors, devices, and systems) to create an integrated platform is always a challenging task. To meet the needs of smart city development, developing that platform to process collected data in real time to support smart city services becomes essential. A robust and scalable framework for integrating big data and analytics solutions thus is required, aimed at providing seamless integration of heterogeneous data to manage city transportation, traffic, energy consumption, schools, hospitals, and other public services in a smart and sustainable manner. This paper extends our preliminary framework studies by discussing how we can implement physical and social sensing using the proposed big data and analytics platform to enable better and smarter services than ever before in great detail. With the support of big data and analytics technologies, we use city mobility services to demonstrate the great potential of the proposed integration and aggregation framework. Specifically, real time data from Citi Bike is collected, processed, and modeled. The developed prototype in support of city mobility management and operations shows a variety of potential benefits of the proposed digital ecosystem platform.

1 Introduction

Today data generated from equipment, devices, social media, mobile apps, and IoT gadgets can be well collected and processed [1, 2]. Big data technologies can facilitate analyzing massive and unstructured data to extract information and discover knowledge [3]. Moreover, open source products and APIs make it cost effective. In this paper, we use New York Citi Bike system [1, 4, 5] as our use case to demonstrate the capabilities of a big data and data analytics ecosystem. This ecosystem in its entirety has been developed using open-source software and commodity hardware.

S. K. Pandey · M. T. Khan · R. G. Qiu (✉)
Big Data Lab, Engineering Division, Penn State University, Malvern, PA 19355, USA
e-mail: robinqiu@psu.edu

H. Yang and R. Qiu (eds.), *Advances in Service Science*, Springer Proceedings in Business and Economics, https://doi.org/10.1007/978-3-030-04726-9_12

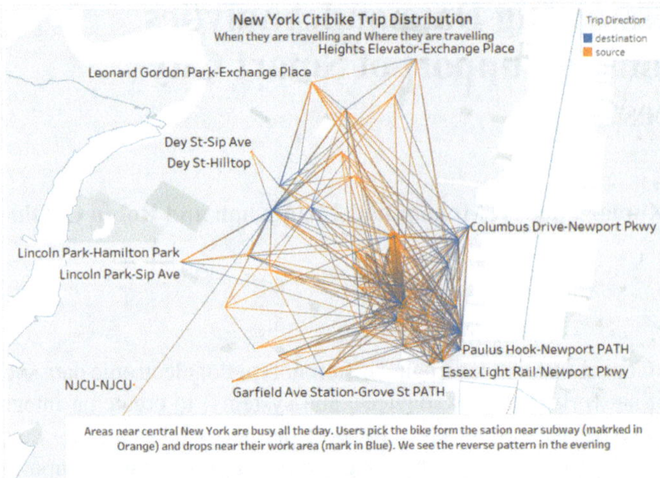

Fig. 1 Observation of Citi Bike riders' morning trips

Citi Bike is the largest America's bike sharing system with 12,000 bikes and 750 stations across New York metropolitan area. A rider can get annual membership or a one-day pass and then take a bike out from any of bike stations and return it to any other in the network. Users use shared bikes to commute to work or school, run errands, get to appointments or social engagements, and do much more [4, 5]. Citi Bike is available for use 24 h/day, 7 days/week, and 365 days/year. It had achieved 50 million trips in Nov 2017 [1]. Uneven demand at time and place results in unbalanced stations in terms of bike availability [6–8]. In morning hours bikes are clustered around the commercial locations, users pick the bikes from residential or subway locations and drop near their workplace (Fig. 1). In the evening hours we observe the reverse trend. Bikes at the stations are balanced manually (Fig. 2). Frequently, trucks are used to transport bikes form one station to another, to cope up with the very dynamic demand pattern on a daily basis. This study aims to minimize this manual redistribution effort [9].

2 Integrating Machine and Human to Enable and Support Smart City Services

There are many sources that predict exponential data growth toward 2020 and beyond [3, 10]. It is predicted that size of the digital universe will double every two years. With this abundance of data, we need special storages and computing platforms that can ingest and analyze data at lower cost. Historically data processing and analysis that has been done by high performance computing machines are no longer cost efficient with the current data explosion. This paper shows how to use commodity

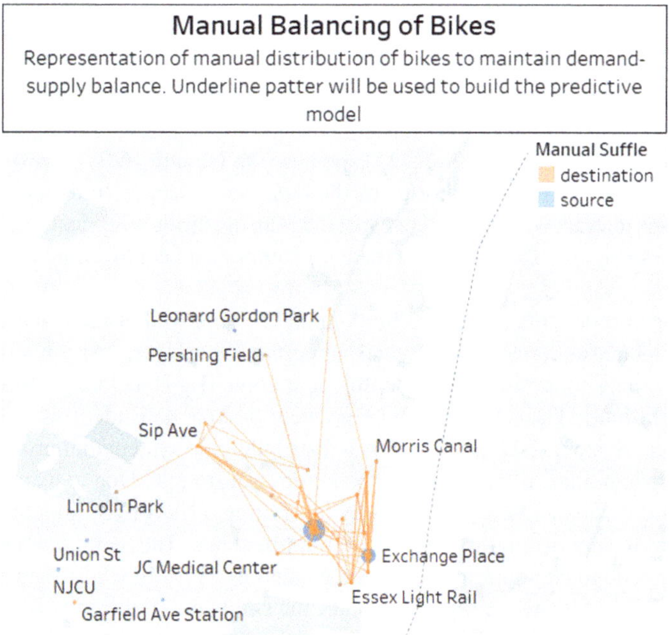

Fig. 2 An example of manual bike rebalancing operation

hardware to establish a big data infrastructure, specifically with open source software realizing operational maturity and established Hadoop data ecosystems, to deploy an enterprise grade solution. Over time, different entities in smart cities can use this platform in the form of service, Big Data as a service (BDaas).

2.1 Big Data as a Service—The Design Principle

The reason to choose a distributed system over a standalone workstation was that an increasing size of data over time implies that a distributed file system must be used to store it and the distributed techniques are required to ensure sufficient scalability. The major advantage of using Hadoop distribution is that all major Hadoop vendors—Cloudera, Hortonworks, IBM, and MapR—offer cloud-based deployments. These vendors allow users to download distributions that can be deployed on-premises or in private clouds on a variety of servers, including Linux and Windows systems. This provides an important advantage of replicating our architecture on cloud and providing global access to data.

2.2 Architecture

The architecture deployed in our big data lab includes 5 workstations running Hadoop Daemons (Fig. 3). We have one Master node and four slave nodes running Hortonworks Data Platform version 2.6.2 [11] and base OS Ubuntu 16.04. Five of the nodes in the cluster act as the repositories of the data we collect from different sources. These nodes interact with each other especially during the data replication and collection process (Fig. 4). Table 1 shows the core components deployed in this discussed cluster.

Hadoop components supported by Ambari are deployed at three service layers, which are named as core, essentials, and supports service layers respectively. Figure 5 provides a snapshot of an Ambari deployment at the Big Data lab, Penn State.

Core Hadoop: The fundamental component of Apache Hadoop is Hadoop Distributed File System (HDFS). HDFS is a distributed file system that provides scalability, fault-tolerance, reliability and economic data storage. Due to its master-slave architecture it enhances computations by leveraging YARN (Yet Another Resource Negotiator) to support multiple data access applications. The rack awareness feature allows for redundancy and minimal loss of data as every block of data resides on multiple racks. High availability eliminates the single point of failure, which can be further enhanced using HDFS federation by logically segregating the contents of the data. Due to operational simplicity, once the cluster is setup it requires minimum intervention. As a result, large clusters of the order of 2000 nodes are thus easily created and manageable.

Essential Hadoop: The following Apache components are deployed at the essential service layer. They are designed to ease working with Core Hadoop.

Apache Pig: This tool is required to design high-level data flow programs that can be compiled into sequences of MapReduce programs. The essential components of Pig are a compiler and scripting language called Pig Latin. For executing a pig script, the data needs to go through three stages: Load, Transform and Dump. Once the data from HDFS is loaded into Pig using Grunt shells, it can be used to execute queries and perform analysis.

Apache Hive: This tool was developed in order to enable SQL features on the data in HDFS. It's often referred to the data warehouse of Hadoop ecosystem since

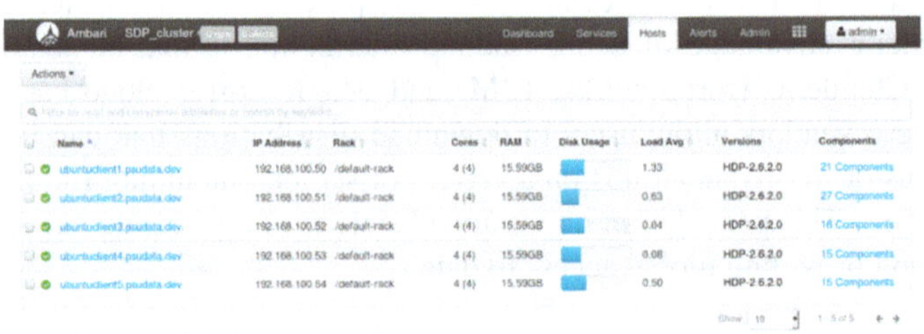

Fig. 3 A cluster of 5 nodes

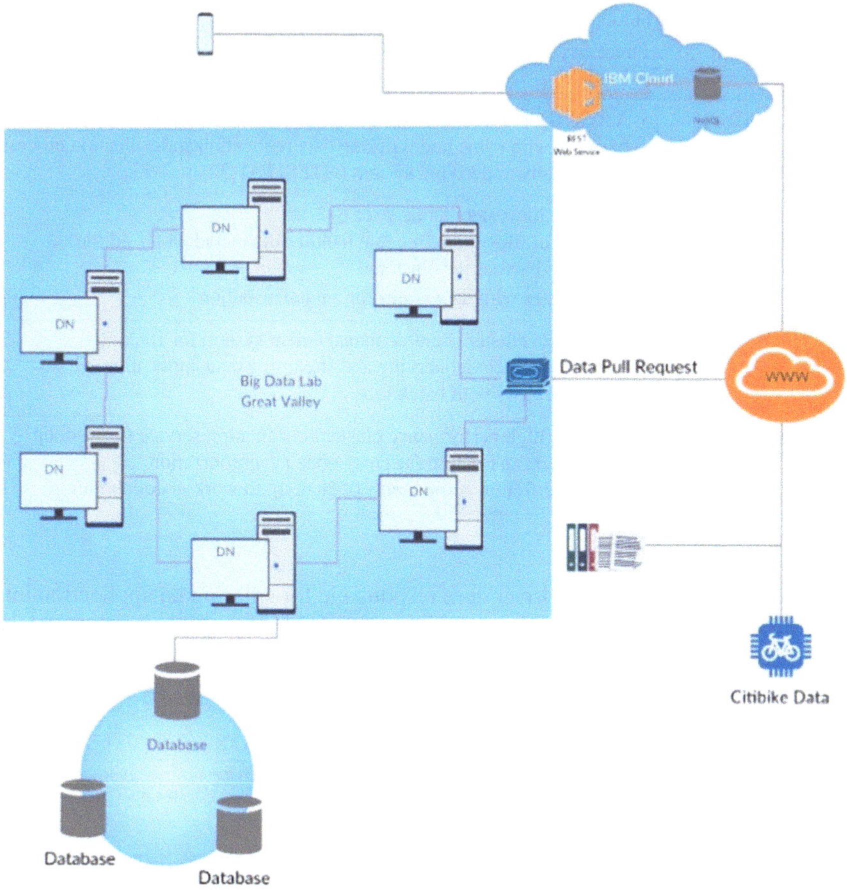

Fig. 4 Overview of the deployed big data ecosystem

it offers processing and analytical capabilities of raw unstructured data stored in HDFS. Since all the queries are converted to map reduce jobs, the performance can be slower yet it provides the facility to write user defined functions (UDF) is highly customizable when it comes to analytical queries.

Apache HBase: A distributed column-oriented database that provides real-time read-write access to files stored in HDFS. The schema-less format of data storage allows the data to be stored in a de-normalized fashion, making databases easily sharable.

Apache Spark: This component is one of the core components of a big data processing framework in a distributed architecture. Spark provides a platform for batch and stream processing and in most cases has replaced Map Reduce jobs due to its in-memory data engine. Resilient distributed dataset (RDD) allows an immutable collection of objects that can be distributed across the cluster. On the processing front,

Table 1 Core cluster Hadoop components

Core cluster components	Functions
Data Node (DN)	• Acts as data repositories for data collected from different source • Interacts with name nodes during data replication and collection process • Runs on any underlying system (NTFS, FAT 32)
Name Node (NN)	• Manages filesystem namespace • Consists of filesystem tree that handles the metadata for all files distributed across the cluster • Coordinates with data nodes for data distribution
YARN	• Serves as a cluster resource management system for Hadoop • Enables interactive querying and streaming data applications simultaneously with batch jobs
Zookeeper	• Manages high performance cluster coordination service for Hadoop • Provides infrastructure for cross-node synchronization • Enables different components of Hadoop to work in collaboration

Spark uses a combination of driver core responsible for splitting an application into individual tasks and individual executors that are assigned to process the workload assigned to them. This feature allows Spark to operate on RDD's in parallel resulting in fast data retrieval with the option of scaling the executors on an application basis.

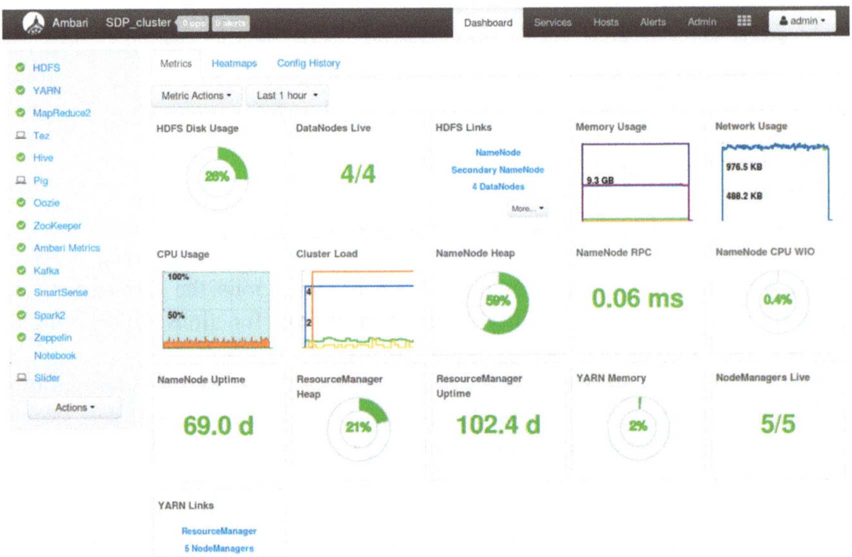

Fig. 5 Snapshot of an Ambari deployment

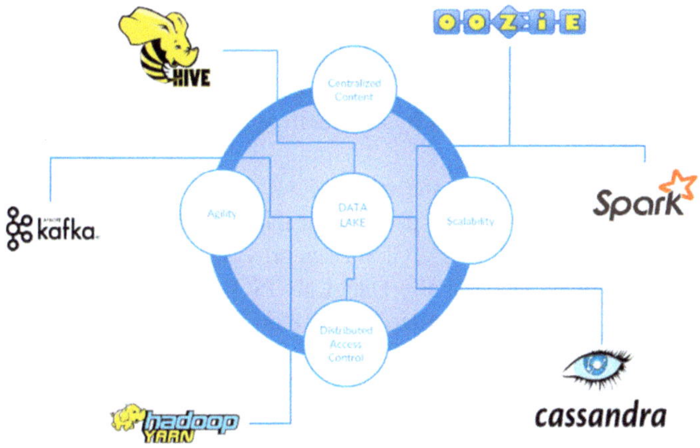

Fig. 6 Data lake architecture overview

Hadoop Support: The following components are deployed at the support service layer. They provide access to monitor Hadoop installation, which ensure all the components are functioning normally.

Apache Oozie: A server-based workflow engine that ensures a collection of actions can be performed seamlessly in the Hadoop ecosystem in order to build application pipelines. It consists of DAC (Direct Acyclic Graph) to represent the sequence execution of various components. It performs this scheduling of jobs using control and action nodes. The entire flow of the execution which includes the start, fail and end nodes as well as the mechanism to control the path of the flow execution, including decision, fork and join nodes are managed by control nodes. The execution of the workflows is triggered at specified intervals, which is performed by the designated action nodes.

Nagios: Nagios is an open Source tool used for monitoring, response, alerting, reporting, maintenance and planning. It provides its services using a dashboard which is very convenient for analysing metrics and identifying any issues before they take the infrastructure down.

Kafka: Kafka is a distributed, resilient and fault tolerant publish-subscribe messaging system, which enables high performance and horizontal scalability. Kafka has been used for de-coupling system dependencies and is the only messaging system in the world to provide exactly once semantics for streaming purposes.

Apache Nifi: It provides a web-based interface to enable seamless integration between different data collecting systems and provides guaranteed delivery. Data provenance can be easily tracked for every possible processor configuration making troubleshooting data pipelines easier.

Modern analytical framework should be capable of seamlessly integrate data from different sources with some intelligence. This architecture creates Data Lake (Fig. 6) where data can be stored value and time sensitivity.

Based on the type of data that is stored in systems, data can be processed in-memory in parallel or stored in less frequent accessed database. Data Lake does not only consolidate all the available data with provided ETL like framework, but also supports Hadoop components by accelerating data analytics processes in many application cases.

Since Penn State does not provide any public IP for this study that can be accessed from outside, IBM Cloud has been used as our intermediate components. A Node.js application has been hosted by IBM Cloud that provides REST endpoints to all those services that want to connect our Big Data Lab. JSON data pushed to a REST URL is saved on a NoSQL database hosted on the cloud, which can be access from any deployed application in the premise infrastructure. Currently an android Citi Bike rider mobile app is pushing rider's geo spatial data through this REST service.

3 Implementation

The Citi Bike API publishes data every five minutes for which we use a cron job to ingest the data into our ecosystem. With Spark, we have the advantage of real time streaming and processing of data. With libraries like machine learning (ML) and ML Lib, now it's easier to apply various machine learning models to perform predictive and descriptive analytics. We used various libraries like urllib, json, requests, pandas in python for data manipulation, cleaning and aggregation in order to pre-process the data and store good quality data in HDFS. Data is collected in HDFS every five minutes and this data is aggregated at the end of a day. The purpose of this data aggregation is to make possible various observations that are based on the timestamp.

Our work for New York Citi Bike is one suitable example to explain the usage of Penn State Big Data Lab infrastructure as a service [4, 5] (Fig. 7). The following items describe all the steps involved in the data processing and analysis of the Citi Bike implementation.

3.1 Data Collection

Citi Bike publishes relevant data in two different formats, Station data and Trip data. Station data includes real time data providing the status of a bike station, such as available bikes for rent, free docking spots, etc. Station data follows General Bikeshare Feed Specification (GBFS) that is refreshed at the interval of five minutes. Oozie manages all the jobs, written in python, in the ecosystem that runs at the interval of five minutes to pull station data from Citi Bike endpoint and then saves it locally in HDFS. A separate block in this script gets the weather data for each station from a separate provider. Citi Bike Trip data includes all riders' trip data and is published every quarter and contains information about trip duration, bike id,

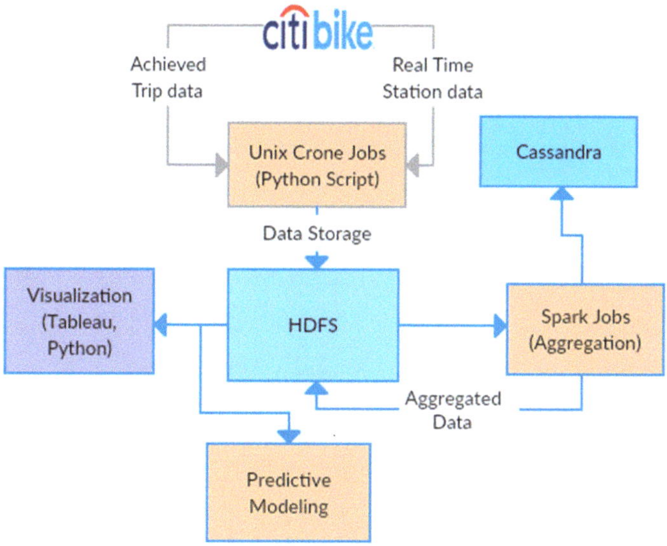

Fig. 7 An example of the usage of Penn State Big Data Lab infrastructure

source station, destination station etc. Currently this data is manually downloaded and saved into HDFS by a python script.

3.2 Data Processing

Once data is accumulated for a complete month in HDFS, Spark job is run for data aggregation. Aggregated data can be stored either in the Hadoop ecosystem (Hive, HBase) or in any NoSQL database. Our system can be easily configured for other data store including RDBMS. We are currently using python wrapper, pyspark, of spark (Fig. 8). Note that wrappers for R can be used interchangeably without any configuration changes.

3.3 Modeling and Visualization

Citi Bike uses Apache Spark to perform exploratory data analysis (EDA), involved in developing machine learning pipelines and using the APIs and algorithms available in the Spark MLlib DataFrames API. The reason behind selecting this data intensive

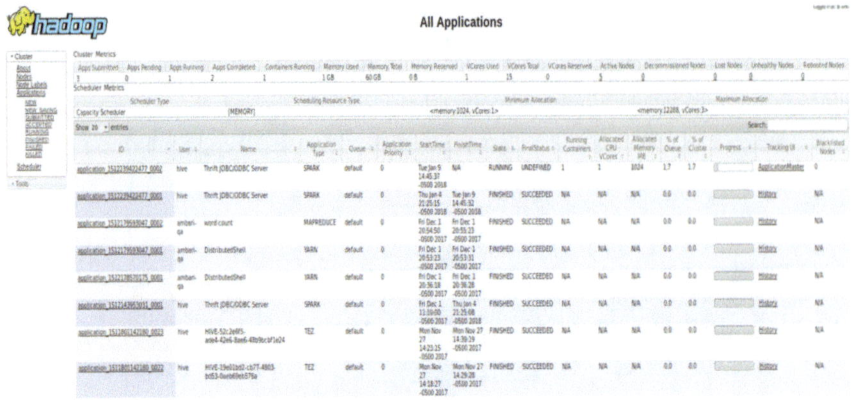

Fig. 8 Examples of task executions for the Citi Bike implementation

computing approach is to rapidly analyze exploding volumes of data at the point of creation and at scale. The adopted framework is mainly used for real time analytics. In addition to supporting real time analysis, data-driven modeling and visualization are also well supported. For example, hypothesis testing using multivariate analysis had been done for hourly trend, daily trend, rain and temperature for the Citi Bike implementation. The best model performance result was recorded when Random Forest was applied (Fig. 9).

4 Conclusions

By simply leveraging Penn State's big lab infrastructure, we have designed and built a big data analytics framework using open source tools. This platform allows us to harness big data and analytics solutions to provide optimum supports for heterogeneous and disparate data processing and accordingly decision making. In addition to Apache Hadoop, Spark and NoSQL, the use of cloud integration in our infrastructure has empowered the customers to remotely access our data and thus facilitate them in providing timely feedback and thereby making it possible work in an agile environment.

In this study we streamed bike station data in real time while retrieving historical riders' trip data from Citi Bike system. We proposed an integration and aggregation framework, aimed at facilitating in building a robust and analytics solution system. Currently, the built big data ecosystem functions as a multi-tenant analytics platform where other projects can be easily deployed on this infrastructure. This platform can also be used to phase out legacy data pipeline systems, resulting in significant cost saving and simplification of how we implement enterprise-grade infrastructures for big data and analytics solution ecosystems.

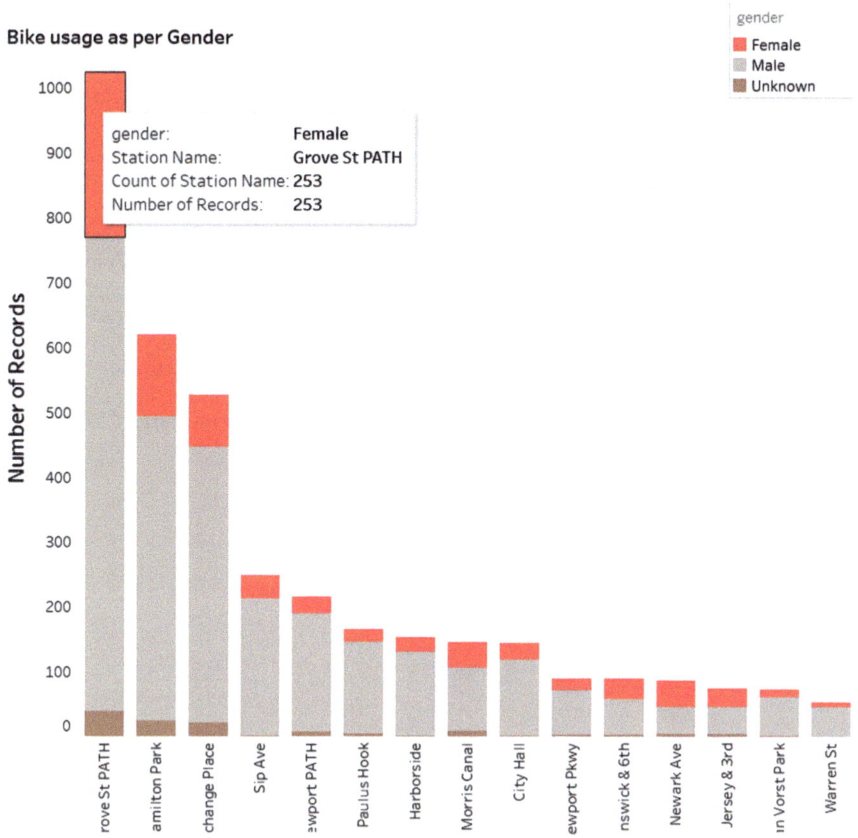

Fig. 9 An exmaple of modeling and visualization

Acknowledgements This work was done with great support and help from the Big Data Lab at Penn State and partially supported by IBM Faculty Awards (RDP-Qiu2016: *Data Analytics in support of City's Smart and Green Mobility Services* and RDP-Qiu2017: *Temporospatial Analytics to Enable Smarter City Mobility Services*).

References

1. Citi Bike. Unlock a bike. Unlock New York. 2018. https://www.citibikenyc.com/.
2. Jin J, Gubbi J, Marusic S, Palaniswami M. An information framework for creating a smart city through internet of things. IEEE Internet Things J. 2014;1(2):112–21.
3. Qiu RG. Service science: the foundations of service engineering and management. Wiley;2014.
4. Qiu RG, Qiu L, Badr Y. Integrating physical and social sensing to enable smart city mobility services. In: Proceedings of 14th IEEE international conference on industrial informatics (INDIN). 2016. p. 909–15.

5. Qiu R, Badr Y, Wang J, Li S. Developing a smart service system to enrich bike riders' experience. In: Proceedings of 2nd International conference on software, multimedia and communication engineering (SMCE2017). 2017. 455–459.
6. Mahony EO, Shmoys DB. Data analysis and optimization for Citi Bike sharing. In: Proceedings of 29th AAAI conference on artificial intelligence. 2015. p 687–94.
7. Preisler T, Dethlefs T, Renz W. Self-organizing redistribution of bicycles in a bike-sharing system based on decentralized control. Fed Conf Comput Sci Inf Syst (FedCSIS). 2016;2016:1471–80.
8. Singhvi D, Singhvi S, Frazier PI, Henderson SG, O'Mahony E, Shmoys DB, Woodard DB. Predicting bike usage for New York City's bike sharing system. AAAI Workshop: Computational Sustainability;2015.
9. Schuijbroek J, Hampshire RC, Van Hoeve WJ. Inventory rebalancing and vehicle routing in bike sharing systems. Eur J Oper Res. 2017;257(3):992–1004.
10. Qiu RG. Computational thinking of service systems: dynamics and adaptiveness modeling. Serv Sci. 2009;1(1):42–55.
11. HTW. Apache hadoop and big data platform for a data driven enterprise. 2018. https://hortonworks.com/.

The Pay Equity Dilemma Women Face Around the World

H. Muge Yayla-Kullu and Lana McMurray

Abstract "Pay discrepancies are bad for business, and yet they are rife in finance" (Morgan 2018). There has been very recent social movements around the world, especially in service industries to question and reduce the gender pay gap. In this research, we examine the pay equity dilemma women face and how it is different in various regions of the world. Our research focuses on the cultural characteristics (such as power distance, individualism, uncertainty avoidance, and masculinity) and how a society's norms affect pay inequality. We also go deeper in our discussions regarding the service industries. By better understanding the underlying reasons of pay inequality, changes can be made that will improve not only a business's bottom-line, but also the quality of lives all around the world.

Keywords Gender gap · Inequality · National culture · Services management

1 Introduction

"The gender pay gap in financial services is astonishing. It is almost 100 years since International Women's Day was first observed, and still we find women at some of the country's [UK's] top financial institutions are paid half as much as men... Barclays International has published a mean gender pay gap of 48%. For bonuses, it's 79%, meaning that for every £100,000 of bonuses handed out to men, women are only getting £21,000. The respective figures are 37 and 64% for RBS, and 33 and 65% for Lloyds" [12].

Women face discrimination in the work place. Report after report show us the grim details on the gender inequality in the workplace especially in management positions

H. M. Yayla-Kullu (✉) · L. McMurray
University of Central Florida, Orlando, FL 32816, USA
e-mail: muge@ucf.edu

L. McMurray
e-mail: lmcmurray@knights.ucf.edu

H. Yang and R. Qiu (eds.), *Advances in Service Science*, Springer Proceedings in Business and Economics, https://doi.org/10.1007/978-3-030-04726-9_13

which are essentially service jobs. The World Economic Forum 2017 index reports that the global gender gap is increasing and is going to take 217 years to close, [3]. Also, "Globally, women are paid less than men. Women in most countries earn on average only 60–75% of men's wages" (Facts and Figures 2017). Moreover, "for every 100 women promoted past entry level positions, 130 men are promoted" [14]. Clearly men and women are not moving at the same promotional pace especially at the higher level managerial positions.

Women all around the world are affected by wage inequality. We have already mentioned the issue in the UK. In the United States, it has been 55 years since the enactment of the Equal Pay Act and women are still earning 82 cents for every dollar a man earns [7]. This law was passed in 1963 and at that time the gender gap was 54 cents for every dollar a man earns. Even with the enforcement of the Equal Pay Act, it is estimated that the pay gap will not close many years to come. In Iceland, thousands of women left work 14 mins early in an orchestrated effort to protest their 14% gender wage gap in 2016. At that time, the wage gap was 72 cents to every man's dollar [7]. Less than a month later France followed suit protesting their 15.1% gender wage gap [1].

Carol Sankar, negotiation trainer, leadership advisor, and founder of The Confidence Factor For Women, reports that she is asked a routine question by women at her numerous training and speaking events. "Don't you think I should wait a few more years before asking for a raise?" Carol explains how this reflects the feeling of doubt and of being under-qualified that permeates the thoughts of women in middle and senior level management roles [13]. A study by Glassdoor showed that 68% of women accept the salary they offered compared to men at 52% per a survey [5].

The benefits of establishing pay equality has been also part of the recent discussions in business circles [15]. "Firstly, gender diversity pays; it's good for the bottom line. Credit Suisse found that companies where women make up at least 15% of senior managers had more than 50% higher profitability than those where female representation was less than 10%. Other benefits include a reduced chance of groupthink, enhanced connection to customers, and access to a wider talent pool. Secondly, the picture so far isn't great. A more gender diverse team brings benefits, but only one in four board members of financial services firms are women. Only 6% of chief executives of financial services firms are women. There is clearly a long way to go. Thirdly, culture is important. Witnesses have told us that the "alpha male" culture in financial services at senior levels is deterring women. Jayne-Anne Gadhia, chief executive of Virgin Money, described this as a culture of winning at all costs, rather than doing the right thing. Recurring cultural themes of our inquiry include sexual comments from male superiors, stereotyping by the "old boys' club" and its arcane recruitment practices, the "motherhood penalty", opaque bonus criteria, and presenteeism, whereby performance is judged by visibility rather than output" (Morgan 2018).

"The gender wage gap has now been intensively investigated for a number of decades, but also remains an area of active and innovative research" [2, p. 789]. Evidence demonstrates that the gender pay gap exists and is a complicated issue that can't be dismissed with the notion that women's personal choices or low self-esteem are the root cause. Wage inequality has many contributing factors, the motherhood penalty, lack of negotiation skills, limited work experience, housework, the treatment of female based jobs compared to male dominated roles, to name a few. Our theory will focus on the fact that most of these factors can be attributed to the societal norms dictated to the people living there. In addition, research shows that "of the current 19-cent gender wage gap, 41% (or about 8 cents) remains unexplained. In other words, 41% of the difference in pay between men and women has no obvious measurable rationale" [4] in the current body of knowledge.

Could national culture be one of under-investigated reasons? It's common knowledge that every nation has its own distinct way of expressing itself. It's a collective display of personality that we call culture. What is not commonly known is the influence culture plays in the gender wage gap. Can women living in different countries expect and accept their salaries to be lower because of the culture they live in? Do some dimensions have more influence on the women's pay gap relative to others?

The goal of this study is to understand the relationships between individual characteristics of national culture (such as power distance, uncertainty avoidance, individualism, and masculinity) and the gender pay gap. We explore the social mechanisms that can help to explain the pay gap inequality. We want to provide an understanding of these social dimensions, increase awareness of cultural barriers that hold women down, and uncover strategies that can aid women in reaching their economic potential.

2 Background Theory and Hypotheses

Social psychologist Prof. Geert Hofstede pioneered research on cross-cultural groups. In his ground breaking study, he developed a theory that organized the behavior of society into a framework known as Hofstede's cultural dimensions' theory. He identified that "Collective mental programming of people in different cultures exists." He referred to culture as the "software of our minds. What we share with those around us" [9, 10]. What Hofstede's [9, 10] research found was that people gathered in the same geographic location share unwritten rules. This group think is the foundation of the mental programming that is passed down from parent to child and forms the society.

Hofstede [9, 10] identifies four dimensions of national culture. More than 116,000 questionnaires were collected from employees at IBM in 40 countries around the world. He devised a cross-country comparison of the assumptions and values and ascribed each country a numerical value on a 100-point scale called an index. The naming convention is specific for each dimension its associated. This scale provides a comparative perspective to examine different cultural dimensions, namely Power Distance Index (PDI), Uncertainty Avoidance Index (UAI), Individualism Index (IDV), and Masculinity Index (MAS).

2.1 Masculinity

We believe masculinity is the most important cultural dimension as it directly relates to how women are seen from the eyes of the society. In a masculine society, gender roles are clearly defined. Men are supposed to be from Mars, women from Venus. Masculine societies are much more openly gendered than feminine societies [9, 10]. Clear discrimination between men and women is readily accepted and expected. A country with high masculinity expresses itself as competitive and showy; and the decisive and assertive traits are necessary for everybody. On the other hand, a feminine society, the genders are emotionally closer. Competing is not so openly endorsed, and there is sympathy for the underdog.

In Global Gender Pay Inequality, Labor markets are explained to have two key margins. (1) pure discrimination where women are paid less for the same work of equal value. (2) differences in value created due to occupational choice (Rockey 2017). Considering Hofstede's dimension of masculinity where women and men roles are clearly defined and enforced, this attitude would suggest the reason why jobs that are predominantly occupied by females are paid less than male centered roles is the cultural perception of "women" and "men".

Claire Cain Miller writes in her article, "Despite generous social policies, women who work full-time are still paid 15–20% less than men, new research shows a gender pay gap similar to that in the United States" [11]. The author further writes, "Children hurt mothers' careers. This is, in large part, because women spend more time on child rearing than men do, whether by choice or not." That's a gendered role in place, a sign of high masculinity. The author notes that men's pay is not affected by the birth of a child. This study also found that Women without children are paid almost 40% more than women with children. This is unfortunate because women overall are already behind the men by 20% [11]. Hence, we posit that masculinity dimension will be highly correlated with gender pay gap differences across the board.

Hypothesis 1. In a country with high masculinity, gender pay gap increases.

2.2 Power Distance

We believe the second most important dimension among all is the power distance. It shows the extent to which the less powerful members of organizations and institutions (like the family) accept and expect that power is distributed unequally. The acceptance of inequality in power lies with the people at the bottom [9, 10]. Each country is different in how people look at power. When different levels of power are evident, some cultures accept that this power is unquestionably right and should go unchallenged while others feel that everyone has a right to speak truth and tame it. Furthermore, to what degree do we expect the powerful to be sovereign and manageable determines our sense of autonomy where we can place controls on that power [9, 10].

In the pay inequity context, we are looking to see to what extent that the attitude of non-questioning of authorities translates into a woman's questioning her boss's decision to give promotion to incompetent-but-male colleagues. Does the female demand what is rightfully theirs or do they give in and "feel grateful that they even have a job"? In a country with low power distance such as Iceland (PDI = 30), a woman would expect to be given a fair wage for a day's work and would not hesitate to protest any discrepancies [7]. Whereas Russia with a high power distance score (PDI = 93) ranks second in Top 20 countries with the biggest percentage pay gaps as reported in MoveHub [8]. Hence, we hypothesize that a nation's power distance characteristics plays an important role in women's pay inequity around the world.

Hypothesis 2. In a country with high power distance, gender pay gap increases.

2.3 Individualism

Next, we discuss individualism. It is the extent to which people feel independent, as opposed to being interdependent as members of larger wholes. It means that individual choices and decisions are expected. On the other end, collectivism means that "one knows one's place" in life, which is determined socially. A self-imposed cultural barrier that women naturally erect that can interfere with economic potential is putting others above themselves. There are many societies that women choose to stay at home and raise children. The in-group association and the society's role for women influences their actions. Women welcome and follow the pre-determined roles. So, majority of these women willingly sacrifice their independence because they feel that the whole society is in better harmony if they stayed home. Hence, we hypothesize that in a collectivistic culture, women may follow the norms of the society where they choose to stay at home or get lower level jobs for the well-being of the male-dominant societies.

Hypothesis 3. In a country with low individualism, gender pay gap increases.

2.4 Uncertainty Avoidance

Lastly, uncertainty avoidance deals with a society's tolerance for uncertainty and ambiguity. It is the extent to which members in a society feel uncomfortable when ambiguity occurs and people try avoiding it. It is about anxiety and distrust in the face of the unknown, people having fixed habits and rituals and a wish to know the truth [9, 10]. An example of uncertainty avoidance in our context would be women staying in jobs that affect their health because it's what they are used to, resigning would create ambiguity and uncertainty of finding another job is overly stressful. For example, some women in France are suffering from an invisible occupational disease. This condition is caused by injury in workers with repetitive duties. It is due to the neglect in their work environment and the dismissal of their injuries by management.

According to Rachel Saada, an expert on labour law in France these are, "stresses the ambiguities" [6]. Labour doesn't recognize the impact to women's health or financial sustainability because they don't have a concrete process for addressing the dangers in "women's work." The avoidance of the term hardship and how it applies to work that women do is causing the problem [6]. Interestingly, France has an UAI score of 86.

This cultural dimension is related to the fear of doing something unknown and it can impact women's decisions in the workplace—maybe more than men. Uncertainty avoidance is another cultural characteristic that women need to learn how to navigate through. Exploring the culture more thoroughly and understanding how to successfully deal with this dimension is an essential skill to learn. We hypothesize that uncertainty avoidance is another cultural characteristic that cause lower wages for women around the world.

Hypothesis 4. In a country with high uncertainty avoidance, gender pay gap increases.

3 Methodology

In order to test our predictions, we use gender gap data published by the World Economic Forum as our dependent variable. We use the most recent data published in The Global Gender Gap 2017 Report. The report is prepared by a joint effort between the World Economic Forum, Harvard Kennedy School of Government, and Institute for Business and Social Impact at the Haas School of Business at UC-Berkeley. "Report benchmarks 144 countries on their progress towards gender parity on a scale from 0 (imparity) to 1 (parity) across four thematic dimensions Economic Participation and Opportunity, Educational Attainment, Health and Survival, and Political Empowerment and provides country rankings that allow for effective comparisons across regions and income groups." We have calculated the "Gender Gap" variable by "1-Global Index" as the Global Index measures equality and we are interested in the "inequality".

Our independent variables come from the well-established national culture literature. Hofstede's dataset has been utilized by thousands of studies. Hofstede's national culture dataset is a product of "a large research project, involving 116,000 questionnaires, about the work-related value patterns of matched samples of industrial employees in 50 countries and three regions at two points in time. Half of the variance in the countries' mean scores can be explained by four basic dimensions, here labeled power distance, uncertainty avoidance, individualism versus collectivism, and masculinity versus femininity. These dimensions are offered as a framework for developing hypotheses in cross-cultural organization studies. Index scores of the countries on the four dimensions correlate significantly with the outcomes of about 40 existing comparative studies" [9]. Since its inaugural release, Prof. Hofstede and his team have continuously updated this dataset expanding it to many more countries. The data is publicly available at https://www.hofstede-insights.com/.

When we merge the two datasets, we end up with 81 countries spanning all parts of the world.

4 Results and Discussions

In this section, we present a summary of our regression results that take gender pay gap as our dependent variable in this paper. We find that masculinity, individualism, and power distance have statistically significant cultural dimensions explaining the gender pay gap confirming our predictions. High masculinity, low individualism (high collectivism), and high power distance adversely affects women's role in society and reduces the amount of pay they receive in return of their fair share of the work (Table 1).

As expected, masculinity is high when a society has distinct definitions for each gender with child-care or household help is expected to be done by women. If women want to work, they are allowed to do menial jobs which pay significantly less. Hence, such societies have the most impact on gender gap ($\beta_{MAS} = 0.083$). Second most impactful dimension turns out to be the individualism ($\beta_{IDV} = -0.071$). In more collectivistic cultures, women's place is at home and not the workplace. In such countries, even the economic participation of women is less than other countries causing an increase in the gender gap. We also find that power distance plays a significant role in gender gap ($\beta_{PDI} = 0.064$). Power distance measures how much inequality is accepted by the powerless. Our finding shows that when women accept to be the "low class citizen" themselves, it hurts their position in the society.

Table 1 Results of the preliminary analysis

National culture dimensions	Gender gap
Power distance	0.064* (0.033)
Individualism	−0.071** (0.031)
Masculinity	0.083*** (0.028)
Uncertainty avoidance	0.038 (0.025)
Constant	21.661*** (3.448)
N	81
R^2	0.3756
adj. R^2	0.3427
F	11.4279

Standard errors in parentheses $^{*}p < 0.10$, $^{**}p < 0.05$, $^{***}p < 0.01$

Our results shows that it is very important that we educate women in the first place. For a more equal world, we should teach women that cultural norms does not determine their destiny and the fight starts with them by not accepting these norms.

5 Concluding Remarks

Gender pay gap and inequality at the workplace especially at the higher level management roles is prevalent around the world. Even the most developed countries like UK and US suffer from widespread inequality practices when it comes to women versus men. There is growing research on the issue and increasing attention to remedy the situation in most parts of the world.

In this research, we aim to look at the problem with a cultural point of view and try to understand the implications of societal norms on the workplace inequity. We use Hofstede's four dimensions and look at their relationship with gender pay gap indices in different parts of the world. We find that masculinity, individualism, and power distance are significant national culture dimensions that help explain the differences in gender gap.

We believe our contribution to the literature is significant since this is the first paper looking at this specific relationship. We hope that our findings help build the body of knowledge on the topic and remedy the situation sooner than later for the good of the whole world.

References

1. BBC News. French firms face gender pay gap fines. BBC News, 7 Mar 2018. www.bbc.com, http://www.bbc.com/news/world-europe-43320041.
2. Blau FD, Kahn LM. The gender wage gap: extent, trends, and explanations. J Econ Lit. 2017;55(3):789–865. https://doi.org/10.1257/jel.20160995.
3. Brinded L. Its going to take 217 years to close the global economic gender gap. Quartz, 1 Nov 2017. https://www.qz.com/1117632/its-going-to-take-217-years-to-close-the-global-economic-gender-gap/.
4. Carnevale AP, et al. Women cant win: despite making educational gains and pursuing high-wage majors, women still earn less than men. 2018. repository.library.georgetown.edu. https://repository.library.georgetown.edu/handle/10822/1049530.
5. Chamberlain et al. New research: demystifying the gender pay gap. Glassdoor Econ Res 23 Mar 2016. https://www.glassdoor.com/research/demystifying-the-gender-pay-gap/.
6. Equal Times. The invisible risks facing working women in France. Equal Times. https://www.equaltimes.org/the-invisible-risks-facing-working. Accessed 26 Mar 2018.
7. Fortune. 3 reasons why the gender pay gap still exists. Fortune. http://fortune.com/2017/04/03/equal-pay-day-2017-wage-gap/. Accessed 19 Mar 2018.
8. MoveHub. Global gender pay gap map|percentage pay gap for men and women. https://www.movehub.com/blog/global-gender-pay-gap-map/. 6 Mar 2014.
9. Hofstede G. Motivation, leadership, and organization: do American theories apply abroad? Organ Dyn. 1980;9(1):42–63. https://doi.org/10.1016/0090-2616(80)90013-3.

10. Hofstede Insights. Country comparison. https://www.hofstede-insights.com/country-comparison. Accessed 11 Mar 2018.
11. Miller CC. Children hurt womens earnings, but not mens (even in Scandinavia). The New York Times. NYTimes.com, https://www.nytimes.com/2018/02/05/upshot/even-in-family-friendly-scandinavia-mothers-are-paid-less.html. Accessed 18 Mar 2018.
12. Morgan N. Finance has a shocking gender pay gap. Shining a light on it is just the start. The Guardian, Thu 8 Mar 2018. https://www.theguardian.com/commentisfree/2018/mar/08/finance-gender-pay-gap-inquiry-nicky-morgan. Last accessed on 31 Oct 2018.
13. Sankar C. Why dont more women negotiate? Forbes, https://www.forbes.com/sites/forbescoachescouncil/2017/07/13/why-dont-more-women-negotiate/. Accessed 19 Mar 2018.
14. Women in the Workplace Study. Getting to gender equality starts with realizing how far we have to go. https://womenintheworkplace.com/. Accessed 7 Apr 2018.
15. Yayla-Kullu HM, et al. Employees national culture and service quality: an integrative review. Serv Sci. 2015;7(1):1–18.

Project and Resource Optimization (PRO) for IT Service Delivery

Haitao Li and Cirpriano A. Santos

Abstract This paper identifies the needs and challenges of IT service project delivery. A hierarchical Project and Resource Optimization (PRO) architecture is presented to provide a comprehensive and systematic roadmap for coping with the decision needs at the strategic, tactical, operational and executional levels. We highlight the data-driven feature of PRO with emphasis on the modeling and algorithmic methdologies to provide dynamic and adaptive decision-support.

Keywords Project management · Resource management · Mathematical programming · Analytics · Data-driven

1 Introduction

Information Technology (IT) is at the heart of any business. While flexible and efficient service delivery is the central goal in managing IT service projects, the gap between flexibility and efficiency has been widening [8]. On one hand, business firms and organizations endeavor to improve productivity, efficiency and speed for service delivery; on the other hand, they would also like to deliver flexible and customized services to meet their clients' unique needs and requirements. While efficiency can be relatively easy to achieve with more standardized and streamlined processes, it is often harder to achieve it with highly flexible and customized services. The tradeoff between flexibility and efficiency in the service industry is analogous to the well-known tradeoff between variety (process-based approach) and volume (product-based or continuous approach) in manufacturing. While mass customization

H. Li (✉)
College of Business Administration, University of Missouri–St. Louis,
St. Louis, MO, USA
e-mail: lihait@umsl.edu

C. A. Santos
Gurobi Optimization, Beaverton, OR, USA
e-mail: santos@gurobi.com

H. Yang and R. Qiu (eds.), *Advances in Service Science*, Springer Proceedings
in Business and Economics, https://doi.org/10.1007/978-3-030-04726-9_14

is a production paradigm that has both advantages [14], it is time for researchers and practitioners to define and create a new paradigm for service delivery.

According to The Future of Corporate IT [9], there are several driving shifts behind this challenge. The first is that a significant proportion of IT projects nowadays is information-based rather than process-based. A process-based project follows the traditional way of managing well-defined tasks/jobs and their relationships in the project. In contrast, an information-based project is largely driven by innovation, analytics and collaboration across organizations, geographical locations, and partners involved in the project. It calls for more flexibility and agility for service delivery. For example, managing the supply chain of a manufacturer often requires collaboration/coordination of suppliers/vendors and logistics providers at different stages in the chain. IT plays an important role in such an integrated supply chain by providing ERP systems to enable data/information to be shared in a seamless way throughout the supply chain. Then business analytics is needed to utilize the available data for better decision-making.

The second shift is that IT must be better integrated in business services for the need of global delivery. Both infrastructure and applications should be developed and built-in to achieve competitive advantage. For example, the cloud technology makes it an efficient infrastructure to store and share data across organization and geographic locations in a seamless way. Thus applications built upon the cloud infrastructure may suit well for the need of global service delivery.

The third is the growing need for externalized service delivery, as a result of knowledge/information sharing and collaboration beyond the boundary of a single firm/organization. Innovation often calls for cross-disciplinary knowledge and collaboration, so that one firm alone will be unlikely to have all the expertise and capabilities required for service delivery. The challenge arises to optimize the mix of internal and external resource utilization and cost for service delivery.

The aforementioned three shifts compel pertinent needs for the transformation of project and resource management to better align an organization's available resources with its strategic directions and operations. Project and resource managers aught to seek answers to the following questions:

- What are the new decision problems to address? Identifying and solving the right problem tailored to the unique needs of an organization is the way to go.
- Is the required data available, sufficient and reliable enough to be employed?
- How to achieve it? From the methodological perspective, what methods or synthesis of multiple methods are needed?

In this positioning paper, we first identify the existing issues and challenges in the current practice and research of the IT service industry. Then an integrated decision-support architecture called Project and Resource Optimization (PRO) is presented to cope with the challenges. We next describe various data-driven optimization schemes to implement the PRO modules. In particular, we present a generic modeling and solution framework to offer dynamic and adaptive decision-support, and elaborate how various analytical methodologies, namely, Descriptive, Predictive and Prescriptive, can be synthesized for providing effective, efficient and reliable solutions.

2 Issues and Challenges

Human Resource (HR) of professionals is the most important asset a service firm can own. Having the right resource at the right place, at the right time and right cost has been the slogan for effective workforce management for decades [23]. HR in IT is often heterogeneous in nature with multiple attributes, e.g., business domain, skill type, job level, location, workforce type and capacity [28]. Some existing issues and challenges in IT HR include: (i) Lack of proactive planning, especially at the strategic level. Resource allocation and assignment are often made as last-minute decisions, so that it may become difficulty to identify internal resources with the right skill on time. (ii) Lack of global view and accessibility of workforce pool, i.e. information about resource capability, capacity and availability is stored and shared locally within an isolated organization; (iii) Resource allocation and assignment decision is made manually and in a decentralized way. These often result in hiring more contingent workforce (CWF) than the regular workforce (RWF) with low internal workforce utilization and morale. Because CWF is usually more costly due to higher direct cost and learning, lower internal workforce utilization directly causes higher service delivery cost and lower profit margin. (iv) Delivery teams are aligned by technologies with siloed mind-sets and poor understanding of end-to-end performance of business impact. (v) Coping with risks and uncertainties reactively by piecemealing change and reorganizing selectively to reduce disruption.

A large body of the existing research in workforce management address the short-term personnel scheduling and assignment decisions, e.g., in airline [45], public transportation [44], healthcare [10], software development and consulting [29] among others. We refer to Ernst et al. [13], Van den Bergh et al. [43] and De Bruecker et al. [12] for systematic and comprehensive reviews in this line of research.

At the strategic and tactical level, a long-term manpower planning model was developed by Gass et al. [20] and Gass [19] to optimize the quantity and skill-mix of military workforce. Anderson [1] considered a strategic level staffing problem to manage the acquisition of knowledge and skills with nonstationary stochastic demand. Gans and Zhou [17] studied a time-varying capacity planning problem with skill levels, learning and turnover. Gresh et al. [22] developed a resource capacity planning (RCP) tool to determine the shortages and excesses of resources over multiple time periods. A simulation application called SimMan was developed by Huang et al. [25] to assess the impact of demand uncertainty on a workforce capacity planning solution. Cao et al. [7] presented a suite of methodologies called OnTheMark, which include various stochastic optimization models and methods, for effective management of human resource supply chains. The recent work of Davis et al. [11] developed a workforce management application for a cohort of individuals with similar attributes under uncertainty.

The existing approaches to workforce resource management tend to address issues and decision needs in a siloed way, although the concept of hierarchical decision framework was advocated by Gass [19] and Pinedo [33] long time ago.

3 Solving the Right Problems: The PRO Architecture

To address the existing issues and challenges in the IT service industry calls for a holistic approach. We present a unified framework called Project and Resource Optimization (PRO) architecture for this purpose. It aims to optimize the strategic alignment, tactical planning and operational utilization of resources, for better project delivery and improved return on investment (ROI). The PRO architecture is also a data-driven optimization framework. We shall elaborate what data will be needed for implementing PRO, and what *data-driven* means in the context of optimization.

A sketch of the PRO architecture for an IT service firm is provided in Fig. 1. Each decision module represented by a rectangular takes input data from its right-hand-side, and provides decision-support to a typical decision-maker at its left-hand-side. At the strategic level on the top, it addresses the Labor Strategy Optimization (LSO) as coined by Li et al. [26] to align the firm's workforce resources with its overall IT budget for the following fiscal year. The LSO module assists executives to optimize the allocation of IT budget in such a way that all the business units get the funding of projects that support the firm's business strategies. It also identifies any gaps in the capacity and capability of resources for the firm to address in a proactive fashion. Then at the tactical level, the Project Portfolio Optimization (PPO) [27], module assists a portfolio manager to optimize the selection and planning of project opportunities that best align with the firms strategies, under limited resources, i.e. budget and workforce resources. Next, after the optimal portfolio of projects has

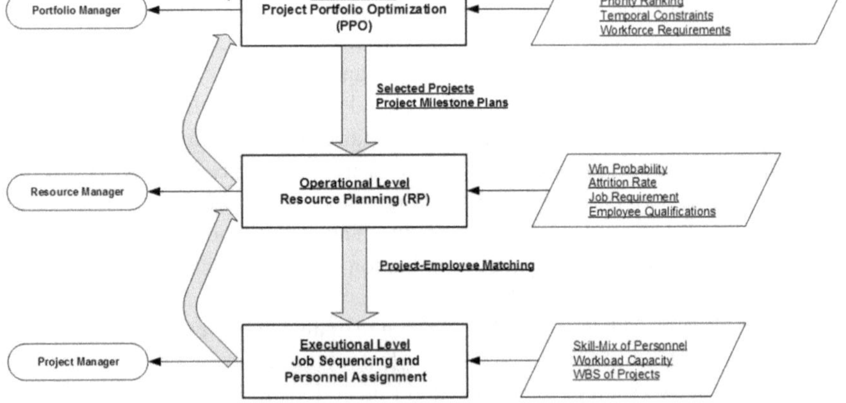

Fig. 1 Sketch of the PRO architecture for an IT service firm

been determined, the Resource Planning (RP) module at the operational level can be employed by a resource manager to optimize the matching between the available resources with the job opportunities with multiple attributes for measuring the matching quality [36]. What follows next is the project scheduling and sequencing decision that can be optimized for a project manager to enhance the efficiency of project execution, e.g., through the project scheduling with multi-purpose resources (PSMPR) [29], in the context of resource-constrained project scheduling [37].

The four modules in the PRO architecture are not isolated but are inter-related through two-way communication. The outputs of a predecessor module provide inputs to its successor; and reversely, the successor sends feedbacks to its predecessor module. For example, the optimized IT budget allocation and workforce capacity from the LSO module serve as the input, and specifically, the resource capacity, to the PPO module. And in turn, the portfolio composition and project milestone prescribed by PPO may re-shape the budget allocation in its upstream LSO module, and also serve as the inputs to its successor RP module.

4 How to Achieve It: Data-Driven Optimization

For one module in the PRO architecture, let the input data be represented by a vector $\mathbf{b}$ of dimension $1 \times n$, $\mathbf{c}$ of dimension $m \times 1$, and a matrix $\mathbf{A}$ of dimension $m \times n$. For now, we assume that $\mathbf{b}$, $\mathbf{c}$ and $\mathbf{A}$ are all deterministic and given as constants. Define $\mathbf{x} \in \mathbb{R}^n$ as an $n \times 1$ vector of decision variables. Then a generic mathematical programming model for the module can be expressed as $Max(Min) : g(\mathbf{b}, \mathbf{x})$, subject to $f(\mathbf{A}, \mathbf{x}) \leq \mathbf{c}$, where $g(\cdot)$ and $f(\cdot)$ is the objective function and a system of constraints, respectively, all being linear. Table 1 conceptually describes the model formulations of the four modules, including the decision variable $\mathbf{x}$, objective function $g(\cdot)$, and constraints $f(\cdot)$.

All of the four models may involve a mixture of continuous and integer (binary) decision variables, thus need the mixed-integer linear programming (MILP, cf. [32]) and/or various metaheuristics [21] to handle. One more note we shall add is that the model formulations presented here serve as examples, but do not mean to be exclusive. Each model component should be customized according to the decision needs and features of an organization. One may also extend the basic constructs to cope with variants and extensions of the problem setting.

4.1 *Paradigms for Data-Driven Optimization*

We now elaborate how the *data-driven* feature of the PRO architecture works. Six paradigms are delineated to implement the data-driven optimization.

Table 1 Conceptual model formulation of the PRO modules

	x	$g(\cdot)$	$f(\cdot)$
LSP	• Amount of each resource • Workforce transformation and cross-training • Gap and idleness of resources	• Maximize the total gross margin	• Meeting target revenues of market offerings • Demand dependency based on bill-of-labor • Available workforce capacities
PPO	• Project selection • Start time of projects and/or project milestones	• Maximize the total NPV • Maximize the priority of selection • Minimize total staffing cost • Minimize overall project completion time	• Portfolio composition requirements • Total budget constraint • Threshold on priority ranking • Temporal constraints among projects and/or project milestones • Resource constraints of workforce, hardware and equipment per period
RP	• Resource-job matching • Job loss and resource idleness	• Minimize the total staffing cost	• Assignment constraints • Constraints to identify job loss and resource idleness
PSMPR	• Project task scheduling and sequencing • Multi-skilled personnel assignment	• Minimize the project makespan • Minimize the total project execution cost	• Assignment of multi-skilled personnel • Assignment of hardware and equipment • Temporal constraints among project tasks

- **Input Data**: The most elementary form of data-driven optimization has its root in math programming where an optimal solution to a linear or integer program varies with the input data [24]. When either $\mathbf{b}$ or $\mathbf{A}$ is uncertain or involves random parameters, it can be replaced by its corresponding point estimate (mean) $\bar{\mathbf{b}}$ or $\bar{\mathbf{A}}$, which is the well-known deterministic or certainty equivalent (CE) optimization approach.
- **Sensitivity Analysis**: It is also known as the *what-if analysis* or *post-optimality analysis* in math programming [24] to examine how the optimal objective value and the optimal solutions respond to the changes of the input parameters. Note that this approach is still deterministic in nature, as it does not consider any uncertainty *prior to* obtaining an optimal solution.
- **Rolling Horizon**: This is a widely applied paradigm in real life deployment of a multi-period optimization. The original decision variable $\mathbf{x}$ is decomposed to $\mathbf{x_1}, \mathbf{x_2}, \ldots, \mathbf{x_T}$, where $\mathbf{x_t}$ is the decision to be made in period t, and T is the total number of periods in the decision horizon. In period t: (i) the point estimates $\bar{\mathbf{b}}$ and $\bar{\mathbf{A}}$ of the input parameters are updated based on the newly observed information/data arrived in t; (ii) then the original model is solved for $\mathbf{x_t}, \mathbf{x_{t+1}}, \ldots, \mathbf{x_T}$;

(iii) only $\mathbf{x_t}$ is implemented for the current period t. The system evolves to the next period and (i)–(iii) repeat.

- **Two-Stage Stochastic Programming**: This paradigm applies for the situation where the decision variable $\mathbf{x}$ consists of the *here-and-now* decision in the first-stage, and the second-stage *recourse* decision only to be made after the actual realization of the random parameters are observed [6]. Here, the model formulation directly uses the probability distribution of random parameters to compute the expected value of certain measure as a function of the recourse variables. The probability distribution can be estimated by various descriptive statistical methods with the available data of the random parameters.
- **Stochastic Dynamic Programming**: When the problem at hand involves decisions to be made over multiple time periods or stages, subject to uncertainty, the stochastic dynamic programming, also known as the Markov decision process (MDP, [35]) approach is a good candidate. The kind of decision-support provided by this paradigm is called a *policy*, which maps a stage-state pair to a decision for the current stage. Because the decision is prescribed based on the current state of the system, it offers a true dynamic and adaptive solution, known as the closed-loop policy [4]. Its data-driven feature works in two-ways: (i) by using the exogenous information to update the state of the system; (ii) by using the historical data to estimate the probability distribution of certain random parameters.
- **Simulation-Optimization**: This paradigm integrates both the simulation and optimization methods for a static (one-stage) problem, or for offering an *open-loop policy* to a multi-stage problem. Among various recipes for integration [15], a successful one is to integrate Monte Carlo (MC) simulation within a metaheuristic framework (cf. [2, 16]), where the metaheuristic is employed to search the solution space and to avoid local optima, and the MC simulation is used to evaluate a candidate solution during the search. The data-driven feature comes into play by estimating the probability distribution in the MC simulation.

The first three paradigms are all *deterministic* in nature in that they assume the input data are either known as constants, or can be replaced by the point estimates via the certainty equivalence assumption. All the last three paradigms explicitly cope with uncertainty. The choice depends on the nature and needs of the problem at hand, e.g., static versus two-stage or multi-period, open-loop or closed-loop policy needed. In the PRO context, the two-stage stochastic programming approach has been applied for the LSO by Li et al. [26] and the stochastic RP (SRP) by Li et al. [28]. The stochastic dynamic programming approaches have been developed for multi-period stochastic resource planning (MPSRP, [41]) and a stochastic resource-constrained project scheduling problem (SRCPSP, [30, 31]).

4.2 Dynamic and Adaptive Decision-Support

Given its apparent generality and flexibility, and the advantage of dynamic and adaptive decision-support, the stochastic dynamic programming or Markov decision process (MDP) approach is further elaborated with its generic modeling framework and solution strategies. An MDP model consists of the following five components.

- **Stage**: A *stage* $t = 1, 2, \ldots, T$ denotes a point of time or process when a decision needs to be made. In resource planning, a stage is often a week; in project scheduling, a stage is the time point when a new task may start.
- **State**: The *state* variable S_t of stage t contains all the information/data needed for making the decision at t. For example, the state variable in resource planning includes all the available resources and job opportunities; the state variable in project scheduling includes the completed tasks, tasks in progress and the current available resource capacity.
- **Decision**: Let $X_t \in \mathbb{D}(S_t)$ denote the set of feasible decisions that can be made at t, where $\mathbb{D}(S_t)$ is the feasible region corresponding with the state S_t in stage t. X_t in resource planning is the resource-job matching decision; and in project scheduling it refers to the set of tasks that can be started at t.
- **State Transition**: The state of the system transits from S_t to S_{t+1} if the decision X_t is made at t, and the exogenous information observed at $t+1$ is $\tilde{W}_{t+1}$ through the transition function $S^M(\cdot)$, i.e. $S_{t+1} = S^M(S_t, X_t, \tilde{W}_{t+1})$. In resource planning, $\tilde{W}_t$ refers to the availability status of resources and job opportunities, which is a random parameter; in project scheduling, $\tilde{W}_t$ may refer to the random disturbance that makes task durations to be uncertain.
- **Value Function**: The value function $g(S_t, X_t, S_{t+1})$ measures the immediate return (or cost) gained (or incurred) when the system is in state S_t at stage t, the decision X_t is made, and the system transits to state S_{t+1}. In resource planning, $g(\cdot)$ is the staffing cost incurred in the current stage; in project scheduling it can be the increment of project makespan in the current stage.

With the above model components well-defined, the overall objective function of the MDP model can be expressed as:

$$Max \quad \mathbb{E}\{\sum_{t=1}^{T} g(S_t, X_t, S_{t+1})\} \tag{1}$$

The goal is to determine the best policy π among the set of all policies Π, such that the objective function (1) is maximized. Let X_t^{π} denote the decision made in stage t following policy π. The *cost-to-go function* to be maximized in t can be written as:

$$J_t(S_t) = \mathbb{E}\sum_{\tau=t}^{T} g(S_\tau, X_\tau^{\pi}, S_{\tau+1}) \tag{2}$$

That is, an optimal policy π^*, given the current state S_t in stage t, should maximize the total expected value for the current and all the remaining stages by following π^*. The optimal closed-loop policy can be obtained by computing the recursive function of Bellman [3]:

$$X_t^{\pi} = \arg \min_{X \subset \mathbb{D}(S_t)} \mathbb{E}\{g(S_t, X_i, S_{t+1}) + J_{t+1}(S^M(S_t, X_t, \tilde{W}_{t+1}))\} \tag{3}$$

Directly solving (3) for the PRO modules suffers curse-of-dimensionality due to three types of high-dimensionality: (i) large state space, (ii) large number of scenarios of random parameters, and (iii) large solution space (often combinatorial in nature). For instance, the number of states and the possible random scenarios in stochastic resource planning, about the availability status of resources and job opportunities, grow exponentially with the number of resources and job opportunities [41]; in stochastic project scheduling with resource constraints, there are numerous states and random scenarios due to all the possible task durations [31], and finding the optimal sequence of tasks under resource constraints is NP-hard [18], even for the deterministic version of the problem.

Therefore, developing computationally tractable algorithms, namely, the approximate dynamic programming (ADP), is a key to success for the MDPs of real life PRO modules. ADP has its root in neuro-dynamic programming (NPD) of Bertsekas et al. [4] and reinforcement learning (RL) [42]. We refer to Si [39] and Powell [34] for a systematic review and treatment on this topic. The essence of ADP is to replace the exact cost-to-go by some form of its approximation. As reviewed by Li and Womer [31], there are two general strategies to achieve this. The first is to devise an *explicit* functional approximation for the cost-to-go. This approach works well for problems having special structure amenable to linear programming or network optimization, and has been applied by Simao et al. [40] for fleet optimization, and by Solomon et al. [41] for the multi-period stochastic resource planning problem.

The second strategy is to replace the exact cost-to-go *implicitly* by some heuristic base policy in the *rollout* framework of Bertsekas et al. [5]. It can be viewed as a heuristic version of the policy iteration algorithm in dynamic programming. This approach is attractive for various combinatorial optimization problems, for which effective and efficient heuristics are available. It has been successfully applied for stochastic vehicle routing [38] and stochastic resource-constrained project scheduling [31].

4.3 Synthesis of Analytical Methods

Either of the two approximation approaches in ADP requires multiple analytical methods to function. It is an excellent arena for designing and implementing hybrid algorithms that integrate various descriptive, predictive and prescriptive techniques. We provide several recipes for integration below.

- **Sample Path via MC Simulation**: An important technique in ADP is the forward iteration through the *sample path* generated by MC simulation. This avoids the need of complete enumeration of states in the backward recursion of the classical dynamic programming.
- **Solving Sub-Problems by Optimization Methods**: The direct value function approximation approach relies on efficiently solving the sub-problem in each iteration through various math programming methods: linear programming, network optimization and integer programming. The key solving technique in the rollout framework is the base policy via some heuristic procedures, i.e. either the custom-built special-purpose heuristics, or the metaheuristic algorithms.
- **Approximating Value Function by Predictive Analytics**: Various predictive techniques including regression, forecasting in statistics, and machine learning in artificial intelligence (AI) can be employed to improve the quality of value function approximation.

5 Journey Ahead

The need for efficient yet flexible service delivery creates opportunities to develop innovative decision-support paradigms for project and resource management in the IT service industry. The Project and Resource Optimization (PRO) architecture presented in this paper delineates a hierarchical system of decision modules ranging from the strategic and tactical levels to the operational and executive levels. We elaborate six data-driven optimization paradigms in the PRO context. Additional exposition is provided on the stochastic dynamic programming approach to provide dynamic and adaptive decision-support in the most general and flexible fashion.

Moving forward, we envision the following directions for researchers and practitioners. It is our belief that innovations originate from real life decision needs and features. It will be a fruitful path for researchers from academia to work with subject matter experts (SMEs) in industry to identify new characteristics of the PRO modules tailored to the needs of a firm, build innovative model formulations, and develop effective, efficient and reliable solution algorithms. The existing research has focused primarily on workforce resources, while the hardware resources (HWR) and software resources (SWR) also play an important role in IT service organizations. One promising direction for future research may incorporate HWR and SWR into the PRO modules to reduce the acquisition cost and improve utilization.

References

1. Anderson EG. The nonstationary staff-planning problem with business cycle and learning effects. Manage Sci. 2001;47(6):817–32.
2. April J, Glover F, Kelly JP, Laguna M. Simulation-based optimization: practical introduction to simulation optimization. In: Proceedings of the 35th conference on Winter simulation: driving innovation. 2003. p. 71–78 (Winter Simulation Conference).

3. Bellman RE. Dynamic programming. New York: Courier Dover Publications; 1957.
4. Bertsekas DP, Bertsekas DP, Bertsekas DP, Bertsekas DP. Dynamic programming and optimal control, vol. 1. MA: Athena scientific Belmont; 1995.
5. Bertsekas DP, Tsitsiklis JN, Wu C. Rollout algorithms for combinatorial optimization. J Heuristics. 1997;3(3):245–62.
6. Birge JR, Louveaux F. Introduction to stochastic programming. New York: Springer Science & Business Media; 2011.
7. Cao H, Hu J, Jiang C, Kumar T, Li TH, Liu Y, Lu Y, Mahatma S, Mojsilović A, Sharma M, et al. Onthemark: integrated stochastic resource planning of human capital supply chains. Interfaces. 2011;41(5):414–35.
8. CEB. The new model for it service delivery, volume i: design, definition and governance. Report, CIO Leadership Council, The Corporate Executive Board Company; 2012.
9. CEB. The future of corporate it: 2013–2017. The Corporate Executive Board. Company: Report; 2013.
10. Cheang B, Li H, Lim A, Rodrigues B. Nurse rostering problems–a bibliographic survey. Eur J Oper Res. 2003;151(3):447–60.
11. Davis MJ, Lu Y, Sharma M, Squillante MS, Zhang B. Stochastic optimization models for workforce planning, operations, and risk management. Serv Sci. 2018;10(1):40–57.
12. De Bruecker P, Van den Bergh J, Beliën J, Demeulemeester E. Workforce planning incorporating skills: state of the art. Eur J Oper Res. 2015;243(1):1–16.
13. Ernst AT, Jiang H, Krishnamoorthy M, Sier D. Staff scheduling and rostering: a review of applications, methods and models. Eur J Oper Res. 2004;153(1):3–27.
14. Fogliatto FS, Da Silveira GJ, Borenstein D. The mass customization decade: an updated review of the literature. Int J Prod Econ. 2012;138(1):14–25.
15. Fu MC. Optimization for simulation: theory vs. practice. INFORMS J Comput. 2002;14(3):192–215.
16. Fu MC, Glover FW, April J. Simulation optimization: a review, new developments, and applications. In: Proceedings of the 37th conference on winter simulation. 2005. p. 83–95 (winter simulation conference).
17. Gans N, Zhou YP. Managing learning and turnover in employee staffing. Oper Res. 2002;50(6):991–1006.
18. Garey MR, Johnson DS. Computers and intractability. A guide to the theory of np-completeness. 1983.
19. Gass SI. Military manpower planning models. Comput Oper Res. 1991;18(1):65–73.
20. Gass SI, Collins RW, Meinhardt CW, Lemon DM, Gillette MD. Or practice the army manpower long-range planning system. Oper Res. 1988;36(1):5–17.
21. Glover F, Kochenberger G. Handbook of metaheuristics. Berlin: Springer; 2005.
22. Gresh DL, Connors DP, Fasano JP, Wittrock RJ. Applying supply chain optimization techniques to workforce planning problems. IBM J Res Dev. 2007;51(3.4):251–61.
23. Grinold R, Marshall K. Manpower planning models. Co: North-Holland Pub; 1977.
24. Hillier FS. Introduction to operations research. New York: Tata McGraw-Hill Education; 2012.
25. Huang HC, Lee LH, Song H, Eck BT. Simmana simulation model for workforce capacity planning. Comput Oper Res. 2009;36(8):2490–7.
26. Li H, Santos CA, Fuciec A, Gonzalez MT, Jain S, Marquez C, Mejia C, Zhang A. Optimizing the labor strategy for professional service firms. IEEE Trans Eng Manage. 2018a (To appear).
27. Li H, Santos CA, Lopez I. New models and methods for project portfolio optimization. Working paper, University of Missouri–St Louis. 2016.
28. Li H, Vargas M, Santos CA, Ramshaw L, Lopez I, Perez S, Valencia C. Optimizing large-scale stochastic resource planning. Working paper, University of Missouri –St. Louis. 2018b.
29. Li H, Womer K. Scheduling projects with multi-skilled personnel by a hybrid milp/cp benders decomposition algorithm. J Sched. 2009;12(3):281–98.
30. Li H, Womer K. Stochastic resource-constrained project scheduling and its military applications. MORS Phalanx. 2011.

31. Li H, Womer K. Solving stochastic resource-constrained project scheduling problems by closed-loop approximate dynamic programming. Eur J Oper Res. 2015;246(1):20–33.
32. Nemhauser G, Wolsey L. Integer and combinatorial optimization. New York: Wiley; 1988.
33. Pinedo M. Planning and scheduling in manufacturing and services. Berlin: Springer; 2005.
34. Powell WB. Introduction to approximate dynamic programming. Approximate dynamic programming: solving the curses of dimensionality; 2011. p . 91–127.
35. Puterman ML. Markov decision processes: discrete stochastic dynamic programming. New york: Wiley; 2014.
36. Santos CA, Gonzalez MT, Li H, Chen KY, Beyer D, Biligi S, Feng Q, Kumar R, Jain S, Ramanujan R, Zhang A. Hp enterprise services uses optimization for resource planning. Interfaces. 2013;43(2):152–69.
37. Schwindt C, Zimmermann J. Handbook on project management and scheduling. International handbooks on information systems, vol. 1&2. Heidelberg: Springer; 2015.
38. Secomandi N. A rollout policy for the vehicle routing problem with stochastic demands. Oper Res. 2001;49(5):796–802.
39. Si J. Handbook of learning and approximate dynamic programming, vol. 2. New York: Wiley; 2004.
40. Simao HP, Day J, George AP, Gifford T, Nienow J, Powell WB. An approximate dynamic programming algorithm for large-scale fleet management: a case application. Transp Sci. 2009;43(2):178–97.
41. Solomon S, Li H, Womer K, Santos CA. Approximate dynamic programming for multi-period stochastic resource planning. Forthcoming in decision sciences, 2018.
42. Sutton RS, Barto AG. Introduction to reinforcement learning, vol. 135. Cambridge: MIT Press; 1998.
43. Van den Bergh J, Beliën J, De Bruecker P, Demeulemeester E, De Boeck L. Personnel scheduling: a literature review. Eur J Oper Res. 2013;226(3):367–85.
44. Wren A. General review of the use of computers in scheduling buses and their crews. Computer scheduling of public transport; 1981. p. 3–16.
45. Yu G, Pachon J, Thengvall B, Chandler D, Wilson A. Optimizing pilot planning and training for continental airlines. Interfaces. 2004;34(4):253–64.

A Unified Framework for Specifying Cost Models of IT Service Offerings

Kugmoorthy Gajananan, Aly Megahed, Shubhi Asthana and Taiga Nakamura

Abstract Information technology (IT) service providers compete to win highly-valued service contracts in a tender-like kind of process. The process starts with clients submitting a request for proposals, for which competing providers prepare a solution that covers the client requirements, and then begin the negotiation with the client trying to win the deal. Traditionally, IT providers design solutions by establishing a laundry list of services that the customer needs. Then, they try to cost and price each of these services individually. The more recent trend is that IT providers identify and design solutions that integrate a set of services bundles, usually called offerings, to allow for standardization and usage of economies of scale. This makes defining cost models for such integrated solution challenging as there is no consistent way to specify these costs for individual offerings which may have their own characteristics. In this work, we provide a unified framework that provides a consistent approach for specifying cost models for different service offerings while being flexible enough to handle individual differences among them.

1 Introduction

Information Technology (IT) service providers often compete via a bidding process to win outsourcing service contracts [1–4]. Such deals consist of complex IT services such as cloud computing, mobile computing, backup, help desk, among others [2, 4–6]. Providers prepare and price solutions and then present such solutions to the clients trying to win the deals.

Obviously, to have a competitive solution, service providers need to design a low-cost solution that fulfills the client's requirements. Traditionally, the approach to define such a solution was to establish a laundry list of services that the customer

K. Gajananan (✉)
IBM Research - Tokyo, IBM, Tokyo, Japan
e-mail: gajan@jp.ibm.com

A. Megahed · S. Asthana · T. Nakamura
IBM Research - Almaden, IBM, San Jose, CA, USA

H. Yang and R. Qiu (eds.), *Advances in Service Science*, Springer Proceedings in Business and Economics, https://doi.org/10.1007/978-3-030-04726-9_15

needs, where each service represents a distinctive element delivering a specific feature and value. A solution can then be defined as a hierarchy of such distinctive services and their costs [7]. Recently, however, an increasing trend for IT services business has been to identify and define solutions based on service offerings. A service offering refers to a set of related products and services grouped together as a bundle [8]. Benefit of implementing service bundling for providers are differentiating them from competitors and decreasing their costs due to standardization and economies of scale.

In an IT services business scenario, one or more service offerings may be required to build integrated solutions that fulfill client requirements. It is challenging for IT providers to come up with cost models for such integrated bundled solutions. Cost models are the parametric equations or formulae used to estimate the costs of a product or a service element. There are three main reasons for this difficulty. The first reason is the lack of consistent approaches to developing cost models for individual offerings. An offering team may have their own custom approaches for building cost models for their respective offerings based on experience and the client requirements their offering fulfills. However, the cost model for an integrated solution should take the overlaps, gaps, similarities and differences of cost models for the multiple offerings that form the solution into consideration. This is especially true when the same service element is included in multiple offerings of an integrated solution. Examples of such latter common services are project management, account management, and account security. The second source of difficulty is that offering teams with the specialized knowledge and expertise in building an application-oriented offering, may not always use the best practices in place which may lead to inconsistencies, as well as knowledge and expertise gaps among multiple offerings. Thirdly, when building a cost model for an integrated solution, solutioners need to understand cost models for different offerings and manually combine them into a single cost model, which is a labor-intensive, time-consuming and error-prone process.

Therefore, IT providers need an efficient, consistent approach to build cost models for different service offerings to make it easier for building cost models for integrated solutions. This approach needs to be simple and unified given the growing portfolio of offerings which are highly technical in nature, and the fewer skilled resources. In this work, we present a unified framework that provide a consistent approach for specifying cost models for different service offerings while accommodating flexibility to handle individual differences among them. We also present a proof-of-concept implementation of our framework showing its effectiveness.

The rest of this paper is organized as follows: In Sect. 2, we review the relevant literature. We then describe our methodology in Sect. 3. In Sect. 4, we provide our proof-of-concept implementation, and lastly, in Sect. 5, we conclude our work and list directions for future work.

2 Literature Review

In this section, we present the state of the art in cost modeling for different services offerings as follows. The services we refer to in our work that are included in the offerings that form integrated bundled solutions follow the taxonomy presented in [9]. Such services follow a hierarchy where the top level of the hierarchy refers to the highest level for the services and each service at that level is further decomposed into lower levels. We refer the readers to the references [10–13] for understanding more details on services in IT service contracts.

Motahari Nezhad and Shwartz [14] described the new set of opportunities and challenges faced by service clients to consume offerings from multi-vendor services. The author presented a conceptual architecture for an open services platform which can allow vendors to offer services together with third party providers seamlessly. There are many ways to build analytical models for predicting costing of services. We refer the reader to [15] for understanding how these analytical models are created and factors that contribute towards it. The studies in [16, 17] present different approaches for evaluating the costs of outsourced IT services based on the written comments from sales personnel pursuing these opportunities.

Different costing models are studied by Li et al. [18] for cloud storage services where they rely on different relevant factors such as storage types, market, and configurations to analyze costs. The authors in [19] studied the different models for IT services for a queueing system. Basu et al. [20] provided another costing model for cloud services, where they developed optimal cost models for cloud providers by using the cloud users as a function of two vectors; a set of parameters directly proportional to the customer utility and ones that have a negative effect on the utility. The relationship between architectural and costing characteristics for different IT services are highlighted in [21] where the authors show that there is cost discrimination between the same products being offered to different customers at different costs. However, their analyses are based on the value proposition of the firm, is focused on Software-as-a-Service (SaaS), and applies directly to characteristics of SaaS. Further, there are also multiple literature studies around the area of costing services and their characteristics. For the sake of conciseness, we refer the readers to the references in [22–25].

As seen from the above literature, there has been a lot of prior studies in the cost modeling of specific IT services included in solutions prepared by IT service providers to respond to clients' requirements. However, to the best of our knowledge, none of these papers study the service bundling in standardized offerings nor present a unified framework for specifying cost models for such offerings in a systematic, consistent manner. This is the objective of our work.

3 A Unified Framework

3.1 Overview

An offering definition refers to the configuration of services bundled in an offering. An offering definition consists of two key constructs: a cost driver and a cost component. Cost drivers refer to factors that drive the cost of a service element in an offering. A cost component captures the cost of a service element in the offering and specifies cost customization options. Each cost component has a set of cost drivers associated with it. This association helps us to model which factors are influencing the cost of a service element. Further, the cost drivers and cost components are all service dependent and are defined by the offerings owner for each offering/service.

We introduce a typical service offering called "virtualization" from an IT service provider, as a running example in this work. In practice, the virtualization offering is aimed at creating a consolidated, virtualized end-user desktop solution for a customer. This offering includes a bundle of services such as hardware and software products, and labor to monitor a virtualized desktop environment. Figure 1 shows hierarchies of cost drivers (a) components (b) and component specification (c) for the virtualization offering definition. For this specific example, we specify the logical grouping of the cost drivers under major categories: "virtualization services" and "infrastructure" required to provide the services. The "virtualization services" have four cost drivers under it whereas "infrastructure" further decomposes into hardware, software, application etc. The leaf nodes cost drivers such as "Number of Point of Deployments XLarge" refer to the actual cost driver factor. In this example, the cost of services driven by this cost driver is determined by the number of deployment points in the whole solution. Intermediate nodes provide a logical grouping referring to the summation of its child nodes. Thus, if an intermediate cost driver node is associated with a cost component of an offering definition, then the formula for computing the cost for the component, would consider the summation of the child nodes of the associated cost drivers as an input.

Similarly, cost components are structured as a hierarchy so that logical grouping of similar components fits together with in an offering definition. In component hierarchy, we consider the leaf nodes corresponding to the actual cost of the services delivered in the offering. For instance, in Fig. 1 the leaf node named "Design and Install Server Monitoring XLarge" is a cost component, and the intermediate nodes "Build Labor" and "XLarge PoDs" are logical groups.

The cost component specification includes the detail information regarding a leaf node of a cost component hierarchy. For example, Fig. 1c shows detailed specification of the cost component 'Design and Install Server Monitoring'. Each leaf node in the cost component hierarchy would have a specification.

(a)
```
Virtualization Offering
    Virtualization Services
        Number of Points of Deployments XLarge
        Number of Points of Deployments Large
        Number of Points of Deployments Medium
        Number of Points of Deployments Small
    Infrastructure
        Application
           Number of Appl. Hosts
        Hardware
            Number of Hosts
            Number of Concurrent Resource
            Amount of Storage in GB
            Number of Network Resources
        Master Images
            Number of Master Images
        Software
            Number of Appl. 'A'
```

(b)
```
Virtualization Offering
    Virtualization Services
        Build Labor
          Large PoDs
            Design and Install Server Monitoring
            Design and Test 'VS' Solution
            Pilot 'VS' Solution
            Prepare Deployment
            Handover to Customer
        Build Scaling Infra.
          Large PoDs
            Install and Config Windows Servers
            Install and Config MS SQL Servers
        Manage Labor
          Large PoDs
             Steady Stage Manage
        Hardware
          Large PoDs
             Compute Resources New and Refresh
             Concurrency Resources New and Refresh
             Storage Resources New and Refresh
             Network Resource New and Refresh
        Software
          Large PoDs
              Application 'A' New and Refresh
```

(c)
```
Component Spec:{
  Name: Design and Install Sever Monitoring
  Type: Labor
  Phase: Transition and Transformation
  Delivery Location From: China (100%)
  Scaling Factor: 'flat'
  Associated Cost Drivers: Number of Points of Deployment Large
}
```

Fig. 1 An example cost driver hierarchy for virtualization offering definition

3.2 *Cost Model Pattern*

In this work, we capture the general concept of how the cost should be computed and customized for each cost component of an offering using a construct that we call "cost model pattern". Each pattern defines the characteristics of cost computation for the component that it would be mapped to, via a set of formulas. The general structure of a formula that represents a cost computation for a component is represented by the following equation:

$$(Component) = \{(baseline_1 * rate_1) + (baseline_2 * rate_2) + \cdots\} * (scaling\,factor) \tag{1}$$

In Eq. 1, the terms baseline 1, baseline 2 etc. refers to the quantities of the cost drivers that are associated with a cost component for which the formula is applied to. Rates, which are parameters, refer to the actual dollar value for the quantities for cost drivers.

The rate value of a cost driver may depend on several factors such as where (country etc.) the underlying service is delivered from and delivered to, exchange rates, tax and local law applied etc. The notion of "scaling factor" represent economies of scale. Overall, the general structure shows how the cost of a component is computed based on its characteristics.

As seen in Fig. 2, a pattern has three subsections: (a) parameters, (b) baselines, and (c) component. The parameters contain place holders that represent information required to customize the cost computation specified in the pattern. A baseline refers to the number of units/quantities of a service element in an offering. Baselines facilitate cost computation of a component by providing constants as well as referring to its associated cost drivers. The formulas in the component section refers to the

(a)

```
{
   Pattern Id: P01,
   Parameters:{
        {
          id: r0,
          name: Base,
          value: ''
        },
        {
          id: r1,
          name: Change Ratio,
          value: ''
        },
        {
          id: g0,
          name: grPercent,
          value: ''
        }
   }
}
```

(b)

```
{
   Pattern Id: P01,
   Baselines:{
        {
          id: b0,
          isInternal: false,
          formula: Sum
        },
        {
          id: b1,
          isInternal: false,
          monthly: [1.0]
        },
        {
          id: t,
          name: Transition Term BL
          isInternal: true,
          monthly: [1.0]
        }
   }
}
```

(c)

```
{
   Pattern Id: P01,
   Component:{
      name: L1,
      type: Labor,
      Phase: TandT
      scale: 'flat',
      globalformula: 'IF((%baseline%)>0; (%t%) * (%r0% + (%baseline%)* %r1%) ;0) * 1',
      localformula: 'IF((%baseline%)>0; (%t%) * (%r0% +(%baseline%)* %r1%) ;0) * 1'
   }
}
```

Fig. 2 An example cost model pattern and its sections **a** parameters, **b** baselines, and **c** component

parameters and baselines defined. The formulas included in the component section of a pattern follows the general structure in Eq. 1.

4 Proof-of-Concept Implementation Approach

In this section, we study the feasibility of using our unified framework to specify offering definitions and generate cost model structures for different service offerings via a proof-of-concept implantation using real data from one of the world's largest IT service providers. We present the overall architecture and detail the functional flow from input definition to output generation as follows.

4.1 Architecture Overview

The primary goal of our approach is to let offering owners specify the offering definition as well as generate a cost model structure based on that definition. Figure 3a illustrates the overall architectural view of our implementation approach with several

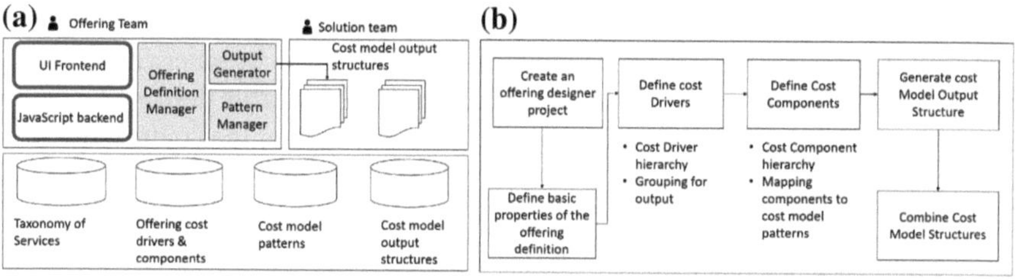

Fig. 3 **a** Architecture of our proof-of-concept implementation, **b** overall functional flow of cost model specification and generation

sub-modules that support different steps of the entire process of the cost model specification and generation.

4.2 *Functional Workflow Overview*

Figure 3b illustrates the functional workflow of our framework. The first step is to create an offering designer project that would contain a single offering definition. Then, the next step is to define the basic properties of the offering definition. Later, the workflow requires one to define the input factors such as cost drivers and cost components, where both are specified in the form of hierarchies. Next, one needs to specify detailed configurations of cost components of interest. Typically, this involves specifying every leaf nodes of the cost component hierarchy of an offering definition. The detailed configuration also includes the specification of service rates depending on the type of service elements. Once the offering definition is specified, we generate the cost model structure as an output for the offering definition specified. The final step is to combine multiple cost structures for different offerings to come up with a single cost model for an integrated solution.

4.3 *Mapping Cost Model Patterns to Cost Component*

In a single offering definition, there may be a set of cost components hierarchically organized. Each of the leaf cost component is mapped to a suitable pattern that describes the cost computation for the component. We briefly describe how our current implementation maps cost components with the cost model patterns in this subsection and generates output structures in the next subsection.

We refer to the cost driver and component hierarchies in the example of the virtualization offering definition as shown in Fig. 1 and describe how each cost component is mapped with a pattern.

In the cost component hierarchy of an offering definition, our approach retrieves the leaf nodes. For each leaf cost component retrieved, it selects a pattern by matching different attributes of the solution such as whether the services are delivered globally or locally, the number of cost parameters required, the type of the contract etc.

4.4 Output Generation

In Fig. 4a, we illustrate the cost model output structure generation process as a sequence of steps. Each leaf cost component of offering definition would be mapped against a pattern and from the pattern, the corresponding formulas for computing the cost would be associated. In addition, parameter values from cost component specifications would customize the formulas. The sequence of steps shown in Fig. 4a would generate a complete output structure of a cost model for an offering definition.

The work flow, for a given offering, assumes that cost component and cost driver hierarchies as well the detail specifications for the leaf cost components in the hierarchy as input. Step A, matches a pattern for each leaf cost component in the hierarchy to a pattern based on its matching criteria. Step B, for each leaf cost component in the hierarchy, generates a cost model specification based on the formulas from the associated pattern as well the component specification, which include associated cost drivers. An example output structure of a cost model specification for an offering is shown in Fig. 4b. The output structure displays sections: (a) components (b) baselines, and (c) rollouts. The components sections would include a set of cost model specifications, one for each leaf node cost component from the hierarchy defined for an offering. For instance, in Fig. 1 the leaf node named "Design and Install Server Monitoring XLarge" is a cost component, which would have cost model specification as shown in Fig. 4b. This specification includes a formula which was derived from the mapping pattern for this cost component. The specification of the cost component

(a)

```
Input:   Cost driver, cost component (with their specification)
         hierarchies for an offering, a set of patterns

Step A: For each leaf cost component in the component hierarchy
        choose a pattern based on a matching criteria based on
        component specification

Step B: Create cost model specification for each pair of cost
        component and a mapped pattern

   Step B.1: Instantiate parameters in the pattern using
             the respective values from cost component

   Step B.2: Instantiate baselines in the pattern using
             the respective values from cost component

   Step B.3: Instantiate component section of output using the
             template in the pattern and specification of the
             component.
      Step B.3.1: Replace placeholders for baselines, and
                  parameters in the formula with actual
                  values (associated cost drivers, parameters
                  baselines etc.)

Step C:  Generate cost model output structure
   Step c.1: Generate rollout for the internal baselines
```

(b)

```
Output Structure: (
  Components:[
    {
      Name: Design and Install - Server Monitoring
      Type: LAB
      Global Formula: IF((Number of Points of
                      Deployment)>0; (Transition Term BL) *
                      (Base + (Number of Points of Deployment) *
                      Change Ratio); 0)*1
        }

      ],
  Baselines:[
        {
          Name: Number of Points of Deployment
          Formula: MVS Services Monthly Qty 1
        },
    ],
  Rollouts: [
        {
          Name: MVS Services Monthly Qty 1
          Starting Qty: 3.0
          Monthly: 3.0, 3.0 .....
         },
    ]
}
```

Fig. 4 **a** Cost model output structure generation process as sequence of steps, **b** ouput structure

"Design and Install Server Monitoring XLarge" includes the cost driver 'Number of Point of Deployments XLarge' as shown in Fig. 1, which is instantiated as baseline in the formula of the cost model specification. Hence, the output structure would include 'Number of Point of Deployments XLarge' as a baseline.

5 Conclusions and Directions for Future Work

In this work, we presented a unified framework that include several constructs for specifying offering definitions for service offerings in IT service contracts. Our framework includes a novel idea of what we called cost model pattern, which is a set of formulas that can have an abstraction of cost computation and can thus be reused for different offerings. Cost model pattern also contain constructs for customizing the formulas for cost computation with respective to a specific offering. To support cost model specification, we also introduce a set of concepts such as cost drivers and cost components.

The conceptualization expressed in our unified framework provides a consistent way for offering teams to specify different offering definitions and thus results in a more efficient preparation of solutions to clients' needs in complex IT service engagements. To support this, we also make use of a service taxonomy. Additionally, we provided a proof-of-concept implementation that shows the effectiveness of our approach.

There are multiple directions for future work. For instance, a current limitation of our framework is that we may need an exhaustive list of patterns that need to be defined to capture cost computations of each cost component in different offerings. Thus, a direction for future research is to come up with an approach that automatically generates a comprehensive cost model structure for an integrated solution that comprises of multiple service offerings. Additionally, another direction for future research is identifying and studying the overlaps and gaps across different offerings and using the results of such study to further enhance and generalize our framework.

References

1. Yin P, Nezhad HRM, Megahed A, Nakamura T. A progress advisor for IT service engagements. In: 2015 IEEE International conference on services computing (SCC). IEEE;2015. p. 592–9.
2. Megahed A, Yin P, Nezhad HRM. An optimization approach to services sales forecasting in a multi-staged sales pipeline. In: 2016 IEEE international conference on services computing (SCC). IEEE;2016. p. 713–9.
3. Megahed A, Gajananan K, Asthana S, Becker V, Smith M, Nakamura T. Top-down pricing of IT services deals with recommendation for missing values of historical and market data. In: Proceedings of the international conference on service-oriented computing (ICSOC). 2016. p. 745–60.
4. Fukuda MA, Gajananan K, Jiang S, Megahed A, Nakamura T, Smith MA. Top-down pricing of a complex service deal. U.S. Patent Application 15/192,884, filed 28 Dec 2017.

5. Gajananan K, Megahed A, Asthana S, Becker V, Nakamura T, Smith M. A method for estimating annual cost reduction of IT service deals. In Proceedings of the IEEE conference on service operations and logistics, and informatics (SOLI). 2014. p. 45–50.
6. Fukuda MA, Gajananan K, Jiang S, Megahed A, Nakamura T, Smith MA. Method and system for determining an optimized service package model for market participation. U.S. Patent Application 15/050,986, filed 24 Aug 2017.
7. Megahed A, Shubhi A, Becker V, Nakamura T, Gajananan K. A method for selecting peer deals in IT service contracts. In: 2017 IEEE international conference on artificial intelligence and mobile services (AIMS). IEEE;2017. p. 1–7.
8. Mine D. Bundling products and service: the new super solution that gives companies and edge. Ivey Bus J. 2013;77(3).
9. Akkiraju R, Smith M, Greenia D, Jiang S, Nakamura T, Mukherjee D. Pusapaty S. On pricing complex IT service solutions. In: Service research and innovation institute global conference. 2014. p. 55–64.
10. Gajananan K, Megahed A, Abe M, Nakamura T, Smith M. A top-down pricing algorithm for IT service contracts using lower level service data. In: 2016 IEEE international conference on services computing (SCC). IEEE;2016. p. 720–7.
11. Megahed A, Gajananan K, Abe M, Jiang S, Smith M, Nakamura T. Pricing IT services deals: a more agile top-down approach. In: International conference on service-oriented computing. Springer Berlin Heidelberg;2015. p. 461–73.
12. Megahed A, Ren GJ, Firth M. Modeling business insights into predictive analytics for the outcome of IT service contracts. In: Proceedings of the IEEE international conference on services computing (SCC). 2015. p. 515–2.
13. Firth MK, Megahed A, Ren G. Assessing probability of winning an in-flight deal for different price points. U.S. Patent Application 15/192,892, filed 28 Dec 2017.
14. Motahari Nezhad HR, Shwartz L. Towards open smart services platform. In: Proceedings of the 50th Hawaii international conference on system sciences. 2017. p. 1103–9.
15. Greenia DB, Qiao M, Akkiraju R. A win prediction model for IT outsourcing bids. In: 2014 Annual SRII global conference (SRII). IEEE;2014. p. 39–42.
16. Carman S, Strong R, Chandra A, Oh S, Spangler S, Anderson L, Bernard JJ. Predictive value of comments in the service engagement process. Proc Am Soc Inf Sci Technol. 2012;49(1):1–6.
17. Nezhad HRM, Greenia DB, Nakamura T, Akkiraju R. Health identification and outcome prediction for outsourcing services based on textual comments. In: 2014 IEEE international conference on services computing (SCC). IEEE;2014. p. 155–62.
18. Li N, Zhang LJ, Xu P, Wang L, Zheng J, Guo Y. Research on pricing model of cloud storage. In: 203 IEEE ninth world congress on services (SERVICES). IEEE;2013. p. 412–9.
19. Li Z, Li M. A hierarchical cloud pricing system. In: 2013 IEEE ninth world congress on services (SERVICES). IEEE;2013. p. 403–11.
20. Basu S, Chakraborty S, Sharma M. Pricing cloud services the impact of broadband quality. Omega. 2015;50:96–114.
21. de Medeiros RW, Rosa NS, Pires LF. Predicting service composition costs with complex cost behavior. In: 2015 IEEE international conference on services computing (SCC). IEEE;2015. p. 419–26.
22. Indounas K, Avlonitis GJ. Pricing objectives and their antecedents in the services sector. J Serv Manage. 2009;20(3):342–74.
23. Xu L, Jennings B. A cost-minimizing service composition selection algorithm supporting time-sensitive discounts. In: 2010 IEEE international conference on services computing (SCC). IEEE;2010. p. 402–8.
24. Fukuda MA, Gajananan K, Jiang S, Megahed A, Nakamura T, Smith MA. Effectiveness of service complexity configurations in top-down complex services design. U.S. Patent Application 14/977,383, filed 22 June 2017.
25. Fukuda MA, Gajananan K, Jiang S, Megahed A, Nakamura T, Smith MA. Augmenting missing values in historical or market data for deals. U.S. Patent Application 15/192,875, filed 28 Dec 2017.

Toward a Context-Aware Serendipitous Recommendation System

Changhun Lee, Gyumin Lee and Chiehyeon Lim

Abstract Recommendation system development has been an important domain in the industrial and academic fields for the past two decades. Recently, the importance of developing a context-aware serendipitous recommendation system has emerged. As such, we investigate the latent features of items that may be recognized by the users of such a system. We assume that users will move from one item to another through the latent features reflected in the sequence of items. Our work specifically focuses on the process of predicting the sequential and changing taste of users. We show the existence of latent features by presenting a topic map and suggest a context-aware serendipitous recommendation system.

1 Introduction

Recommendation system development has been an important domain in the industrial and academic fields for the past two decades. The primary goal of this system is to provide users with personalized items based on past records to improve their satisfaction. However, as techniques improve and research on recommendation systems increases, researchers are facing the question of how well a user is satisfied with the recommendation (i.e., the value of the recommendation). For example, the importance of developing a "serendipitous" recommendation system has emerged as part of improving the value of the recommendation system.

Thus far, accuracy is considered a representative measure for estimating the value of a recommendation system. Accuracy indicates the probability that the user will appreciate the item recommended [1]. Although using this measure sounds simple and logical, a few researchers argue that other things should be taken into account [2–4]. As a result, the necessity of measuring "diversity" and the concept of serendipitous recommendation has emerged. Diversity is a literal indicator of how diverse

C. Lee · G. Lee · C. Lim (✉)
School of Management Engineering, Ulsan National Institute of Science and Technology, Ulsan 44919, Republic of Korea
e-mail: chlim@unist.ac.kr

H. Yang and R. Qiu (eds.), *Advances in Service Science*, Springer Proceedings in Business and Economics, https://doi.org/10.1007/978-3-030-04726-9_16

items are included in the recommended set and is therefore inevitably associated with the concept of serendipitous recommendation. However, diversity alone does not secure a serendipitous recommendation, because serendipity means both unintended and useful discovery (i.e., unexpectedly satisfactory discovery) [2, 5]. In other words, a recommendation system built specifically to increase diversity will recommend novel items regardless of the user's satisfaction. For example, given a recommendation system that recommends unfamiliar movies to the user on purpose, the user will probably be dissatisfied with the recommended movies because his/her taste is sacrificed for diversity. On the other hand, for a system that recommends familiar movies only, it is likely that the user's taste will be over-reflected such that the system will recommend accurate but obvious movies (i.e., movies that the user would have discovered by himself/herself). This is not a serendipitous recommendation system because the recommended movies are not unexpected. This situation is a well-known trade-off relation between diversity and accuracy. Hence, many studies focused on increasing the diversity while minimizing the accuracy loss for a serendipitous recommendation [6, 7], and our work likewise is conducted in a similar way.

Seeing that serendipity alleviates the trade-off between diversity and accuracy, we need to define the concept of diversity and figure out how to measure it. To define and measure diversity, we paid attention to the latent features of items. We assumed that people would move from one item to another through a latent feature reflected in a sequence of items. To sum up, we considered that there are latent features that link each item and the variety of the topics measures the diversity.

Specifically, in this study, we developed a movie recommendation system with user rating data. Our data included who the users were and how and when they rated movies with a certain score. The ratings were recorded chronologically and could thus be regarded as a list of consecutive movies that reflect the latent features. Our work specifically focused on the impact of the latent features by treating a recommendation as a process of predicting the sequential and changing taste of users. We indirectly show the existence of latent features by presenting a topic map and suggest a context-aware serendipitous recommendation system.

This paper is organized as follows. First, Sect. 2 addresses the main idea on which our recommendation system is built based on a concept from cognitive psychology field. Then, Sect. 3 presents a detailed description of the algorithm used in our work. Sections 4 and 5 explain the metrics and data, respectively. Finally, Sects. 6 and 7 cover the evaluation and conclusion respectively.

2 Context: Spreading Activation Model

In cognitive psychology, one of the most famous theories regarding human semantic processing is spreading-activation theory. This theory is based on Quillian's theory of semantic memory [8]. Qullian viewed memory search as an activation spreading from the concept nodes through a semantic network [9]. For example, when people are asked to state everything about machines (stimuli), they start off with clear facts (reaction), such as "it is manmade," "it has moving parts," and so on. Soon, however, they begin to give less clear facts, such as "a typewriter is machine." This stimuli–reaction mechanism is represented by a spreading-activation model, which consists of nodes and links. Each node corresponds to the concept that people can recall as a reaction against the stimuli, and the relational links indicate how strongly the concepts are related. Nodes preferentially activate the peripheral nodes, which are strongly related, and the activation spreads in a way that the activated nodes activate other linked nodes, and so on.

Motivated by the spreading-activation model, we assumed that people semantically link movies on the basis of latent features. For instance, a user logs into a rating website (e.g., IMDB or MovieLens) to rate a movie he/she just watched. After the user rates the first movie (stimuli), the movies he/she thinks are relevant (reaction) will pop up in his/her head and he/she will probably rate one of them. This process will continue until a user log out. What makes the user rate consecutively from one movie to the next is the latent features.[1] We call this concept "context."

3 Algorithm

Two popular techniques of information retrieval are used to extract and train the latent features, namely, Latent Dirichlet Allocation (LDA) and Word2Vector (W2V). LDA provides a global picture of latent features (e.g., the number and type of latent features in data) by allocating every movie into every topic, while W2V vectorizes each movie with local information related to its latent feature.

3.1 Latent Dirichlet Allocation

LDA is a topic modeling technique [10]. Basically, it was developed to extract latent features (i.e., topics) from a corpus using the overall structure of the words in documents.[2] This is why we used LDA. We wanted to know which and how many topics exist in the overall data. As a result, a total of 13 topics were derived. Figure 2a shows the change in topics per user by time. Specifically, it visualizes the route by which a user passes through his/her own semantic network, such as shown in Fig. 1.

LDA is a soft clustering technique. In LDA, each movie is non-exclusively allocated to each topic with a specific probability, but what we want is to exclusively allocate each movie to each topic. Therefore, additional work was carried out to transform the result of LDA into a hard clustering format. First, we made a

[1]Those features could be the actor, director, genre, series (e.g., Marvel comics), animation, japanimation, atmosphere and so on.

[2]LDA allocates the words into every topic based on the Dirichlet distribution.

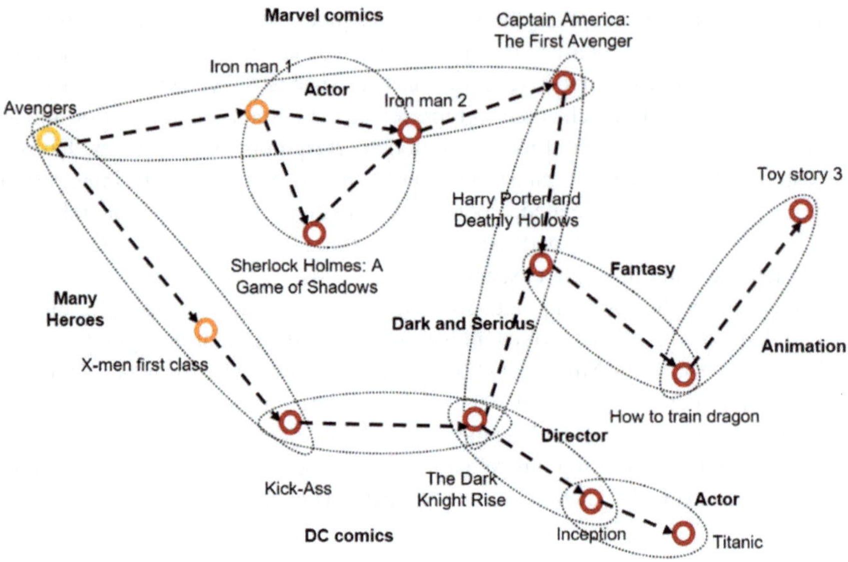

Fig. 1 Example of movie semantic network

movie–topic matrix. The values in matrix are the probabilities of being allocated to each topic per movie. Then we applied k-means clustering to exclusively allocate each movie to each topic. The proper number of k was determined by the elbow method, as shown in Fig. 2b. In our case, $k = 6$.

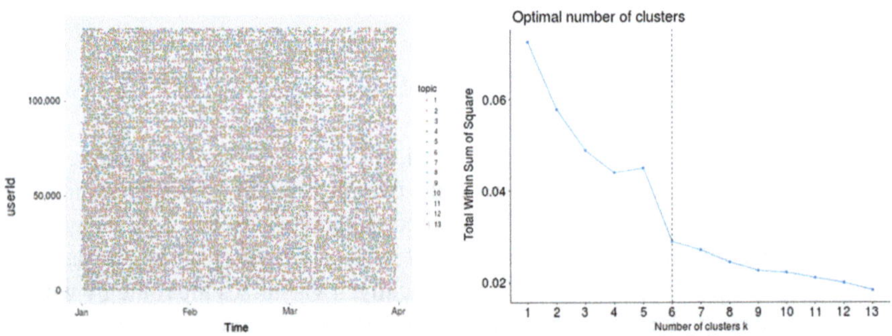

(a) Topic Map (b) Finding k

Fig. 2 Topic map and finding *k*

3.2 Movie2Vector

Our recommendation algorithm, Movie2Vector, is an application of W2V, which is a famous word embedding technique proposed in [11]. When a bundle of corpus is given, W2V technique trains local information per word by a moving training window. This technique is based on the concept that the meaning of a word is locally defined in a relationship with the neighborhood words. We assumed that the movie sequence per user has a similar property with the word sequence in the document and therefore applied W2V in our recommendation system.

4 Metrics

We used two metrics in our work, namely, the accuracy and diversity defined in this section. Before proceeding to the metrics, let us briefly explain our notations. $Q_u = \{Q_1, \ldots, Q_n\}$ notates a set of query movies per user and is used as input. $R_u = \{R_1, \ldots, R_n\}$ denotes a set of recommended movies, that is, the output. Test set per user $T_u = \{T_1, \ldots, T_n\}$ is the sequence of the movies after Q_u, where Tu indicates a set of target movies following a query set. *TP* is a set of topics. A mapping function $M(*)$ is also utilized to map a set of movies into a set of topics to which each movie is allocated.

4.1 Accuracy

Hit rate indicates how many items in T_u are hit. In other words, it shows how many items in R_u matches with items in T_u. This is a simple and popular metric to measure accuracy and was therefore employed in this study.

$$\boldsymbol{Accuracy} = \frac{\boldsymbol{n}(\boldsymbol{T}_u \cap \boldsymbol{R}_u)}{\boldsymbol{n}(\boldsymbol{T}_u)} \tag{1}$$

4.2 Diversity

We defined our own diversity metric. We assumed that the diversity originates from the diverse latent features and thus measured the diversity on basis of a variety of topics. The metric is defined as a ratio of new topics that appeared in R_u over the total number of T.

$$\boldsymbol{Diversity} = \frac{\boldsymbol{n}(\boldsymbol{M}(\boldsymbol{R}_u)) - \boldsymbol{n}\big(\boldsymbol{M}(\boldsymbol{Q}_u) \cap \boldsymbol{M}(\boldsymbol{R}_u)\big)}{\boldsymbol{n}(\boldsymbol{TP})} \tag{2}$$

5 Data

Choosing an appropriate dataset is crucial in evaluating an algorithm. Our work uses the rating sequence of each user, meaning that we need user rating, timestamp, and the identifier of the corresponding movies. The confidence of dataset needs to be considered as well. We assumed that the more referred the dataset was, the more confident it would be. The two criteria (appropriateness and confidence) led us to use the MovieLens dataset. MovieLens is a website that provides a movie rating service[3] run by GroupLens, a research laboratory at the University of Minnesota. It has offered various types of datasets since 1998, and the datasets are basically given by size, from 100 K to 20 M, each of which contains genome scores, genome tag links, movies, ratings, and tag files. The size and file construction of the MovieLens dataset makes it worthy of use and popular in the recommendation system research field, such that there is even a paper which covers the history of the dataset [12]. As of May 28, 2018, when you search "movielens" in GoogleScholar, you can retrieve about 14 K results. This figure means a fairly large number of articles are referring to the dataset, which proves its confidence. Among the files in each dataset, the ratings file includes information about users' preference. This file takes the form of ⟨user id, movie id, rating, timestamp⟩. Overall, the MovieLens dataset achieves both appropriateness and confidence and is, therefore, the best dataset for applying our algorithm. Our dataset from MovieLens includes 20 M ratings and 46.5 K tag applications across 27 K movies by 13.8 K users between January 09, 1995 and March 31, 2015. It was first released on April 2015 and most recently updated on October 2016. According to the data provider, the dataset was constructed by random extraction only for users who had rated at least 20 movies.

6 Evaluation

We argue that latent features exist and move users from one movie to next, which represents the change in context. Although the latent features are invisible, we considered them definitely hidden behind the sequential records of the rated movies. Our recommendation system was evaluated if it reflected this perspective.

The result of recommendation is plotted in Fig. 3. We compared our algorithm (W2V) to a user-based collaborative filtering (CF) and the random recommendation (RD) algorithm. Every point in Fig. 3 indicates a result of a recommendation per user. We ran each algorithm on randomly selected 300 users. If we carried out random sampling once per user, the result of recommendations can do over-fitting to the specific query set of the users. Thus, a cross-validation was performed. The number of validations was set to 20, and it described that every point in Fig. 3 represents the per user mean diversity and mean accuracy of the recommendations derived from 20 times of random sampled queries. Figure 3 and Table 1 clearly show that W2V

[3]https://movielens.org/.

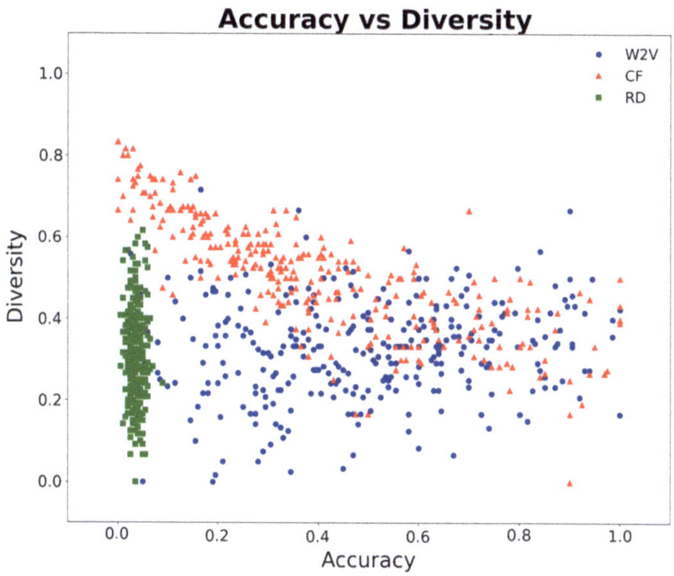

Fig. 3 Accuracy versus diversity

Table 1 Performance comparison

Algorithm	Accuracy	Diversity
W2V	0.499	0.324
CF	0.397	0.510
RD	0.035	0.314

performs better than CF and RD on the accuracy perspective. This finding suggests that context-awareness is a noteworthy factor in recommending an item.

Additionally, the way CF observations are distributed shows a negative slope, which explains the trade-off relation between accuracy and diversity. On the contrary, W2V displays a relatively flat distribution. The weakening trade-off in W2V once again supports our argument that context-awareness is important in serendipitous recommendation. However, our algorithm is limited in terms of the average diversity. This work is an ongoing effort, and the present conference paper shows the in-progress result.

7 Concluding Remarks

One cognitive psychology theory, the so-called spreading-activation theory explaining about human semantic processing, motivated us to develop the proposed recommendation system. We assumed that the cognitive factor affects the rating behavior of users and must therefore be considered in the recommendation system designed for her/him. In the proposed system, the semantic links in spreading-activation theory are mapped into the latent features (i.e., topics), and we termed them as context. We applied the LDA and W2V for our algorithm to be context-aware. Then, we showed that our context-aware recommendation system performs better on the accuracy perspective, which proved the importance of context-awareness.

Although our algorithm alleviates the traditional trade-off problem in recommendation systems, thus far, our interim result only seems to be context-aware-accurate not serendipitous. Nonetheless, we think the resulting non-pattern (circle dots in Fig. 3) at least shows a possibility of being serendipitous. We are continuing to improve our algorithm for it to be serendipitous by increasing diversity while minimizing accuracy loss.

We conclude this paper by emphasizing its relevance to the service science. Nowadays, recommendation systems have become indispensable elements in many service systems, from item recommendation in e-commerce services to friend recommendation in social network services. The development of high-performance recommendation algorithms will soon become the core competitiveness of many service firms.

References

1. Ziegler CN, et al. Improving recommendation lists through topic diversification. In: Proceeding WWW'05 proceedings of the 14th international conference on world wide web. 2005. p. 22–32.
2. Herlocker JL, et al. Evaluating collaborative filtering recommender systems. ACM Trans Inf Syst (TOIS). 2004;22(1):5–53.
3. Ge M, et al. Beyond accuracy: evaluating recommender systems by coverage and serendipity. In: RecSys'10 proceedings of the fourth ACM conference on recommender systems. 2010. p. 257–60.
4. McNee SM. Being accurate is not enough: how accuracy metrics have hurt recommender systems. In: Proceeding CHI EA'06 CHI'06 extended abstracts on human factors in computing systems. 2006. p. 1097–1101.
5. Toms EG. Serendipitous information retrieva. In: DELOS workshop: information seeking, searching and querying in digital libraries. 2000. p. 17–20.
6. Lathia N, et al. Temporal diversity in recommender systems. In: SIGIR'10 proceedings of the 33rd international ACM SIGIR conference on research and development in information retrieval. 2010. p. 210–7.
7. Zhang Y, et al. Auralist: introducing serendipity into music recommendation. In: Proceedings of the fifth ACM international conference on Web search and data mining. 2012. p. 13–22.
8. Collins AM, Loftus EF. A spreading-activation theory of semantic processing. Psychol Rev. 1975;82(6):407–28.
9. Quillian MR. Word concepts: a theory and simulation of some basic semantic capabilities. Behav Sci. 1967;12.
10. Blei DM, et al. Latent Dirichlet allocation. J Mach Learn Res. 2003;3:993–1022.
11. Mikolov T, et al. Efficient estimation of word representations in vector space, arXiv preprint arXiv:1301.3781. 2013.
12. Harper FM, Konstan JA. The movielens datasets: history and context. ACM Trans Interact Intell Syst. 2015; 5(4):1–19, Article 19.

Analysis of Service Execution in the On-line Sports Gambling Industry

James Roche, Pezhman Ghadimi and Vincent Hargaden

Abstract The on-line sports gambling industry has experienced significant growth in the last decade as well as a high-level of merger and acquisition activity among the leading multi-national companies that make up the sector. The companies in this sector also exhibit some of the unique challenges in on-line service supply chain management, including co-production process design, demand variability and resource utilization. Our research focuses on the post-merger service process design of one of these multi-national sports gambling organizations, particularly, on the efficient use of human resources within one of its service offerings (betting on soccer matches). Using discrete event simulation, a baseline model was developed to capture the "as-is" service process. The analysis identified number of bottlenecks, long queue times and capacity utilization issues. A revised "to-be" process was designed and when implemented, resulted in an increase of fourteen percent in overall utilization as well as removing variability in employee workload across the seven-day week.

Keywords On-line sports gambling · Service supply chain management
Discrete event simulation

1 Introduction

The gambling industry, especially on-line betting, has experienced significant growth over the past decade. Current figures show annual yields of £13.7 billion for 2016–2017 in the United Kingdom alone [1] and an estimated £490 billion to £700 billion wagered worldwide [2]. However, the management of sports betting by large corporations is a relatively new phenomenon. Historically, sports betting took place within bookmaking shops or at race-tracks, commonly referred to as 'bookies', and wagers were placed on a small range of events such as daily horse racing meetings

J. Roche · P. Ghadimi · V. Hargaden (✉)
Laboratory for Advanced Manufacturing Simulation, School of Mechanical & Materials Engineering, University College Dublin, Belfield, Dublin 4, Ireland
e-mail: vincent.hargaden@ucd.ie

H. Yang and R. Qiu (eds.), *Advances in Service Science*, Springer Proceedings in Business and Economics, https://doi.org/10.1007/978-3-030-04726-9_17

and high-profile sporting events. By the late 1990s, the first website enabling on-line play appeared. The development of the internet has had a transformative effect on the industry; there has been an increase in the number of on-line gambling web sites, magazines related to on-line gambling, and TV shows that involve gambling. In 2015, the search engine 'Google' received 11.1 million hits a month for the term 'Bet365' compared to 9 million hits per month for the world renowned brand 'Nike' [3].

The sector has experienced a high-level of mergers and acquisitions (M&A) activity recently. For instance, GVC holdings publicly announced their acquisition of Ladbrokes and Gala Coral, who themselves had previously merged in 2015 [4] having followed the path of other major players such as Paddy Power and Betfair's merger in 2016 [5]. As the industry becomes mature, the trend towards consolidation will see an increased focus on the design and efficiency of these service-based companies supply chains.

Generally, a review of the design of a supply chain occurs at the following times; the opening of a new business/business division, to improve growth and revenue, and after the completion of a merger of two firms [6]. It is the latter that provides the context for the research in this paper. M&As are renowned for being difficult processes across all levels of a company. This is amplified within service supply chains as the need for quick integration of the best of both firms offers opportunities to reap the benefits of undertaking the M&A.

The research to date on this sector is limited and has tended to focus on the science of customer behavior [7] and the effects of the 'Internet Gambling Enforcement Act' in the United States [8]. The research described in this paper focuses on the relatively unexplored domain of the on-line gambling industry through a service supply chain lens, and in particular the application of discrete event simulation (DES) as an analytical approach to evaluate and improve the service co-production processes following a merger of two companies. The current research activity sits in the SSCM literature by contributing a characterization of a unique e-business industry supply chain into a service supply chain context. Moreover, a DES approach has been utilized in modelling and analysis of the efficient use of resources within the newly merged company. The base model currently implemented (as-is) in the case company was revised to a CONstant Work-In-Process (CONWIP) flow system (to-be). The comparison results of the current and revised models stood out to management as an indicator to why there is such a high level of staff turnover historically in this role, an insight that was known but never fully quantified in a similar manner.

The reminder of this paper is as follows. Section 2 is dedicated to a literature review on service supply chain management. Section 3 provides the developed research design and methodology used within the research. Following this, the Sect. 4 explores deeper into the system and the implementation details of the simulation application is presented in this section. Section 5 presents the results discussion and comparisons between base line and revised models. Some concluding remarks are also included in this section.

2 Literature Review

Service supply chain design and management (SSCM) presents unique challenges when compared with conventional manufacturing supply chains [9–12]. These difficulties originate primarily from the peculiarities of service exchanges, the complexity of the processes for the co-design and delivery of services and the difficulty in visualizing the supply chain [13]. They are diverse in nature and highly contextual and more importantly service procurement is not done in a centralized fashion [14]. In most cases, a distinction is made between the supply of tangible goods and intangible goods which focuses on the supply chains of service offerings [13].

SSCM processes are different from those of a product supply chain in various aspects. For instance, the distinctive feature of SSCM is based on customer-supplier duality [10]. Emerging from the work of [9, 14] a Service Supply Chain (SSC) model was developed (Fig. 1). Here, the 'core service' provides benefit to the customer and the combination of supporting services and is the focal subject to the transaction. As depicted, the model covers three basic units in the chain: the supplier, the service provider and the customer. The service provider is the focal company in the supply chain that performs the service and is analogous to the manufacturer's role in the traditional supply chain literature. The supplier is the company which supplies additional services to the service provider, where these additional services contribute directly to the production of the core service in the chain. Due to the intangibility in service supply chain processes, the traditional logistical and functional operations existing in manufacturing-based supply chains cannot be realized in service supply chains [9]. Moreover, the presence of customers in SSCM presents challenges, such as the efficient utilization of employees [15, 16]. Unused labor-hour capacities tend to be lost without the presence of customers [10]. Therefore, managing these capacities plays critical role in effective SSCM which provides the analytical focus of the DES approach in this paper.

3 Research Methodology

3.1 Research Scope

Due to the scale of a multinational sports betting company's service supply chain, the scope of the research presented in this paper focuses on the soccer segment of the company's on-line sportsbook. The sportsbook is an e-commerce platform that offers customers the service of placing a wager on the outcome of an event (i.e. a soccer game). Each soccer game (an event) will have a calculated probability of occurrence and this probability assists in the generation of the game's 'odds', otherwise known as the 'price'. The service offered on the customer interface is the ability to place wagers on these generated 'odds'. For many, the only distinguishable difference in the product is the 'price' offered, meaning that each change in a game's probability

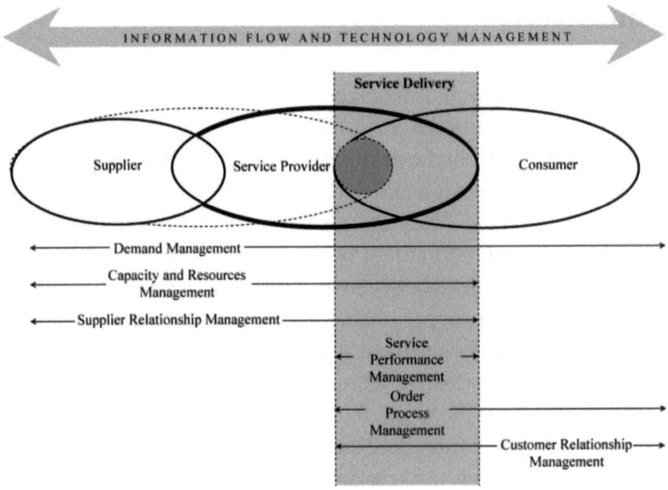

Fig. 1 Service supply chain (SSC) model [9]

represents a new product offering. Across the on-line sports betting sector, companies offer hundreds of differing products within each event, that undergo 'price changes' as the event probability changes. A single company therefore offers hundreds of thousands of various products daily.

Using the SSC model [9], a high level service supply chain map of the soccer sportsbook was developed (Fig. 2). Based on the mapped service supply chain structure, each stage in the chain was expanded into detailed individual processes; connecting links were generated and validated with senior management in the company. The complete process map provides visual representation of the service supply chain and uses conventional supply chain terminology along each tier of the chain e.g. loading bay, initial manufacturing, trading inventory, distribution, sportsbook inventory and customer interface. The basic operation of the current supply chain is as follows; generate an event (product) information from 3rd party entities (suppliers), by initial refinement of the incoming data (initial manufacturing) before running it through the trading inventory (value-added process). The product is then introduced to the destination management (distributor) where product segmentation for different customer groups (B2B and B2C) is distributed to the final warehouse (inventory). The next process sees the product as a service and offered to customers within the customer interface (consumer), which is the e-commerce platform. Following this, real-time information (transactions and wager agreement) within this unit of the chain is reciprocated downwards to the central service provider.

One of the principle concerns for management in the newly merged company is the efficient use of resources in the process. The 'soccer' product dedicates 190 h per week to manual tasks, some of which are a direct result of the merging of the two separate systems.

3.2 Research Approach

Discrete Event Simulation (DES) was used as the approach to carry out the resource management analysis. The specific process task analyzed was the 'Mapping of Feed Providers'. This process occurs after an event has undergone its necessary 'initial manufacturing' (refer to Fig. 2). A connection is made between each event and its corresponding event from the 'Feeds Platform'—that represents a platform for 3rd party information to be translated into appropriate language for the Soccer Division's internal models. This task allows for an event to reach optimal potential, as currently the 'In-Play' spectrum of the company's on-line football offering represents 60% of total volume.

DES deals exclusively with dynamic, stochastic system and is appropriate for systems for which changes in system occur only at discrete points in time. The application of DES in SSCM is mainly focused on health-care industry where various scholars developed DES models for applications such as accident and emergency department (A&E), Inpatient facilities, Outpatient clinics and other hospital units [17]. A DES model was developed for identifying operating strategies that lead to better resource utilization of a case hospital without degrading the service quality [18]. The application of DES in other service industries such as financial services, consulting and software development has also been reported [19]. To our knowledge, the application of DES in resource management of any specific processes of a e-service supply chain such as an on-line gambling firm has not been addressed. The

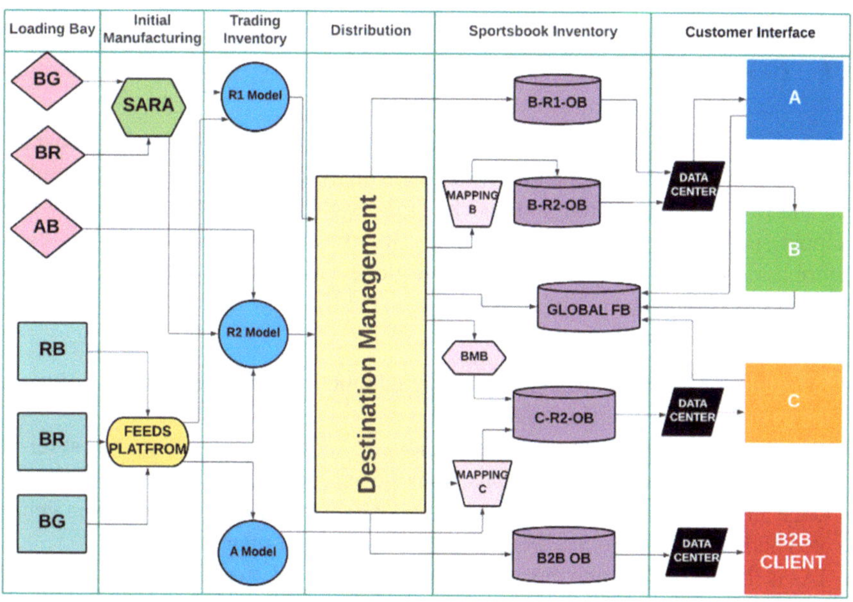

Fig. 2 Process map of soccer supply chain

research steps pursued in this paper have been adopted from [20]. They discussed that prior to developing the simulation model that the process of defining how a system works provides great insight into what changes need to be made, part of this is due to the fact that rarely there is one individual responsible for understanding how an entire system works. The aim of the analysis is to identify potential ways to reduce inefficiencies within the company's resource management by developing an improved configuration of processes and isolating bottlenecks. The steps followed to develop the simulation is as follows:

(1) Development of the flow of the intrinsic tasks conducted within each process,
(2) Initial collection of historical data of selected entities (people, parts, processes machines, servers, etc.) and associated variables, with structured query language of appropriate databases and time studies,
(3) Followed by restructuring and refining of data to assist in formulation of model representation,
(4) Appropriate entities, events and variables translated into the modelling software (Arena),
(5) The simulation verified by senior stakeholders, any issues resolved and rendered to confirm the model is correctly translating the data,
(6) The final step is the validation of the model to develop an acceptable level of confidence.

4 Implementation and Results

4.1 Baseline Model

A baseline model was developed of the flow of work conducted by trading assistants (TAs) within the soccer segment. This enables analysis of the utilization of employees and to identify potential bottlenecks in the system. Current scheduling of feed mapping is derived from the schedule of the 3rd party companies that offer live event information from football events across the globe. Events begin to become available for booking 72 h in advance of transitioning to the 'In-Play' spectrum of the product. The process flow is as follows:

- Schedule is generated through an automated function. This is the arrival point into the system for all events created.
- The resources in the system are the employees, known as TA's, they are first seized to manually check and book events from the appropriate feed providers.
- Once an event is booked the resource is released. A connection link is then generated and streamed into the 'Feeds Platform' ready for mapping.
- TAs is then seized to map the concurring event and link to each other before being released back into the system.

Figure 3 depicts the developed conceptual model for further DES analysis. Data were collected by several means. Historical data of events created within the Soccer supply chain was gathered through a relatively straight-forward process of using structured query language on several company internal databases. This data, when refined, allowed for generation of the probability of events going 'In-Play' and the occurrences of events entering the system. Feed providers provided data on events that were booked within their system. Data in relation into the human element was obtained through a time-study. Due to confidentiality reasons, the actual data points for various building blocks of the model are not presented in this paper. Table 1 provides detail of the building blocks utilized within the developed simulation model together with the generated distributions and probabilities input data.

The baseline simulation was developed in Arena simulation software and run over a six-month period, using historical data to review the performance of actual events to an alternative configuration. The six-month period from July 2017–December 2017 was chosen as it offered an opportunity to review the performance of the manual tasks following the merger. A small proportion of the work is conducted in the first 24 h as the number of the events available to be booked peaks in the twelve hours prior to the game going 'In-Play'. This leads a cascading effect and a bottleneck forms within the system. The baseline configuration of resources is a pure push system where the system behaves as an open queuing network, in which jobs enter the line and depart after one pass, meaning that the number of jobs available can vary over time. The baseline model was developed to represent an event being pushed through each separate task as they become available and subsequently releasing the resource.

The results of the simulation runs are tabulated in Table 2. A screen shot of the developed model in Arena is depicted in Fig. 4. As expected, to achieve the system output needed, the utilization rate of an employee spikes in the latter part of the week (peak demand) as events steadily become backlogged and workload increases.

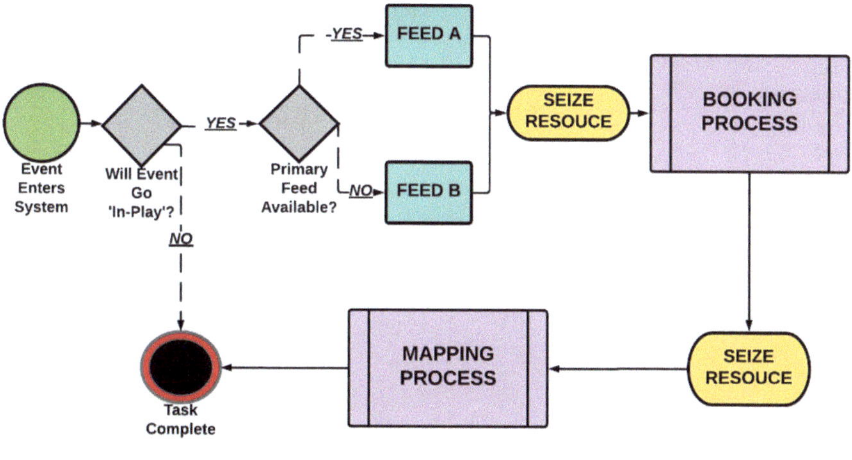

Fig. 3 DES conceptual model

Table 1 Input data utilized in various components of the developed DES model

Module name (type)	Data type	Distribution/probability input
Feed A "process"	Normal	NORM (3.48, 0.23)
Feed B "process"	Normal	NORM (4.18, 0.31)
Book event "delay"	Triangular	TRIA (1.02, 1.26, 1.51)
Unavailable to book "delay"	Triangular	TRIA (0.46, 1.04, 1.14)
Wait for availability "delay"	Normal	NORM (15, 0.53)
Will event go in-play "decision"	2-Way chance	76%
Is primary feed available "decision"	2-Way chance	71%
Is booking available "decision"	2-Way chance	44%
Event created "process"	Log normal	LOGN (2.59, 2.35)

Table 2 Baseline model summary results

Base line	Monday	Tuesday	Wednesday	Thursday	Friday	Saturday	Sunday
No. of resources (TAs)	1	2	2	4	4	4	4
Avg time primary feed booking (min/event)	1.43	1.33	1.34	1.23	1.17	1.16	1.23
Avg time secondary feed booking (min/event)	2.22	2.12	2.12	1.99	1.81	1.84	1.86
Avg time resource idle (mins/event)	0.74	0.66	0.66	0.37	0.14	0.05	0.08
Avg time feed mapping (mins/event)	1.93	1.91	1.91	1.81	1.81	1.84	1.84
Avg. system output	117	197	197	391	547	841	786
Utilization (%)	90	83	83	82	115	177	166

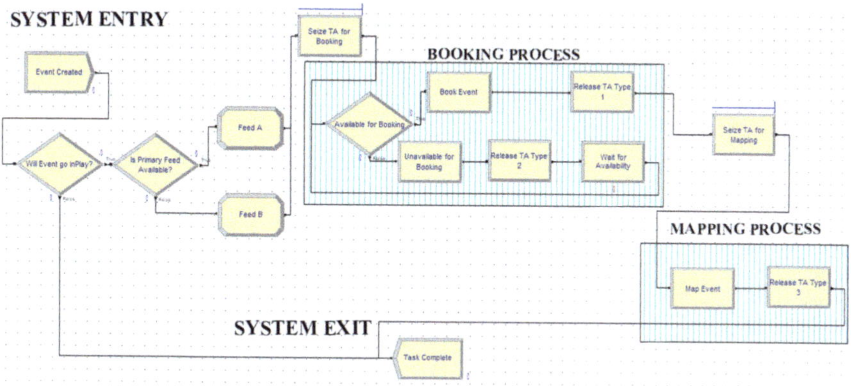

Fig. 4 Interface of Arena for the developed DES model

4.2 *Model Verification and Validation*

Verification aims at testing the computer-based model against the conceptual model [21]. Verification is performed to show that all parts of the model are able to work correctly. The model was verified through taking small steps, one at a time and making sure that the completed step was working well before continuing with the next step. The validation process was conducted to examine the accuracy of the model as compared to the real system. To do so, the average number of jobs was selected as the measure of model validity. In this respect, the actual average number of jobs arrived at system during 7 working days (2,750 jobs) was compared to the average number of customers obtained in seven iterations of the simulation model (2,554 jobs). The traditional comparison between these two values shows an acceptable difference (7.1%) which establishes the validity of the simulated model.

4.3 *Process Improvement*

An alternative configuration was proposed and is representative of a 'Just-in-Time' pull system. The tasks within the overall process would remain the same, schedule is generated as previously, events are firstly booked within 3rd party entities and mapped accordingly. However, the revised configuration would see all events booked and mapped in a 3-h window prior to an event transitioning to 'In-Play'. This system envisions a CONstant Work-In-Process (CONWIP), where from a modeling perspective the system looks like a closed queuing network, in which resources never leave the system, but instead circulate around the network indefinitely.

The results of the simulation runs are tabulated in Table 3. Time taken to conduct each task reduced, and overall the time spent idle was close to none existent in the peak times. The cumulative effect of the alternative system saw a reduction in the

Table 3 Revised model refined results

Revised model	Monday	Tuesday	Wednesday	Thursday	Friday	Saturday	Sunday
No. of resources (TAs)	1	1	1.5	2.5	4.5	5.5	5
Avg time primary feed booking (min/event)	1.03	0.94	0.91	0.88	0.94	0.91	0.92
Avg time secondary feed booking (min/event)	2.04	1.88	1.81	1.83	1.78	1.77	1.79
Avg time resource idle (mins/event)	0.22	0.16	0.15	0.11	0.06	0.04	0.05
Avg time feed mapping (mins/event)	1.86	1.79	1.8	1.74	1.74	1.77	1.77
Avg. system output	188	231	234	413	638	745	716
Utilization (%)	97	111	113	119	115	120	115

utilization needed to produce the appropriate output on weekly basis and therefore increasing the efficiency of work.

5 Discussion and Conclusion

Management at the company suspected that an extensive bottleneck was occurring in the booking and mapping processes. This was confirmed in the baseline simulation, as queue times for mapping were relatively large and, in some cases, where the event count peaked, were unacceptable when the queue time represented twenty percent of the total time in the system. By isolating this, the root cause analysis determined that there was never a definitive schedule or time when events are due to become available from the current 3rd party providers. This tended to occur more by approximation. This lack of a strict and definitive timeframe led to under and over-scheduling of resources in the baseline model, as illustrated through the utilizations levels—too few employees near the end of the week.

Assessment of these shortcomings in the baseline model enabled the revised model to be designed to reduce the queueing time and improve utilization levels to a steady level throughout the seven-day week. The revised and improved configuration uses a CONWIP flow system, which prevents overloading of the system. The premise used in developing this system requires the employees book and map events in a period prior to the transition to 'In-Play' in which ninety-nine percent of events are available, essentially removing the variability. After analyzing the historical booking calendar in the same period, three hours prior to transition was confirmed as the ideal window. The results show a steadier system, where employees' utilization levels were consistent and queue times near peak times plunged to six percent of the total time in the system in comparison to the baseline model.

One of the issues that this service provider has to contend with is the variability of demand from the beginning of the week compared to the latter part of the week, which coincides with when many of the soccer games are scheduled. The simulation illustrated that while resources increased as the week unfolded, the backlog of work conducted on Friday, Saturday and Sunday was far greater than the proposed one-hundred percent utilization rate. This figure stood out to management as an indicator of why the company experiences such a high level of turnover of TA staff. While results from any simulation represent a compromise between exact reality and what the computer and the modeler can comprehend and construct, the figures from the baseline to the revised configuration show that reduction of schedule variability (i.e. waiting time for booking availability) allowed the TAs to perform at an improved rate over all tasks. Queues at peak times reduced from twenty percent to six percent of the total time in the system and previous bottleneck issues were resolved. Average utilization rate increased by fourteen percent. Utilization levels in the revised model, while still higher than one-hundred percent, remained relatively steady throughout the seven-day week. The main reason for experiencing utilization rates more than one hundred percent is due to the higher arrival rates than the service rates, requiring additional servers.

In conclusion, SSCM presents unique challenges when compared with conventional manufacturing supply chains. Following a merger of two of the largest brands in on-line sports betting, opportunities existed to realize the benefits of such an integration. Through the development of discrete event simulation models, specific issues in the co-production process were identified and improvements developed. Moreover, the approach used for the unit of analysis in this paper can be applied to other segments of the on-line sportsbook within the company's service supply chain.

References

1. Commission G. Annual report and accounts. Retrieved 20 Nov 2018. https://www.gov.uk/government/publications.
2. Statista. Sports betting—statistics and facts. https://www.statista.com/topics/1740/sports-betting/.
3. Hunt G. The growth in online betting. https://www.siliconrepublic.com/play/the-growth-in-online-betting-infographic.
4. Ahmed M. How UK beat the odds to win at online gambling. https://www.ft.com/content/044a3d9e-7d1a-11e7-9108-edda0bcbc928.
5. O'Halloran B. Paddy power merger with betfair clears final hurdle. https://www.irishtimes.com/business/retail-and-services/paddy-power-merger-with-betfair-clears-final-hurdle-1.2498353.
6. Lee HL. The triple-a supply chain. Harvard Bus Rev. 2004;82(10):102–13.
7. Shaffer HJ, Peller AJ, LaPlante DA, Nelson SE, LaBrie RA. Toward a paradigm shift in Internet gambling research: from opinion and self-report to actual behavior. Addict Res Theory. 2010;18(3):270–83.
8. McBurney JJ. To regulate or to prohibit: an analysis of the internet gambling industry and the need for a decision on the industry's future in the United States. Conn J Int'l L. 2005;21:337.
9. Baltacioglu T, Ada E, Kaplan MD, Yurt And O, Cem Kaplan Y. A new framework for service supply chains. Serv Ind J 27(2):105–124; 2007.
10. Boon-itt S, Wong CY, Wong CW. Service supply chain management process capabilities: measurement development. Int J Prod Econ. 2017;193:1–11.
11. Chen Z. Service supply chain management: generalised and applied framework perspective. Int J Serv Econ Manage. 2014;6(1):63–96.
12. Cho DW, Lee YH, Ahn SH, Hwang MK. A framework for measuring the performance of service supply chain management. Comput Ind Eng. 2012;62(3):801–18.
13. Yuen KF, Thai VV. The influence of supply chain integration on operational performance: a comparison between product and service supply chains. Int J Logist Manage. 2017;28(2):444–63.
14. Ellram LM, Tate WL, Billington C. Understanding and managing the services supply chain. J Supply Chain Manage. 2004;40(3):17–32.
15. Sampson SE, Spring M. Customer roles in service supply chains and opportunities for innovation. J Supply Chain Manage. 2012;48(4):30–50.
16. Skaggs BC, Galli-Debicella A. The effects of customer contact on organizational structure and performance in service firms. Serv Ind J. 2012;32(3):337–52.
17. Günal MM, Pidd M. Discrete event simulation for performance modelling in health care: a review of the literature. J Simul. 2010;4(1):42–51.
18. Romano E, Iuliano D. A simulation/optimisation approach to support the resource allocation in service firms. Int J Procure Manage. 2018;11(1):53–75.
19. Jahangirian M, Eldabi T, Naseer A, Stergioulas LK, Young T. Simulation in manufacturing and business: a review. Eur J Oper Res. 2010;203(1):1–13.
20. Law AM, Kelton WD, Kelton WD. Simulation modeling and analysis. New York: McGraw-Hill;2007.
21. Persson F, Olhager J. Performance simulation of supply chain designs. Int J Prod Econ. 2002;77(3):231–45.

Decision Modeling in Service Science

Ralph D. Badinelli

Abstract The purpose of this paper is to highlight the need for innovative decision modeling in the field of service science. Each step of the service journey through a service ecosystem is initiated by a decision to integrate resources among actors and engage in a service activity. Consequently, engagement decisions are the driving force of any service journey and decision models are the foundation of service-system models. Each engagement decision must be modeled and executed as joint, adaptive, stochastic and perhaps fuzzy decisions among all actors who are involved in the associated service activity. However, such models are sparse in the research literature, and the current emphasis on predictive analytics and data science seems to distract attention from their development. Three examples of service systems are provided in this paper to illustrate this conclusion.

1 Introduction

The purpose of this paper is to stimulate initiatives for innovative decision modeling in the field of service science. One motivation for this purpose arises from my experience in forming sessions for the Service Science cluster at annual INFORMS meetings and as a reviewer on the editorial staff of the journal *Service Science*. In these efforts, I find a dearth of research that embodies decision modeling that is truly relevant to service. It seems that the service science community risks the attraction towards, on the one hand, a regression to the pre-Service-Dominant-Logic (SDL) perspective on service and, on the other hand, the modern fascination with the new technology suite of big data analytics (BDA), Internet of Things (IoT), machine learning (ML), artificial/augmented intelligence (AI) and data science. Related to this suite, one of the more confusing debates within the community in recent years is over the question, what is smartness in service systems? On this questions, we can make one essential point. Systems that sense, record, translate, measure, estimate and predict fall short

R. D. Badinelli (✉)
Business Information Technology Department, Virginia Tech, Blacksburg, VA 24061, USA
e-mail: ralphb@vt.edu

H. Yang and R. Qiu (eds.), *Advances in Service Science*, Springer Proceedings in Business and Economics, https://doi.org/10.1007/978-3-030-04726-9_18

of complete intelligence, as the latter capability includes the synthesis of the cause-effect relationships that are relevant to an actor's decisions. Therefore, there is an essential element of a smart service system that is not captured by the technology suite. In this paper we expose and describe this element as the descriptive model of the engagement decision and assert that service innovation requires sophisticated models of this type of decision.

To put this argument concisely, Fig. 1 summarizes the forms of analytics that are necessary to a fully functional DSS as well as the key role of the decision maker. It is apparent that many researchers and practitioners lack a clear understanding of this structure. Therefore, the contribution of this paper is an exposition of key insights into the following basic points:

- Decision modelling lies at the core of a DSS,
- Decision models are co-created with the decision makers,
- Predictive modelling exists to *support* decision modelling,
- Optimization *capitalizes* on decision modelling.

A central assertion of this bullet list and of Fig. 1 is that a descriptive model of the engagement decision is the foundation of all of the other forms of analytics that may be involved in a model of a service system. The paper will pursue this assertion as

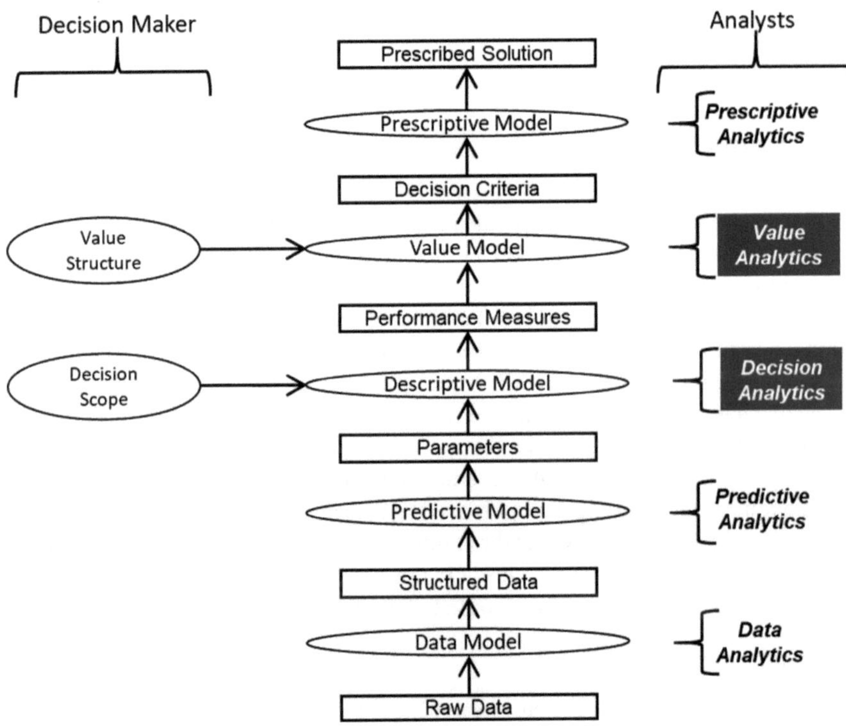

Fig. 1 Model relationships (adapted from [9])

follows. Section 2 reviews the history of decision modeling in service science from the origins in SDL, Sect. 3 presents two examples of important service systems and the models of the core decisions that guide the service journeys and Sect. 4 derives key conclusions about the proper application of the popular technology suite for predictive modeling in support of engagement decisions.

2 Perspectives on Service Science

The history of service science is revealing of the role of decision models in service systems. The modern definition of service begins with Service Dominant Logic (SDL) originating with marketing researchers [1, 2]. From SDL, a new and richly multi-disciplinary science was born. Disciplines such as psychology, economics, sociology even philosophy joined the movement before OR/MS. The Service Science Section of INFORMS was formed in 2007.

We should remind ourselves of the definition of service that gave rise to the Service Science Section of INFORMS. To wit, service is founded on co-creation of value, a concept that upsets the conventional IHIP (intangible, heterogeneous, instantaneous, perishable) definition of service [1, 3]. Co-creation as a process and value as a performance measure inspire new forms of decision modeling.

As researchers in a still fledgling science, the OR/MS community came late to the party but has much to contribute. Other disciplines have illuminated the field by recognizing the complicatedness and complexity of service. Consequently, the phenomenon of emergence has become very popular in service research. I suggest that the perspectives of other disciplines have over-emphasized the emergence phenomenon and that OR/MS has the opportunity to de-mystify service through mathematical modeling. Much progress in this direction has already been made [4, 5] and a lack of scientific precision in much service literature has been clarified [6–9]. However, the integration of decision modeling into the ontological framework of service that has been built remains wide open for innovative development.

Our focus on decision modeling requires a review of the core mechanisms of service. Following a reductionist approach, co-creation of value can be seen to occur through activities that integrate resources from service actors (the distinction between service providers and service recipients is impossible to reconcile with SDL so all participants are identified as actors) and transform them into output resources from which value is extracted. The service activity is the analogue of a process of a conventional supply chain model. However, unlike conventional supply chains that are managed by central authorities, service systems are dynamic, flexible and innovated on the fly. Consequently, a personalized service journey through a chain of co-creative activities spanning a hypernetwork of service systems is the way that service is realized [10]. See Fig. 2.

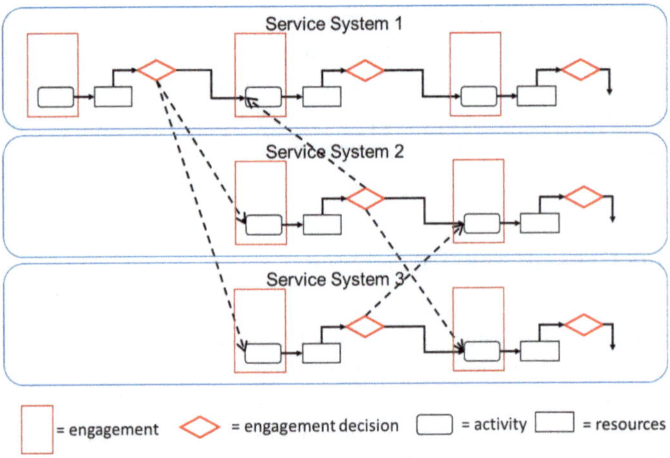

Fig. 2 A service journey through a service ecosystem [9]

3 Decisions Models in Service

Decisions determine the trajectory of the service journey. As engagement decisions are the driving force of any service, decision models are the foundation of service system models. Each step of the service journey is initiated by a decision to integrate resources in a service activity. The actors who authorize these decisions and extract value from the resources that are co-generated by the associated activity are represented by the performance measures and parameters of these decisions [11]. Furthermore, engagement decisions must be made jointly by all actors that participate in a service activity. Therefore, building models of engagement decision in service systems is much more challenging than building models of typical decisions in managing supply chains, where many researchers find inspiration.

Engagement decisions are complicated and complex [12, 13]. Decisions are complicated when they contain overwhelming number of variables and parameters. Indeterminacy in well-defined parameters is represented by stochastic properties of decision models [14]. Imprecision in parameter specification introduces complexity and is represented by fuzziness of decision models [15]. Accurate representations of these phenomena and useful heuristic solutions are waiting to be achieved.

Engagement decisions are also adaptive [16]. Engagement decisions take place within a multi-step service journey (Fig. 2) and learning is a fundamental part of an actor's service journey. Learning is followed by adaptation of the engagement decisions of actors. There are few truly adaptive decision models in the literature.

Finally, and most fundamentally, engagement decisions are joint decisions [17]. By definition, co-creation requires value generation for all actors in a service exchange. Engagement decisions commit resources from all agents that take part in the related service activity. Without a balanced commitment of resources, the co-creation of value will not succeed. This aspect of engagement decisions is absent in most service models, which leads to an asymmetry in the engagement decisions. Ignoring this asymmetry is one of the impediments to fully exploiting the capabilities of the new technology suite. Furthermore, OR/MS researchers have not made much progress in providing decision models that address this problem. We can illustrate this condition with a few examples of important service systems.

3.1 Healthcare Example

The conventional view of service provider and service recipient leads to asymmetry in making engagement decisions. Consider a simple healthcare service system (HSS) that consists of a patient and a medical clinic. Using all available data and a predictive model, the system generates a prediction about the presence of a particular ailment in the patient. This type of prediction is produced by a model with a binary dependent variable, such as the very commonly used logistic regression model. Based on the prediction, the patient or the medical clinic (depending on who receives the first signal from the predictive model) must decide whether to make an appointment at the clinic. This engagement decision for the patients would imply tradeoffs among cost, time, pain and discomfort, long-term health and other personal factors. This decision for the clinic would include performance measures of direct costs, capacity utilization and service level. In either case, the decision maker must weigh the potential outcomes of the decision by the likelihood that the prediction is correct. The linchpin of the decision is the probability of the ailment, which depends on the availability of data and the accuracy of the predictive model.

First, we will take the point of view of the patient in making the engagement decision. We assume that the patient and the HSS update the database of relevant data and re-generate the predictive model at certain time intervals, which we call epochs. At each epoch, the engagement decision is made by the patient. As the outcomes of the decision depend on the correctness of a prediction, the rational decision maker will base the decision on a risk analysis, which follows the familiar newsvendor problem. For the patient (P), the parameters of this model are as follows.

c_{PD} cost to the patient of an accurate diagnosis performed at the clinic. This cost includes the monetary expense to the patient of performing tests at the clinic in order to ascertain the correct diagnosis as well as the "cost" of the inconvenience, discomfort and anxiety that is associated with the visit to the clinic.

c_{PT} cost to the patient of treatment of the ailment after a true positive diagnosis. This cost includes the monetary expense to the patient of the treatment as well as the subjective evaluation of the inconvenience, discomfort and anxiety that is associated with the treatment.

c_{PI} cost of ignoring the presence of the ailment until a more reliable prediction occurs. This cost includes the subjective evaluation of the pain, discomfort and worsened medical condition of the patient that results from delaying treatment [18].

Each of these "costs" must be estimated in consideration of the actual monetary expenses of the patient and the HSS as well as the pain, discomfort, time spent, anxiety and other subjective dimensions of value. Consequently, the relative scale of these parameters can vary greatly across different conditions, diseases and individuals.

Each decision maker is faced with a rather simple choice, which we illustrate with the decision tree in Fig. 3. Define, p = Precision = the probability of a true positive prediction, given a positive prediction = $TP/P = TP/(TP + FP)$.

This condition implies the following decision rule for accepting the engagement proposal:

$$\text{If } p > p_c = \frac{c_{PD}}{c_{PI} - c_{PT}} \text{ then engage, otherwise wait until the next epoch.} \quad (1)$$

It is a well-established fact that the distribution of waiting times of a service system is a function of the capacity of the service system, the average rate of demand for service and the volatility of the demand rate and of the service process times. Service systems that experience higher unpredictability in demand rates and/or in service process times experience higher waiting times. To model this phenomenon in what follows, we will express the cost parameters of the HSS as functions of the uncertainty of the demand process. Given that M stands for medical clinic, we define,

λ the average rate of arrivals of patients to the clinic
σ standard deviation of the time between arrivals of patients to the clinic.

The corresponding cost parameters for the medical clinic are as follows:

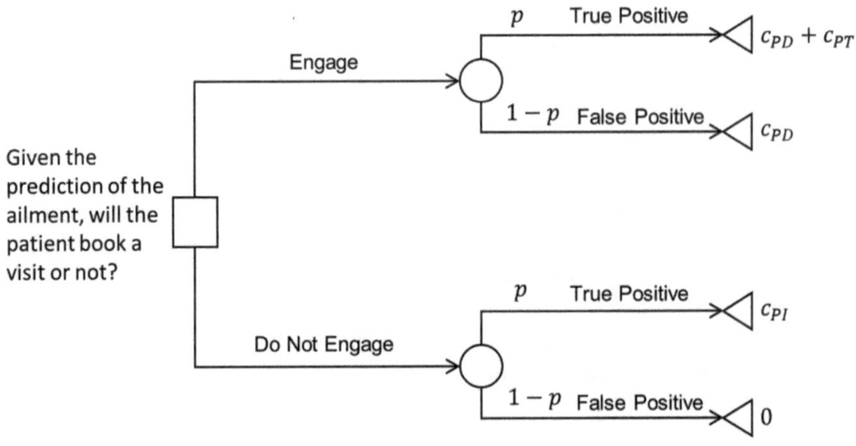

Fig. 3 Engagement decision tree

$c_{MD}(\lambda, \sigma)$	cost to the medical clinic of diagnosis in the form of the expense of performing tests at the clinic to ascertain the correct diagnosis
$c_{MT}(\sigma)$	cost to the medical clinic of treatment of the ailment
$c_{MI}(\sigma)$	cost to the medical clinic of ignoring the prediction and delaying treatment of the ailment
$c_{MT}(\sigma), c_{MI}(\sigma)$	are increasing functions of the randomness in patient arrivals
$c_{MD}(\lambda, \sigma)$	is an increasing function of frequency of visits and randomness in patient arrivals.

A final note about the medical clinic's decision is how the difference in the parameters of the decision for the medical clinic lead to a different optimal intervention plan from the one that the patient would choose. A similar analysis of the engagement decision of the medical clinic produces the following decision rule:

$$\text{If } p > p_c = \frac{c_{MD}(\lambda, \sigma)}{c_{MI}(\sigma) - c_{MT}(\sigma)} \text{ then engage, otherwise wait until the next epoch.} \tag{2}$$

The asymmetry in the cost parameters between the patient's decision and the medical clinic decision imply that the patient and the medical clinic have different thresholds for deciding when a patient should visit the clinic. This conflict will manifest itself in one participant or the other failing to optimize its costs.

3.2 *Retail Example*

Perhaps the most popular use of BDA is the ubiquitous application of predictive classification models to ascertain the propensity of individual consumers to purchase a product. The encouraging outcome of this use of BDA is the influence it has on returning retailing to its roots as a service. Before retailing was mass-produced by department stores, this industry provided personalized products and information to customers. However, the aspiration of modern retailers to serve "a market segment of one" through individualized BDA models leads to the same asymmetry in decisions that is faced by the HSS. The retailer's engagement decision is represented by the well-recognized assortment problem, which has the following, simplified general form:

$$\max_{\{r_{knt}, a_{knt}\}} P$$

Subject to:

$$L_{nt} - w_{nt} \geq 0, \forall n \in N_t, 1 \leq t \leq T$$

$$C_{nt} - \sum_{k \in K_t} v_k i_{knt} \geq 0, \forall n \in N_t, 1 \leq t \leq T$$

where,

r_{knt}	price of sku k for location n in period t,
a_{knt}	binary variable to indicate sku k can be in the assortment for location n in period t,
P	Total system-wide profit over the planning horizon, including inventory holding costs, employee training costs, transportation costs, ordering costs, re-stocking costs, vendor discounts,
i_{knt}	inventory of sku k location n in period t,
L_{nt}	stockout risk limit for location n in period t,
$w_{nt} = \int_{i_{nkt}}^{\infty} \psi_{nkt}(x)dx$	(Value-at-Risk measure),
ψ_{nkt}	pdf of the total demand for sku k location n in period t. This pdf is constructed as a convolution of the ψ_{knt},
C_{nt}	total capacity of location n in period t for holding inventory,
v_k	the "footprint" of sku k in terms of the per-unit number of cubic feet of space or number of square feet of shelf space that the item requires.

One large retailer is known to have used its binary classification models for demand prediction to determine directly the stocking of sku's in its stores without regard to the tradeoffs among the company costs and constraints inherent in this problem statement. This totally customer-centric view leaves the retail actor in the service exchange exposed to potentially costly inefficiencies.

3.3 Maintenance, Repair and Overhaul Service (MRO)

The use of predictive modeling in the MRO industry parallels that of retailing. Instead of predicting individual consumer demand, the technology suite predicts failure of individual components in complicated systems such as jet engines, automobiles, power plants, etc. Coupled with numerous IoT sensors that acquire massive amounts of raw data, BDA provides predictions of component failures throughout each system. Unfortunately, the engagement decision of the owner of the asset is much like that of the patient in the HCSS example. In contrast, is the engagement decision of the MRO value chain, illustrated in Fig. 4, which is plagued with long lead times, tight capacity constraints of highly specialized personnel and equipment and an overwhelming number of sku's for spare parts. Hence, the engagement decision of the MRO operator involves many tradeoffs that are not presented to the asset owner.

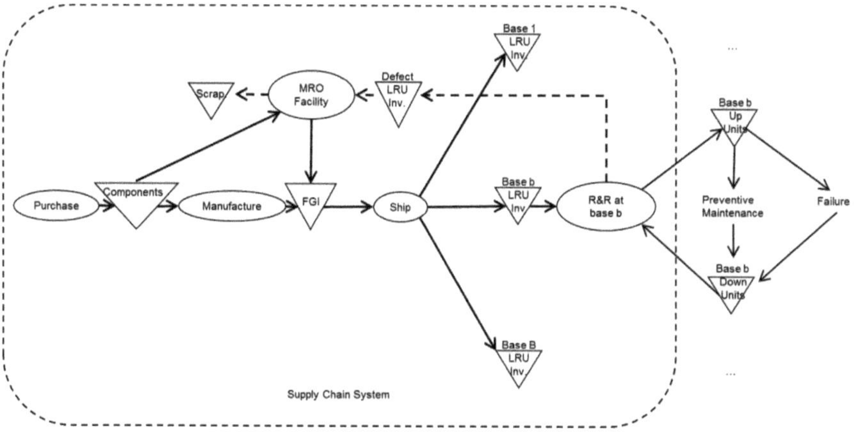

Fig. 4 MRO value chain, *FGI* finished goods inventory; *MRO* maintenance, repair and overhaul; *LRU* line replaceable unit; *R&R* remove and replace

4 Conclusions

The examples in Sect. 3 of the models for the engagement decisions reveals some interesting insights, which we now present. In particular, we can offer a critique of the effectiveness of the technology suite for predictive modeling in support of engagement decisions.

- On the asymmetric effects of decisions: It is generally believed that the use of the technology suite for predictive modelling will enable a highly responsive, personalized service. However, the use of these technologies is not a win-win proposition for all of the actors. Personalized predictive analytics in support of the engagement decisions can have asymmetric effects. For example, the benefits that accrue to the patient from a visit to a doctor can put stress on the resources of the medical care system, and if the prediction of an ailment turns out to be false, these resources are wasted. Furthermore, as the examples above illustrate, this phenomenon is general, in the sense that it applies to most service systems. The root cause of the asymmetry in the benefits of a service engagement is the difference in value structures of different actors. Therefore, an engagement decision that can be authorized by only one actor, such as a patient, a retail customer or an owner of a repairable asset does not represent the joint, co-creative process that service requires.
- On the effective use of predictive analytics: The precision and sensitivity of predictive models has a profound effect on the performance of service systems when these models are applied inappropriately. Many efforts to apply the technology suite have involved exciting uses of newly available datasets for predictive modelling. All too often however, prediction leads directly to an engagement decision. That is, predictive modelling is directed at recommending solutions instead of

supporting decision models. Predictive models are useful when they are used to estimate parameters of decision models. Unfortunately, trendy applications of the technology suite often attempt to allow the outcome of a predictive model to determine whether to engage. For example, the prediction of a component failure in a large assembly such as an engine can be allowed to instigate a visit to a repair shop, without any consideration of available resources or priorities of other cases.

- On the flexibility of resource capacity and scheduling of service systems: The ability of a medical clinic, retailer and MRO operator to respond to individual service journeys is limited by the natural inertia of capacitated service systems to respond to demand. Therefore, as the adoption of the technology suite raises expectations of prompt intervention, there will be increasing pressure on service systems to implement lean practices, process improvements and finite capacity scheduling that will enhance the ability of the systems to respond to demand volatility. Furthermore, the owners of service systems must adapt to operating within a service ecosystem that enables actors to enter and leave any single service system as their individual engagement decisions dictate (see Fig. 2). However, the inertia of the service system constrains the rate of change of capability, capacity and inventory. These constraints exist in all service systems. Different actors in a service engagement experience different levels of agility in resource allocation, which implies different priorities for scheduling the engagements.

Given the limitations and expense of process improvements, we can assert that ultimately, patient and medical clinic, customer and retailer, asset owner and MRO facility will have to cooperate to adjust the scheduling of engagements in consideration of the values and constraints of all actors. This will require combinations of scheduling advances or delays of some interventions to compromise among the performance measures of all actors. In this way, the service systems becomes truly co-creative. All service systems need to co-decide the engagement decisions with actors, jointly making use of the technology suite. In summary, each engagement decision must be modeled and executed as joint, adaptive, stochastic and perhaps fuzzy decisions among all actors who are involved in the associated service activity.

References

1. Vargo S, Lusch R. Evolving to a new dominant logic for marketing. J Mark. 2004;68:1–17.
2. Vargo S, Akaka M. Service-dominant logic as a foundation for service science: clarifications. Serv Sci. 2009;1(1):32–41.
3. Sampson SE, Froehle CM. Foundations and implications of a proposed unified services theory. Prod Oper Manage. 2006;15(2):329–43.
4. Alter S. Making a science of service systems practical: seeking usefulness and understandability while avoiding unnecessary assumptions and restrictions. In: Demirkan H, Spohrer JC, Krishna V, editors. The Science of service systems. New York: Springer; 2011. P. 61–72.
5. Ferrario R, Guarino N, Janiesch C, Kiemes T, Oberle D, Probst F. Towards an ontological foundation of service science: the general service model. In: 10th international conference on Wirtschaftsinformatik. Zurich, Switzerland.

6. Maglio PP, Spohrer J. Fundamentals of service science. J Acad Mark Sci. 2008;36(1):18–20.
7. OMG (Object Management Group). Value delivery modeling language. Accessed 15 June 2018. http://www.omg.org/spec/VDML/1.0.
8. Qiu R. Service science: the foundations of service engineering and management. New York: Wiley; 2014.
9. Badinelli R. Modeling service systems. Business Expert Press; 2015.
10. Chan W, Hsu C. Service scaling on hyper-networks. Serv Science. 2009;1(1):17–21.
11. Lessard L. Modeling value cocreation processes and outcomes in knowledge-intensive business service engagements. Serv Sci. 2015;7(3):181–95.
12. Barile S. Management sistemico vitale. Torino: G. Giappichelli; 2009.
13. Ng I, Badinelli R, Dinauta P, Halliday S, Lobler H, Polese F. S-D logic: research directions and opportunities: the perspective of systems, complexity and engineering. Mark Theor. 2012;12(2):213–7.
14. Badinelli RD. A stochastic model of resource allocation for service systems. Serv Sci. 2010;2(1):68–83.
15. Badinelli R. Fuzzy modeling of service system engagements. Serv Sci. 2012;4(2):135–46.
16. Qiu RG. Computational thinking of service systems: dynamics and adaptiveness modeling. Serv Sci. 2009;1(1):42–55.
17. Qiu R. We must re-think service encounters. Serv Sci. 2013;5(1):1–3.
18. Paulussen TO, Zöller A, Heinzl A, Braubach L, Pokahr A, Lamersdorf W. Agent-based patient scheduling in hospitals. In: Kirn S, Herzog O, Lockemann PC, Spaniol, O editors. Multiagent Engineering Theory and Applications in Enterprises. Heidelberg, Germany: Springer; 2006. PP. 255–276.

Predicting Call Center Performance with Machine Learning

Siqiao Li, Qingchen Wang and Ger Koole

Abstract In this paper we present a simulation-based machine learning framework to evaluate the performance of call centers having heterogeneous sets of servers and multiple types of demand. We first develop a simulation model for a call center with multi-skill agents and multi-class customers to sample quality of service (QoS) outcomes as measured by service level (SL). We then train a machine learning algorithm on a small number of simulation samples to quickly produce a look-up table of QoS for all candidate schedules. The machine learning algorithm is agnostic to the simulation and only uses information from the staff schedules. This allows our method to generalize across different real-life conditions and scenarios. Through two numerical examples using real-life call center scenarios we show that our method works surprisingly well, with out-of-sample fit (R-squared) of over 0.95 when comparing the machine learning prediction of SL to that of the ground truth from the simulation.

1 Introduction

Call centers, as a typical large-scale service system, has attracted attention from the operations research (OR) community for more than 20 years. An important problem in OR is workforce scheduling, which in general can be categorized into 4 steps [1]:

1. **Forecasting**: predicting the arrival rates of each customer type over the planning horizon based on historical data.
2. **Staffing**: determining the minimum number of agents needed to meet a required quality of service (QoS) in each period (e.g., intervals). In practice, QoS can be measured in various ways such as the abandon rates, average response time (i.e., waiting time), etc.

S. Li · Q. Wang (✉) · G. Koole
Department of Mathematics, Vrije Universiteit Amsterdam, 1083 HV Amsterdam, The Netherlands
e-mail: Q.Wang@uva.nl

Q. Wang
Amsterdam Business School, Plantage Muidergracht 12, 1018 TV Amsterdam, The Netherlands

H. Yang and R. Qiu (eds.), *Advances in Service Science*, Springer Proceedings in Business and Economics, https://doi.org/10.1007/978-3-030-04726-9_19

3. **Shift scheduling**: determining the number of agents assigned to each shift for each agent group.
4. **Rostering**: assigning employees to the shifts.

In this paper we focus on steps 2 and 3 and introduce a simulation-based machine learning framework to predict QoS measurements. Our method can lead to an efficient way of solving the staffing and shift scheduling problem integrally.

A fundamental challenge in staffing and scheduling service systems is reaching certain QoS targets at minimum costs. This challenge becomes particularly complicated when considering call centers that have multi-skill agents and multi-class customers with heterogeneous arrival rates, as we need to decide the staffing level with the best configuration of multi-skill agents over the planning horizon with consideration of shift types. To exploit the flexibility to choose staffing levels in different time periods (e.g., intervals) so that the scheduling costs can be minimized, we consider the staffing and shift scheduling problem integrally.

Due to the complexity, few papers have studied the integrated problem in a multi-skill multi-class setting. The key difficulty is the lack of closed-form expressions for QoS measurements. To cope with this, the papers that study the integrated problem either rely on simulation [2, 3] or approximate the QoS by modeling the system into certain existing mathematical models such as fluid models or queuing models [4]. The advantages of using simulation are multi-folded: it tends to produce more reliable results [4, 5], takes into consideration the transient effects between intervals, allows for more diversified QoS measurements, and makes fewer assumptions so as to be more realistic. On the contrary, LP approximations which mostly based on fluid models and queueing models, often need strict conditions such as heavy-traffic systems, short service times, impatient customers, and specific routing policies. Thus it can be less robust and is difficult to apply in practice. Moreover, modern call centers are not dealing only with calls, but also include other channels such as email and chat. Unfortunately, simulation methods can be very time consuming, which is a significant drawback in practice as it is common for practitioners to iterate multiple times between shift scheduling and rostering. In these cases, it is important to have a reliable staffing and shift scheduling method with short computation times.

As a result, we wonder whether there is an approach that can efficiently (i.e., quickly with acceptable accuracy) estimate the QoS measurements without strict assumptions. This paper proposes the following approach: introducing a machine learning algorithm that is trained on simulation results to predict the QoS measurements. More specifically, we first develop a general simulation model for call centers, and under any given scenario, we randomly generate a number of possible schedules with their corresponding QoS. The computational burden in this stage is not heavy as it does not need to be in real time. Subsequently, we train a machine learning algorithm on these schedules and are then able to quickly produce a "look-up table" with the QoS of all possible schedules. In this way, we can take advantage of the simulation method without having to bear the costs of long computation times.

2 Model

In this section, we provide a general description of our problem, based on which a simulation model is built.

We consider a call center where arriving customers are categorized in I types of service: $\{1, \ldots, I\}$. These customers can be calls, emails or chats that are served by agents divided into G groups: $\{1, \ldots, G\}$. Agents within the same group have the same skill set, which gives the subset of service types that this agent can serve. Finally, there are K different shifts: $\{1, \ldots, K\}$. The schedule then can be represented by $n_{g,k}$ which is the number of agents staffed to group g, and shift k. With respect to the QoS measurement, we choose the service level (SL), which is most commonly used in call centers. It is defined as the proportion of customers who wait less than a given time threshold, also known as the acceptable waiting time (AWT), over a time period. In practice, managers often pay attention to daily or weekly SL of each call type. Sometimes, they also evaluate several service types together as a set. In future work, other QoS measurements can be included easily if needed.

We do not assume any particular arrival process, service time distribution, or patience distribution, instead we only need to be able to simulate them. In practice, the arrival rates are normally derived from the forecast results, and redials and reconnects can also be included. We only assume that service within the same type will be served in a First Come First Served (FCFS) basis.

In the multi-skill multi-class environment, the choice of routing rules is also important for achieving good system performance. Again, we do not need to restrict our model to a specific routing policy. Later, we present the performance of our proposed approach by comparing two scenarios derived from real call centers, where static priority is used and preemption is allowed for some call types: once an agent starts serving a call, there can be an interruption by other calls with higher priorities.

The goal of this paper is to approximate the simulation with a machine learning algorithm so that a near optimal schedule can be found efficiently. To do so, once the simulation model is validated, we can model the simulation outcomes as a prediction problem, and then train a machine learning algorithm to predict SL outcomes for each scenario and schedule.

3 Machine Learning Approach

In order for the formulation to be generalizable to different scenarios and be invariant to simulation details, we only use information relating to the scheduling decision variables. Since SL is a continuous variable between 0 and 1, we model this as a regression problem and minimize the sum of squared error (SSE) between the true and predicted SL values of simulated samples during the training process. The general formulation of our problem can be written as:

$$arg_f min \sum_{s=1}^{S}(f(n^s_{1,1}, n^s_{1,2}, \ldots, n^s_{1,k}, n^s_{2,1}, n^s_{2,2}, \ldots, n^s_{2,k}, \ldots$$
$$n^s_{G,1}, n^s_{G,2}, \ldots, n^s_{G,k}) - SL^s_{true})^2 \tag{1}$$

where S is the total number of simulated samples, and $n^s_{g,k}$ is the number of staff for skill group g scheduled in interval k.

Note that $n^s_{g,k}$ can be derived easily from the agent schedules.

f is a function that produces a SL prediction for inputs of a given staffing policy. An example of a function f that can minimize the SSE is least squares regression, where optimal weights for a linear combination of the staffing variables are computed. However, although we expect changes in SL to be monotonic with changes in the number of staff scheduled, least squares regression is not a suitable function to approximate the call center staffing problem due to non-linearity of SL in response to the number of staff.

We use Gradient Boosted Decision Trees (GBDT) for this problem [6]. It is also known by other names such as Multiple Additive Regression Trees (MART), Gradient Boosting Machine (GBM) or Tree Boosting. They all refer to the same technique which applies an algorithm called Gradient Boosting that uses classification or regression trees (CART) as base learners [7]. GBDT is an iterative algorithm and it works by training a new regression tree for every iteration to minimize the residual of predictions made by the previous iteration. The predictions of the new iteration are then the sum of the predictions made by the previous iteration and the prediction of the residual made by the newly trained regression tree in the new iteration.

GBDT is one of the most powerful machine learning algorithms and has been used to win most of the recent predictive analytics competitions [8]. It is also well suited for the problem of predicting SL from staff schedules. Unlike least squares regression, GBDT is both non-linear and non-parametric, and is able to exploit interactions between input variables. In this paper we use a fast and accurate implementation of GBDT called Light GBM, which is currently used in most solutions to data science challenges [9].

4 Numerical Experiments

We perform numerical experiments using two simulation scenarios to test the performance of machine learning predictions of SL in approximating the SL outcomes from the simulations. Both scenarios are derived from real-life call centers. The first scenario is fairly simple and is meant to be an easy case for the machine learning approximation. The second scenario is complex and is intended to test the capabilities of the machine learning solution in large-scale and highly complicated scenarios. In our case, we are trying to evaluate the SL of given schedules so that useful training data is expected to have well-distributed SL outputs and a good coverage of all shift types. To avoid simulation runs with poor coverage, we need to decide some scheduling parameters based on the given scenarios before generating random schedules: the upper/lower bound (U, L) of the total assigned agents, and the maximal/minimal (u, l) number of agents assigned to each combination of shifts and agent groups. Moreover, the agents are assigned to the combinations in a random order to avoid bias.

Scenario 1 is from a mid-sized English call center that has only inbound calls, in five different languages. Five corresponding agent groups are considered, one for each language, but the non-English groups are also able to handle English calls but with a lower priority. No interruptions are allowed. We randomly simulate 10,000 schedules within the constraints of $U = 150$, $L = 20$, $u = 10$, and $l = 0$. Weekly SLs are measured by the proportion of the customers who wait less than their corresponding AWTs among all customers of a week.

Scenario 2 is designed by combining several mid-sized call centers. In total, we have 29 service types and 23 agent groups. Service types include 16 inbound call channels, 6 chat channels, and 7 email channels. The attributes of these service types vary greatly, including workload, arrival profiles, average service time, and patience. This results in highly complex relationships between staff schedules and SL. Chats, calls, and emails are differentiated as follows.

Chats and calls are real time, but multiple chats can be handled in parallel by a single agent. When the maximum number of parallel chats is limited (e.g., to 2 or 3), the service time distributions will not depend much on the current level of concurrency. The reason is that customers also need time to reply and this time can be used by the agent to respond to other customers. Emails are not answered in real time so customers do not abandon. Usually the AWT of emails is much longer than calls so they do not need to be handled within the interval of arrival. However, the waiting time of an email customer also includes the service time since they "wait" until they receive the answer.

Among the 23 agent groups, most of the groups' skill sets are overlapping and 4 groups are independent from others. The routing rule we used is still a static priority policy, but some emails can be interrupted by calls. Since this scenario is much bigger than the first one, we generated 20,000 random schedules with $U = 500$, $L = 20$, $u = 100$, $l = 0$.

5 Results

We present the results of the numerical experiments. For each scenario, we randomly selected 70% of the simulated samples to train the machine learning algorithm and test its performance on the other 30% of the samples. GBDT is able to perfectly fit any dataset it is trained on, so we must hold out a subset of the original dataset to test its performance. The goal of this paper is not to assess the ability of GBDT to reproduce simulation outcomes for samples it has already seen, but for samples it has not seen. The 30% of the samples that are held out can also be interpreted as how well GBDT can approximate the simulation in practice.

Table 1 GBDT performance on simulation data

Evaluation metric	Scenario 1	Scenario 2
MAE	0.035	0.013
MAPE (%)	12.46	34.51
WAPE (%)	4.72	11.67
R^2	0.977	0.955%
N	3000	6000

In order to improve the performance of GBDT we also included two additional sets of variables, $v_1^s = \sum_{k=1}^{K} n_k^s$ for each skill group g and $v_2^s = \sum_{g=1}^{G} n_g^s$ for each shift k to function f in Eq. (1). These variables are aggregates of the staffing numbers in Eq. (1) and helps to guide GBDT to better solutions with fewer training iterations.

Table 1 presents the performance of GBDT on predicting SL for the numerical experiment. For the 30% of the simulated samples that we use to evaluate performance, we compute the mean absolute error (MAE), mean average percent error (MAPE), weighted average percent error (WAPE), and the coefficient of determination R^2 between the predicted and actual SL outcomes.

The error rate in Scenario 1 is quite low with MAE of 0.035, while the MAPE is substantially higher at 12.46%. This is due to the samples with low SL which have higher MAPE values as compared to MAE. The 4.72% WAPE is more telling as it weighs samples by their SL values, and the predictions have an R^2 value of 0.977 which shows a strong fit. Overall this suggests that GBDT is able to perform very well on predicting SL for small-to-mid-sized call centers that are similar to Scenario 1. GBDT's performance on Scenario 2 is more complicated. Although MAE is low at only 0.013, the MAPE is very high at 34.51% due to large number of samples with very low SL, and even the WAPE is very high at 11.67%. However, the low SL is a result of how complex Scenario 2 is, and even though GBDT's performance is worse, it is still a better option than relying on simulations in real time to find acceptable solutions due to the large parameter space of 29 service types and 23 agent groups.

6 Conclusion

This paper presents a machine learning approach to evaluating call center staffing schedules for the integral staffing and shift scheduling problem. This is a challenging problem due to the lack of closed-form QoS measurements which leaves simulation as the best existing method to evaluate staffing schedules. However, simulation is slow and therefore cannot be used for real-time optimization of staffing schedules for large call centers, leaving faster methods to be desired. Our approach uses Gradient Boosted Decision Trees (GBDT) to train a prediction model on simulation samples that are generated offline for the purpose of evaluating staffing schedules online. We show that GBDT performs very well on a numerical example of a small-to-mid-sized call center and moderately well on a numerical example of a large and complex call center. Both are promising and can potentially allow for real-time optimization of staffing and shift schedules.

References

1. Atlason J, Epelman MA, Henderson SG. Optimizing call center staffing using simulation and analytic center cutting-plane methods. Manage Sci. 2008;54(2):295–309.
2. Cezik MT, L'Ecuyer P. Staffing multiskill call centers via linear programming and simulation. Manage Sci. 2008;54(2):310–23.
3. Avramidis AN, Chan W, Gendreau M, LEcuyer P, Pisacane O. Optimizing daily agent scheduling in a multiskill call center. Eur J Oper Res. 2010;200(3):822–32. ISSN 0377-2217.
4. Bodur M, Luedtke JR. Mixed-integer rounding enhanced benders decomposition for multiclass servicesystem staffing and scheduling with arrival rate uncertainty. Manage Sci. 2017;63(7):2073–91.
5. Ingolfsson A, Campello F, Wu X, Cabral E. Combining integer programming and the randomization method to schedule employees. Eur J Oper Res. 2010;202(1):153–63. ISSN 0377-2217.
6. Friedman JH. Greedy function approximation: a gradient boosting machine. Ann Stat. 2001;29(5):1189–232.
7. Breiman L, Friedman J, Stone CJ, Olshen RA. Classification and regression trees. CRC press; 1984.
8. Chen T, Guestrin C. Xgboost: a scalable tree boosting system. In: Proceedings of the 22nd acm sigkdd, international conference on knowledge discovery and data mining. ACM;2016. p. 785–94.
9. Ke G, Meng Q, Finley T, Wang T, Chen W, Ma W, Ye Q, Liu TY. Lightgbm: a highly efficient gradient boosting decision tree. Adv Neural Inf Proc Syst. 2017:3149–3157.

Information Directed Policy Sampling for Partially Observable Markov Decision Processes with Parametric Uncertainty

Peeyush Kumar and Archis Ghate

Abstract This paper formulates partially observable Markov decision processes, where state-transition probabilities and measurement outcome probabilities are characterized by unknown parameters. An information theoretic solution method that adaptively manages the resulting exploitation-exploration trade-off is proposed. Numerical experiments for response guided dosing in healthcare are presented.

1 Background and Problem Statement

A Markov decision process (MDP) is a tuple $\mathcal{M} = (S, A, T, R, N)$. Here, S is a finite set of states; A is a finite set of actions; T denotes the transition probability function $T(s'|s, a)$, for $s, s' \in S$ and $a \in A$; R denotes the reward function $R(s'|s, a)$, for $s, s' \in S$ and $a \in A$; and $N < \infty$ is the planning horizon [7]. A decision-maker observes the state $s_t \in S$ of a system at stage $t \in \{1, 2, \dots, N\}$ and then chooses an action $a_t \in A$. The system then evolves to a state $s_{t+1} \in S$ with probability $T(s_{t+1}|s_t, a_t)$. The decision-maker collects a reward $R(s_{t+1}|s_t, a_t)$. This process repeats until the end of stage N. A policy trajectory $\pi = (\pi_1, \pi_2, \dots, \pi_N)$ is a decision-rule that assigns actions $\pi_t(s_t) \in A$ to states $s_t \in S$, for $t = 1, 2, \dots, N$. The decision-maker's objective is to find a policy trajectory $\pi = (\pi_1, \pi_2, \dots, \pi_N)$ that maximizes the expected reward $J_\pi(s_1) = E[\sum_{t=1}^{N} R(s_{t+1}|s_t, \pi_t(s_t))]$. MDP models of problems-of-interest to the Service Science community in health care, transportation, energy, supply chain management, inventory control, and revenue management are discussed in [1, 6, 7]. For instance, in inventory control, the state often corresponds to the inventory level, the decision is the order quantity, the transition probabilities are defined by the uncertainty in demand, and the rewards correspond to revenue minus inventory holding and shortage penalty costs.

In some MDPs, the state is not observable; the decision-maker instead has access to imperfect measurements. A typical example is medical treatment planning, where the doctor makes measurements that provide information about a patient's health

---

P. Kumar · A. Ghate (✉)
Industrial & Systems Engineering, University of Washington, Seattle, WA 98195, USA
e-mail: archis@uw.edu

H. Yang and R. Qiu (eds.), *Advances in Service Science*, Springer Proceedings in Business and Economics, https://doi.org/10.1007/978-3-030-04726-9_20

state. Such systems can be modeled as Partially Observable Markov decision processes (POMDPs) [2], which are described by a tuple $\mathcal{Q} = (S, A, T, R, N, O, Z)$. Here, O is the finite set of possible measurements and Z represents the measurement probability function $Z(o_{t+1}|s_{t+1}, a_t)$. This equals the probability that the measurement outcome at the beginning of stage $t+1$ is o_{t+1}, given that the system state is s_{t+1} after choosing action a_t. This POMDP can be reformulated into an MDP whose state is defined as $x_t = (x_t^1, \ldots, x_t^{|S|})$, where x_t^i is the probability that the actual system state is $i \in S$ at stage t. These so-called "belief states" belong to the $|S|$-dimensional probability simplex $X = \{x \in \mathbb{R}_+^{|S|} | \sum_i x^i = 1\}$, and their evolution can be described as follows. Let $P(o_{t+1}|x_t, a_t) = \sum_{s_{t+1} \in S} Z(o_{t+1}|s_{t+1}, a_t) \sum_{s_t \in S} T(s_{t+1}|s_t, a_t) x_t(s_t)$ denote the probability that the next observation will be o_{t+1} given that action a_t was chosen in belief state x_t. Moreover, let $\phi_{s_{t+1}}(x_t, a_t, o_{t+1})$ denote the probability that the next state is s_{t+1} given that action a_t was chosen in belief state x_t and observation o_{t+1} was made. Standard algebraic manipulations using the Markov property and Bayes' theorem yield $\phi_{s_{t+1}}(x_t, a_t, o_{t+1}) = \Big(Z(o_{t+1}|s_{t+1}, a_t) \sum_{s_t \in S} T(s_{t+1}$ $|s_t, a_t) x_t(s_t)\Big) / P(o_{t+1}|x_t, a_t)$. Let $\phi(x_t, a_t, o_{t+1}) \in X$ denote the $|S|$-dimensional vector formed by collating probabilities $\phi_{s_{t+1}}(x_t, a_t, o_{t+1})$ across $s_{t+1} \in S$. In other words, $\phi(x_t, a_t, o_{t+1})$ is the new belief state when the decision-maker observes o_{t+1} after choosing action a_t in belief state x_t. Bellman's backward recursive equations of dynamic programming for this belief state-based MDP reformulation of the POMDP, and standard algorithms for their solution are available in the literature [2]. POMDP models of problems-of-interest to the Service Science community in health care, military, scheduling, communications, and quality control are reviewed in [5].

This paper focuses on POMDPs wherein the decision-maker does not know the transition probability function T and the measurement probability function Z. Specifically, let Λ be a finite set whose elements λ index the possible transition probability functions T_λ. Similarly, let Θ be a finite set whose elements θ index the possible measurement probability functions Z_θ. The family of possible POMDPs thus is $\mathcal{Q}_{\lambda,\theta} = \{S, A, T_\lambda, R, N, O, Z_\theta\}$. The "true" POMDP is $\mathcal{Q}_{\lambda^*,\theta^*}$, where $\lambda^* \in \Lambda$ and $\theta^* \in \Theta$. The decision-maker does not know λ^*, θ^*. The decision-maker begins with prior belief probability mass functions (pmfs) $\alpha_1(\cdot)$ on λ and $\beta_1(\cdot)$ on θ. This induces a joint belief pmf $\delta_1(\cdot, \cdot)$ on the pair (λ, θ). This joint pmf is updated via Bayes' Theorem as measurements drawn from the true probability function Z_{θ^*} are observed starting from an initial belief state $x_1 \in X$. The decision-maker wishes to learn λ^*, θ^* while maximizing expected reward. This problem is termed a POMDP with parametric uncertainty in this paper.

We recently proposed Information Directed Policy Sampling (IDPS) for MDPs with parametric uncertainty [4]. IDPS is based in information theory. At each stage, it requires the decision-maker to solve a convex problem to minimize a so-called information ratio. The numerator of this ratio equals the square of the expected regret of distributions over policy trajectories. The denominator equals the expected mutual information between the resulting system-state trajectory and the parameter's posterior. Thus, IDPS attempts to explicitly manage the exploration (denominator) versus exploitation (numerator) trade-off in MDPs with parametric uncertainty. IDPS for MDPs was an extension of Information Directed Sampling, which was originally proposed for bandit problems [8]. Here, we further generalize IDPS to POMDPs with parametric uncertainty.

2 Information Directed Policy Sampling

To generalize IDPS to POMDPs, we need new notation. A policy trajectory $\pi = (\pi_1, \pi_2, \ldots, \pi_N)$ is a sequence of functions $\pi_t : X \to A$; function π_t assigns an action from A to each possible belief state x_t in stage t. The uncountable set of all such policy trajectories is denoted $\mathcal{P}$. Let $\pi^t = (\pi_t, \pi_{t+1}, \ldots, \pi_N)$ denote the tail of a policy trajectory π. The set of tail policy trajectories is denoted by $\mathcal{P}^t$. Also let $\pi^*_{\lambda,\theta} = (\pi^*_{1,\lambda,\theta}, \ldots, \pi^*_{N,\lambda,\theta})$ denote an optimal policy trajectory for POMDP $\mathcal{Q}_{\lambda,\theta}$. Similar to Sect. 1, define

$$\phi_{s_{t+1},\lambda,\theta}(x_t, a_t, o_{t+1}) = \Big(Z_\theta(o_{t+1}|s_{t+1}, a_t) \sum_{s_t \in S} T_\lambda(s_{t+1}|s_t, a_t)x_t(s_t)\Big)/P_{\lambda,\theta}(o_{t+1}|x_t, a_t) \tag{1}$$

as the probability that the next state is s_{t+1} given, x_t, a_t, o_{t+1} in POMDP $\mathcal{Q}_{\lambda,\theta}$. Here, we use subscripts λ, θ to denote quantities arising from POMDP $\mathcal{Q}_{\lambda,\theta}$.

The information gain between two random variables W and Y is defined by $I(W; Y) = \sum_{w,y} P(w, y) \ln \frac{P(y|w)}{P(y)}$, where the letter P denotes the appropriate joint, conditional, and marginal distributions. In our POMDP context, W takes values in the parameter set $\Lambda \times \Theta$, while Y takes values from observation tuples $(o_{t+1}, \ldots, o_{N+1})$ when policy π^t is implemented starting in belief state x_t. The pmf of Y is denoted by

$$\begin{aligned}
P^{\pi^t}_{\lambda,\theta}(o_{t+1}, \ldots, o_{N+1}|x_t) &= P^{\pi^t}_{\lambda,\theta}(o_{t+2}, \ldots, o_{N+1}|o_{t+1}, x_t)P^{\pi^t}_{\lambda,\theta}(o_{t+1}|x_t) \\
&= P^{\pi^t}_{\lambda,\theta}(o_{t+2}, \ldots, o_{N+1}|o_{t+1}, x_t)P_{\lambda,\theta}(o_{t+1}|x_t, \pi_t(x_t)) \\
&= P^{\pi^t}_{\lambda,\theta}(o_{t+2}, \ldots, o_{N+1}|x_{t+1}, o_{t+1}, x_t)P_{\lambda,\theta}(o_{t+1}|x_t, \pi_t(x_t)) \\
&= P^{\pi^{t+1}}_{\lambda,\theta}(o_{t+2}, \ldots, o_{N+1}|x_{t+1})P_{\lambda,\theta}(o_{t+1}|x_t, \pi_t(x_t)) \\
&\ \ \vdots \\
&= \prod_{\ell=t}^{N} P_{\lambda,\theta}(o_{\ell+1}|x_\ell, \pi_\ell(x_\ell)).
\end{aligned}$$

The information gain is thus given by

$$g_t(\pi^t|x_t,\delta_t) = \sum_{\substack{\lambda\in\Lambda\\ \theta\in\Theta}} \sum_{o_{t+1},\ldots,o_N} \left(\prod_{\ell=t}^{N} P_{\lambda,\theta}(o_{\ell+1}|x_\ell, \pi_\ell(x_\ell)) \right)$$

$$\delta_t(\lambda,\theta) \ln \left[\frac{\prod_{\ell=t}^{N} P_{\lambda,\theta}(o_{\ell+1}|x_\ell, \pi_\ell(x_\ell))}{\sum_{\substack{\lambda\in\Lambda\\ \theta\in\Theta}} \prod_{\ell=t}^{N} P_{\lambda,\theta}(o_{\ell+1}|x_\ell, \pi_\ell(x_\ell))\delta_t(\lambda,\theta)} \right]. \tag{2}$$

Define $\tilde{R}_{\lambda,\theta}(o_{t+1}|x_t, a_t)$ as the expected reward earned in stage t upon choosing action a_t in belief state x_t and observing o_{t+1} in POMDP $\mathcal{Q}_{\lambda,\theta}$. That is,

$$\begin{aligned}
\tilde{R}_{\lambda,\theta}(o_{t+1}|x_t,a_t) &= \sum_{s_{t+1}}\sum_{s_t} \tilde{R}_{\lambda,\theta}(o_{t+1}|x_t,a_t,s_{t+1},s_t) P_{\lambda,\theta}(o_{t+1},s_{t+1},s_t|x_t,a_t)\\
&= \sum_{s_{t+1}}\sum_{s_t} \tilde{R}_{\lambda,\theta}(o_{t+1}|a_t,s_{t+1},s_t) P_{\lambda,\theta}(o_{t+1},s_{t+1}|s_t,x_t,a_t) P_{\lambda,\theta}(s_t|x_t,a_t)\\
&= \sum_{s_{t+1}}\sum_{s_t} R(s_{t+1}|s_t,a_t) P_{\lambda,\theta}(o_{t+1}|s_{t+1},s_t,x_t,a_t) P_{\lambda,\theta}(s_{t+1}|s_t,x_t,a_t) P_{\lambda,\theta}(s_t|x_t,a_t)\\
&= \sum_{s_{t+1}}\sum_{s_t} R(s_{t+1}|s_t,a_t) Z_\theta(o_{t+1}|s_{t+1},a_t) T_\lambda(s_{t+1}|s_t,a_t) x_t(s_t)\\
&= \sum_{s_{t+1}} Z_\theta(o_{t+1}|s_{t+1},a_t) \Big(\sum_{s_t} R(s_{t+1}|s_t,a_t) T_\lambda(s_{t+1}|s_t,a_t) x_t(s_t) \Big).
\end{aligned}$$

Now let $V_t^*(\lambda,\theta|x_t) = \sum_{\ell=t}^{N} \tilde{R}(o_{\ell+1}|x_\ell, \pi_\ell^*(x_\ell))$ denote the expected tail reward accumulated on implementing an optimal policy in POMDP $\mathcal{Q}_{\lambda,\theta}$ starting in belief state x_t, if the realized measurement outcomes trajectory equals $o_{t+1}, o_{t+2}, \ldots, o_{N+1}$. Let $U_t^*(\lambda,\theta|x_t)$ denote the expected value of $V_t^*(\lambda,\theta|x_t)$ with respect to the stochastic trajectory $o_{t+1}, o_{t+2}, \ldots, o_{N+1}$. Similarly, $V_t(\lambda,\theta,\pi^t|x_t) = \sum_{\ell=t}^{N} \tilde{R}(o_{\ell+1}|x_\ell, \pi_\ell(x_\ell))$ for any tail policy π^t, and $U_t(\lambda,\theta,\pi^t|x_t)$ is its expected value; $U_t(\lambda,\theta,\pi^t|x_t) = E_{(o_{t+1}:o_{N+1})\sim P^{\pi^t}_{\lambda,\theta}(o_{t+1}:o_{N+1}|x_t)}[V_t(\lambda,\theta,\pi^t|x_t)]$, and $U_t^*(\lambda,\theta|x_t)$ is defined similarly. The expected regret of implementing tail policy trajectory π^t starting in belief state x_t and posterior δ_t is defined as

$$\Delta_t(\pi^t|x_t,\delta_t) = \mathop{E}_{(\lambda,\theta)\sim\delta_t} \left[U_t^*(\lambda,\theta|x_t) - U_t(\lambda,\theta,\pi^t|x_t) \right]. \tag{3}$$

Let D^t denote the set of all probability distributions ν^t over tail policies in $\mathcal{P}^t$. Then, given the pair (x_t, δ_t) at the beginning of stage t, the expected regret and expected information gain are given by

$$\Delta_t(\nu^t|x_t,\delta_t) = \mathop{E}_{\pi^t\sim\nu^t}[\Delta_t(\pi^t|x_t,\delta_t)], \tag{4}$$

and

$$g_t(\nu^t|x_t,\delta_t) = \mathop{E}_{\pi^t\sim\nu^t}[g_t(\pi^t|x_t,\delta_t)]. \tag{5}$$

The information ratio is defined as

$$\Psi_t(\nu^t|x_t, \delta_t) = \frac{(\Delta_t(\nu^t))^2}{g_t(\nu^t)}. \tag{6}$$

The decision-maker then solves the convex problem

$$\nu_*^t \in \underset{\nu^t \in D^t}{\operatorname{argmin}}\, \Psi_t(\nu^t|x_t, \delta_t). \tag{7}$$

The resulting procedure is summarized in Algorithm 1. A worst-case regret bound for this algorithm is derived in the first author's doctoral dissertation [3] by extending our proof of a similar bound for MDPs [4].

IDPS is a forward approximate solution procedure that only computes decisions in belief states encountered during run-time. It does not compute an entire policy. This is common in POMDP solution procedures because the belief state space is uncountable and hence the full policy cannot be stored. This procedure is implemented next on a medical treatment planning problem.

Algorithm 1 Information Directed Policy Sampling

Require: POMDPs $\mathcal{Q}_{\lambda,\theta} = \{S, A, T_\lambda, R, N, O, Z_\theta\}$ for $\lambda \in \Lambda$ and $\theta \in \Theta$. Prior pmf $\delta_1(\cdot,\cdot) = \alpha(\cdot)\beta(\cdot)$. Initial belief state x_1.

function IDPS
  for episode $k = 1, 2, 3, \ldots$ **do**
    Set $t = 1$.
    Initialize x_1; and prior $\delta_1(\cdot,\cdot) \leftarrow \delta_{N+1}(\cdot,\cdot)$ if $k > 1$.
    repeat
      Compute distribution $\nu_*^t = \underset{\nu^t \in D^t}{\operatorname{argmin}}\, \Psi_t(\nu^t|x_t, \delta_t)$.
      Sample $\pi^t = (\pi_t, \ldots, \pi_N) \sim \nu_*^t$.
      Implement action $\pi_t(x_t)$.
      Observe o_{t+1} drawn from $P_{\lambda^*,\theta^*}(\cdot|x_t, \pi_t(x_t))$.
      Estimate x_{t+1} using

$$\underset{(\lambda,\theta)\sim\delta_t}{E}\left[\frac{Z_\theta(o_{t+1}|s_{t+1}, \pi_t(x_t)) \sum_{s_t \in S} T_\lambda(s_{t+1}|s_t, \pi_t(x_t))x_t(s_t)}{P_{\lambda,\theta}(o_{t+1}|x_t, \pi_t(x_t))}\right], \forall s_{t+1}.$$

      Update $\delta_{t+1}(\lambda, \theta) \propto P_{\lambda,\theta}(o_{t+1}|x_t, \pi_t(x_t))\delta_t(\lambda, \theta)$, for $\lambda \in \Lambda$ and $\theta \in \Theta$.
      t $\leftarrow$ t+1
    until end of horizon N
  end for
end function

3 Numerical Experiments on Response-Guided Dosing

The partially observable response-guided dosing problem is recalled from the first author's doctoral dissertation. Consider a treatment course with N sessions indexed by t. The disease state in session t is X_t, and the treatment's side effect is Y_t. Disease state X_t is an integer from the interval $[0, m]$; larger values represent worse disease states. Side effect states Y_t are also integers from the interval $[0, n]$; larger values represent worse side effects. A dose d_t is chosen for a session after measuring $o_t = (o_{X,t}, o_{Y,t})$. The measured disease scores $o_{X,t}$ take integer values in $[0, m_o]$. The measured side effect $o_{Y,t}$ takes integer values from $[0, n_o]$. Doses d_t are integers from $[0, \bar{d}]$, where $\bar{d} < \infty$ is the maximum permissible dose. The disease state and side effects evolve according to a transition probability distribution shown in Table 1. The measurement outcome probability for the disease scores is $Z^X(o_{x,t+1}|x_{t+1}, d_t) = \mathcal{B}(o_{x,t+1})$, where $\mathcal{B}(o_{x,t+1})$ is the Binomial distribution $B_{o_{x,t+1}}(n', 0.5)$ with $n' = 2(\lceil \frac{m_o}{m} \rceil x_{t+1} + 0.05(\bar{d} - d_t))$. Similarly, the measurement outcome probability for side effects is $Z^Y(o_{y,t+1}|y_{t+1}, d_t) = \mathcal{B}'(o_{y,t+1})$, where $\mathcal{B}'(o_{y,t+1})$ is a Binomial distribution $B_{o_{y,t+1}}(n'', 0.5)$ with $n'' = 2(\lceil \frac{n_o}{n} \rceil y_{t+1} + \theta d_t)$. The joint measurement outcome probability is given by $Z(o_{t+1}|(x_{t+1}, y_{t+1}), d_t) = Z^X(o_{x,t+1}|x_{t+1}, d_t) \times Z^Y(o_{y,t+1}|y_{t+1}, d_t)$. The patient's utility function was assumed to take the $R(s_{t+1}|s_t, d_t) = c_X(X_{t+1}, q_X) + c_Y(Y_{t+1}, q_Y) - cd_t$, where

$$c_X(X_t, q_X) = \frac{1}{m^{q_X}}(m^{q_X} - X_t^{q_X}),$$

and

$$c_Y(Y_t, q_Y) = \frac{1}{n^{q_Y}}(n^{q_Y} - Y_t^{q_Y}).$$

The function $c_X(X_t, q_x)$ represents the patients utility while being in disease state X_t; the function $c_Y(Y_t, q_Y)$ represents the patient utility with side effect Y_t. The higher disease/side effect states imply lower utility for the patient, as can be observed in the function forms. The last term cd_t represents the patient's disutility on receiving higher doses.

For our numerical simulations, we employed $m_o = n_o = 6$, $m = n = 3$, $c = 6$, $q_X = q_Y = 2$, $k = 0.05$, and $\bar{d} = 3$. The decision-maker knows that $\theta \in \Theta = \{0.05, 0.1, \ldots, 0.4, 0.45\}$, and $\lambda \in \Lambda = \{0.05, 0.1, \ldots, 0.45\}$. Figure 1 plots the posterior of the true parameter averaged over 50 runs for IDPS and Thompson Sampling with 25 sessions. The figure shows that IDPS learned faster than Thompson Sampling. Figure 2 plots the cumulative reward for IDPS and Thompson Sampling. The figure shows that IDPS earned higher rewards than Thompson Sampling.

Table 1 State transition probabilities $P^X(X_{t+1}|x_t, d_t)$ and $P^Y(Y_{t+1}|Y_t, d_t)$. The latter is parametrized by λ

Dose d_t	Probability distribution over disease states X_{t+1} $P^X(X_{t+1}\|X_t, d_t)$			Probability distribution over side effects Y_{t+1} $P^Y(Y_{t+1}\|Y_t, d_t)$		
	$X_{t+1} = X_t - 1$	$X_{t+1} = X_t$	$X_{t+1} = X_t + 1$	$Y_{t+1} = Y_t - 1$	$Y_{t+1} = Y_t$	$Y_{t+1} = Y_t + 1$
$d_t > 0$	$0.7 + 0.3\frac{d_t}{d}$	$0.3 - 0.3\frac{d_t}{d}$	0	0	$1 - \lambda\frac{d_t}{d}$	$\lambda\frac{d_t}{d}$
$d_t = 0$	0	0.7	0.3	0.7	0.3	0

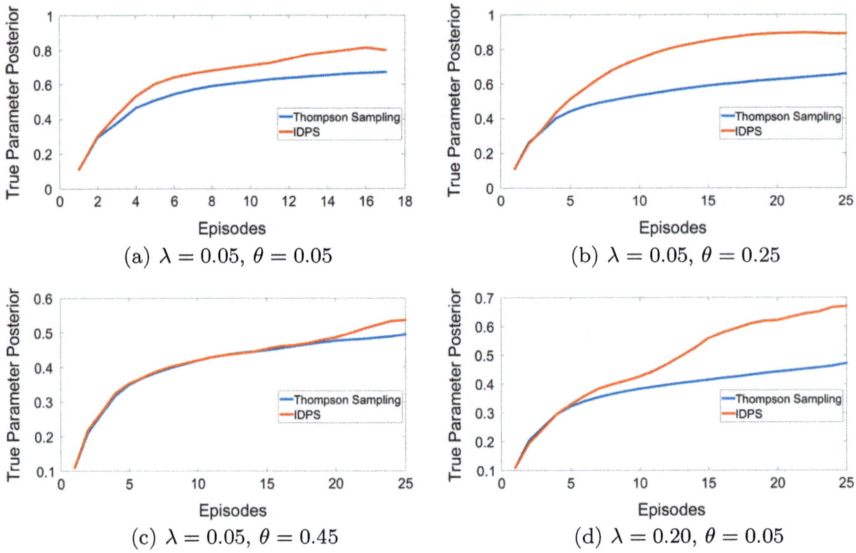

Fig. 1 Unknown transition and measurement outcome distributions: posterior for true parameter values

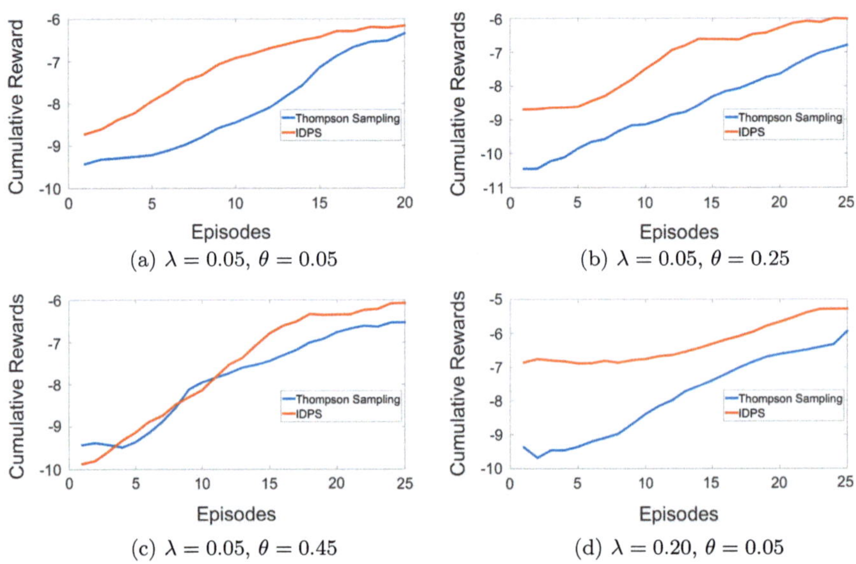

Fig. 2 Unknown transition and measurement outcome distributions: averaged cumulative reward

4 Conclusions

We presented an extension to POMDPs of our recent work on an information theoretic approach to learning in MDPs with parametric uncertainty. Our framework models a decision-maker who is uncertain about parameters that characterize the transition and measurement outcome probabilities in a POMDP. The decision-maker begins with a prior on these parameters, and, at each time-step, implemented an action prescribed by a policy that minimizes the information ratio. This calls for solving a convex program in an attempt to optimize the exploration versus exploitation trade-off that is at the heart of sequential learning problems. The system then stochastically evolves to a new state, a new measurement is acquired, and the prior is updated using Bayes' formula. We presented an idealized version of the algorithm where all calculations were assumed to be performed exactly. This is implementable for small problems. Future work could focus on designing computationally efficient, approximate versions of our approach, which could scale to large problems.

Acknowledgements This research was funded in part by the National Science Foundation via grant CMMI #1536717.

References

1. Boucherie R, van Dijk NM. Markov decision processes in practice. Basel, Switzerland: Springer; 2017.
2. Krishnamurthy V. Partially observed Markov decision processes. Cambridge, United Kingdom: Cambridge University Press; 2016.
3. Kumar P. Information theoretic learning methods for Markov decision processes with parametric uncertainty. Ph.D. thesis, University of Washington, Seattle; 2018.
4. Kumar P, Ghate A. Information directed policy sampling for Markov decision processes with parameteric uncertaint. unpublished; 2018.
5. Lovejoy WS. A survey of algorithmic methods for partially observed Markov decision processes. Ann Oper Res. 1991;28(1):47–65.
6. Powell WB. Approximate dynamic programming: solving the curse of dimensionality. Hoboken, NJ, USA: Wiley; 2007.
7. Puterman ML. Markov decision processes: discrete stochastic dynamic programming. New York, NY, USA: Wiley; 1994.
8. Russo D, Van Roy B. Learning to optimize via information directed sampling. Oper Res. 2017;66(1):230–52.

Buffered Probability of Exceedance (bPOE) Ratings for Synthetic Instruments

Giorgi Pertaia and Stan Uryasev

Abstract Credit Rating is an important characteristic of company in financial market. Investors determine the appropriate yields (required return) for the assets such as Bonds and CDO tranches, based on credit rating. Current methodology for measuring credit rating for synthetic instruments is based on probability of exceedance concept. The probability of exceedance has several drawbacks as a measure of risk. The most important is that it does not measure the magnitude of loss in the event of default. Therefore, financial instruments with very different exposures in the event of default may have the same rating. This paper illustrates, how the new measure called Buffered Probability of Exceedance (bPOE) can be used to calculate the credit ratings. The bPOE has exceptional qualitative and quantitative characteristics, compared to the probability of exceedance. bPOE is sensitive to the thickness of the tail of the loss distribution. Therefore, the exposure in the event of default impacts the ratings based on bPOE.

Keywords Buffered probability of exceedance · bPOE
Probability of exceedance · POE · Conditional Value-at-Risk · CVaR · Ratings
Collateralized debt obligation · CD

1 Introduction

Credit ratings are widely used by investors to assess the credit risk of a security. Currently there are three major credit rating providers (known as "Big Three"): Moody's, Standard and Poor's and Fitch Group. These agencies rate various financial instruments, including so called synthetic instruments. In finance, a synthetic instrument or position, is a way to create the payoff of a financial instrument using other

G. Pertaia (✉) · S. Uryasev
Industrial and Systems Engineering, University of Florida, Gainesville, FL 32611-6595, USA
e-mail: gpertaia@ufl.edu

S. Uryasev
e-mail: uryasev@ufl.edu

H. Yang and R. Qiu (eds.), *Advances in Service Science*, Springer Proceedings in Business and Economics, https://doi.org/10.1007/978-3-030-04726-9_21

financial instruments. The financial crisis of 2008 showed that credit ratings might not measure the risk appropriately for synthetic instruments such as Collateralized Debt Obligations (CDOs). One such case was the American International Group (AIG), a financial institution that purchased a significant number of CDOs tranches. The U.S. Government had to bailout AIG for \$85 billion and bought 80% of its equity. Rating agencies use an approach based on risk measure called Probability of Exceedance (POE). For a random variable X and some threshold x POE is defined as $\mathbb{P}(X > x)$.

This paper suggests a new rating model based on the Buffered Probability of Exceedance (bPOE), that is an improvement compared to the current POE based model. POE based rating do not measure the exposure in the case of default. If one instrument has a heavy-tailed loss distribution, while other one has light-tailed distribution, under POE based rating, they can have the same rating (in some cases, instrument with heavy-tailed loss might even have higher rating). bPOE based rating model will assign lower rating to the instruments with the heavy tails, thus removing incentive to accumulate low default probability but high exposure assets, as it happened with AIG.

2 Current Rating Models

Rating agencies have been collecting the default statistic of the rated companies for decades. The agencies publish the tables of default probabilities for each rating class over a given time horizon. Table 1 gives the Standard and Poor's default probability table. For example, BBB rating corresponds to a financial instrument with 1 year probability of default (PD) satisfying the inequality $0.08\% < \text{PD} < 0.23\%$.

Ratings models for synthetic instruments are quite complicated because of various assumptions and approaches involved in modeling of underlying instruments. However, the approach for issuing the rating, when simulation model is built, is quite simple. We will explain this approach with an example based on the Merton model. Suppose that a firm finances its operation by issuing a single zero-coupon bond with face value B_T payable at time moment T. It is assumed that at every time moment $t \in [0, T]$, the company has total assets A_t, following Geometric Browning motion dynamics.

Merton model assumes that the default of the company occurs when the firm has no capital (equity) to pay back the debt holders. Because the only payment the zero-coupon bond makes is at time T, that is the only moment when the default can occur. It is straightforward to calculate the probability of default (PD) for a given firm. The probability of default at time T is

$$\mathbb{P}(default) = \mathbb{P}(A_T < B_T) \tag{1}$$

Formula (1) can be rewritten in terms of POE by changing the sign of assets and liabilities,

Table 1 PD as a function of rating (published by Standard and Poor's). The PD is measured over a given time horizon

Average cumulative default rates for corporates by region (1981–2015) (%)

Time horizon (years)

Rating	1	2	3	4	5	6	7	8	9	10	11	12	13	14	15
U.S.															
AAA	0.00	0.04	0.17	0.29	0.42	0.54	0.59	0.67	0.76	0.86	0.90	0.95	1.00	1.10	1.21
AA	0.04	0.08	0.18	0.32	0.46	0.61	0.76	0.88	0.98	1.09	1.19	1.28	1.37	1.45	1.55
A	0.08	0.21	0.37	0.56	0.75	0.97	1.22	1.45	1.70	1.95	2.18	2.38	2.58	2.75	2.95
BBB	0.23	0.61	1.02	1.54	2.10	2.65	3.15	3.65	4.15	4.64	5.12	5.50	5.86	6.23	6.60
BB	0.81	2.51	4.58	6.60	8.38	10.14	11.61	12.96	14.17	15.27	16.16	16.94	17.60	18.16	18.75
B	3.93	8.99	13.39	16.81	19.50	21.71	23.55	25.01	26.29	27.46	28.44	29.22	29.94	30.57	31.19
CCC/C	28.21	38.67	44.55	48.32	51.13	52.19	53.32	54.15	55.18	55.84	56.47	57.15	57.92	58.54	58.54
Investment grade	0.12	0.33	0.57	0.88	1.19	1.52	1.83	2.13	2.42	2.72	3.00	3.23	3.45	3.66	3.89
Speculative grade	4.13	8.18	11.72	14.58	16.90	18.84	20.47	21.84	23.07	24.17	25.08	25.85	26.54	27.13	27.70
All rated	1.76	3.52	5.07	6.37	7.45	8.39	9.18	9.87	10.50	11.08	11.57	11.98	12.35	12.68	13.01

$$\mathbb{P}(default) = \mathbb{P}(A_T < B_T) = \mathbb{P}(-A_T > -B_T)$$

Thus, PD is a POE of random variable $-A_T$ with threshold $-B_T$. Table 1 can be used to convert the PD calculated using the Merton model into a rating (e.g., if 1 year PD satisfies inequality 0.08% < PD < 0.23%, then the company has BBB rating). Despite unrealistic assumptions, the Merton model provides the base for more complex models which are widely used in the industry.

3 bPOE Ratings

We propose a new methodology for assigning ratings to synthetic instruments, based on the bPOE concept. bPOE with threshold v for a random variable X is equal to the probability of the right tail of the distribution of X such that the average of this tail is equal to v. Formally, bPOE can be defined as follows (see, [1]),

$$bPOE(v) = \min_{a \geq 0} \mathbb{E}[a(X - v) + 1]^+ \tag{2}$$

where $[x]^+ = max\{x, 0\}$. bPOE is equal to one minus inverse of Conditional Value-at-Risk (CVaR), where CVaR in the average of the tail having the probability $1 - \alpha$. Formally, CVaR is defined as follows (see, [2]),

$$CVaR(\alpha) = \min_C \left(C + \frac{1}{1-\alpha} \mathbb{E}[X - C]^+ \right)$$

By definition bPOE equals POE = $1 - \alpha$ of the right tail with $CVaR(\alpha) = v$, see Fig. 1.

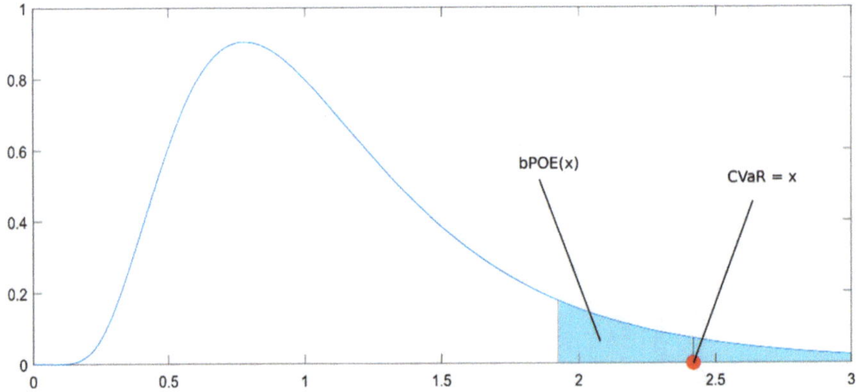

Fig. 1 Relationship of bPOE and CVaR (the shaded region area is equal to bPOE)

For more information about properties of bPOE see [1]. Note, the formula (2) is considered a property of bPOE in paper [1], however, it is convenient to use it as a definition.

For evaluation of ratings, we suggest to replace POE with bPOE, calculated for the same threshold. bPOE, by construction, is always greater than the POE with the same threshold. For example, for the standard normal distribution, bPOE is roughly 2.4 times higher than the POE with the same threshold. For the log-normal distribution with parameters $\mu = 0$ and $\sigma = 1$, bPOE is roughly 3.2 time higher than POE. We propose to rescale the probabilities in the rating tables, by bPOE/POE ratio calculated for the exponential distribution. bPOE ratings will be calculated using the intervals from the new table. There are two reasons why exponential distribution is a good candidate for rescaling:

1. Exponential distribution is the "demarcation line" between heavy-tailed and light-tailed distributions. The distribution is called heavy-tailed if

$$\lim_{v \to \infty} e^{\lambda v} \mathbb{P}(X \geq v) = \infty, \quad \forall \lambda > 0.$$

   i.e., heavy-tailed distribution has heavier tails than the exponential distribution with arbitrary parameter λ.
2. The bPOE(v)/POE(v) ratio for the exponential distribution with arbitrary parameter $\lambda > 0$ and arbitrary threshold value $v > EX$, is constant and equal to e $= 2.718\ldots$. bPOE for the exponential distribution (see [3]) equals $e^{1-\lambda v}$. The POE for the exponential distribution equals $e^{-\lambda v}$, thus, the ratio of bPOE to POE equals e.

4 Case Study

This section illustrates bPOE ratings for a CDO's loss distribution. The data comes from paper [4]; the dataset can be downloaded from (http://www.ise.ufl.edu/uryasev/research/testproblems/financialengineering/structuring-step-up-cdo). This data represents a loss distribution of the underlying assets of the CDO over the period of one year, generated by Standard and Poor's CDO Evaluator. Figure 2 shows the histogram of the loss distribution.

We take threshold values of 0.12 and 0.15 to illustrate bPOE ratings. The POEs for the threshold values 0.12 and 0.15 are 2.73% and 0.8% respectively. Based on the Standard and Poor's ratings, 2.73% falls into the B rating bracket, because it is within the following interval 0.81%<2.73%<3.93% and 0.8% fall into the BB rating bracket, because it is within the following interval 0.23%<0.8%<0.81%.

The bPOEs for 0.12 and 0.15 threshold values are 7.24% and 2.06%. By scaling the probabilities in the Table 1, using the exponential distribution coefficient e, we get that the ratings are not changed for this case. The bPOE corresponding to the 0.12 threshold value falls in the interval 2.718*0.81%<7.24%<2.718*3.93% and bPOE

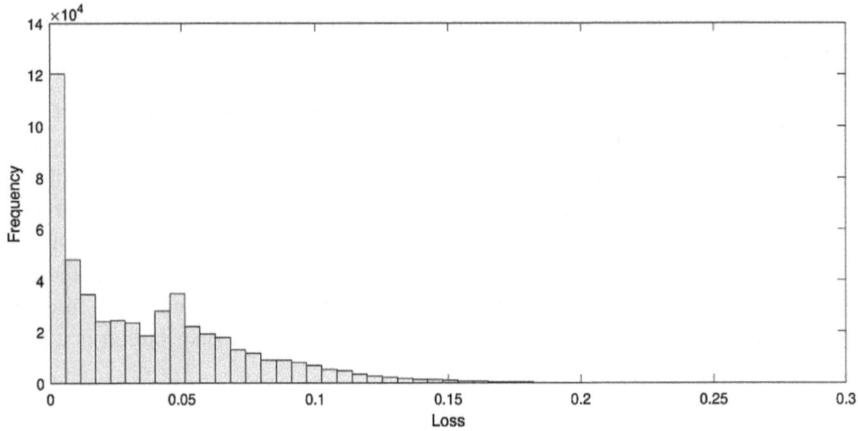

Fig. 2 The histogram of the loss distribution

for 0.15 falls in the interval 2.718*0.23% < 2.06% < 2.718*0.81%, corresponding to B and BB ratings respectively. The fact that ratings are unchanged means that the loss distributions tail is similar to the tail of the exponential distribution. Therefore, the bPOE ratings are close to the POE ratings for this dataset.

5 Summary

The paper presented the application of bPOE for defining the credit ratings. bPOE accounts for information about magnitude of the losses in the tail of the distribution. The paper proposed coefficient equal to e for converting existing rating probability bounds into the bPOE rating probability bounds. The conversion factor is based on the assumption that losses are distributed exponentially. Any loss distribution that has heavier tails than the exponential distribution (and thus is a heavy tailed distribution), will have lower bPOE rating than the POE based rating.

References

1. Mafusalov A, Uryasev S. Buffered probability of exceedance: mathematical properties and optimization. SIAM J Optim. 2018;28(2):1077–103.
2. Rockafellar RT, Uryasev S. Optimization of conditional value-at-risk. J Risk. 2000;2(3):21–41.
3. Mafusalov A, Shapiro A, Uryasev S. Estimation and asymptotics for buffered probability of exceedance. Europ J Op Res. 2018.
4. Veremyev A, Tsyurmasto P, Uryasev S. Optimal structuring of CDO contracts: optimization approach. J Credit Risk. 2012;8(4), Winter 2012/13.
5. Case study: structuring step up CDO. Data and Codes: http://www.ise.ufl.edu/uryasev/research/testproblems/financialengineering/structuring-step-up-cdo.

Service Quality Assessment via Enhanced Data-Driven MCDM Model

Vahab Vahdat, Seyedmohammad Salehi and Nima Ahmadi

Abstract Tourism and hospitality industry has brought large economical revenue for both developing and developed countries. However, with the increase in tourists' diversity, needs, and expectations, the need for hotels with higher quality of services has emerged. This research evaluates and compares the quality of service in two different types of hotels that exist in the historic cities: first, hotels that are located in the historic sites of the city offering mostly the city architecture, culture, life style, and local cuisines second, modern hotels that are outside the buffer zone of the historic site, equipped with modern technology and offer more standardized services and international cuisines. In this research, a stylized multi-phase framework is used to assess the quality of service from a modified-SERVQUAL model. Two sets of surveys are distributed among the hotel administrators and travelers. Using Analytic Hierarchy Process (AHP), fuzzy set theory, and Technique for Order Preference by Similarity to the Ideal Solution (TOPSIS), the relative importance of each SERVQUAL dimension in the hotel industry is investigated and the hotel types are ranked accordingly. Our results indicate that hotels that are located in historic sites are more favorable for the tourists.

V. Vahdat (✉)
Department of Mechanical and Industrial Engineering, Northeastern University,
Boston, MA 02114, USA
e-mail: vahdatzad.v@husky.neu.edu

S. Salehi
Computer and Information Sciences, University of Delaware, Newark, DE 19716, USA
e-mail: salehi@udel.edu

N. Ahmadi
Department of Industrial Engineering and Engineering Management, Western New England University,
Springfield, MA 01119, USA
e-mail: nima.ahmadi@wne.edu

H. Yang and R. Qiu (eds.), *Advances in Service Science*, Springer Proceedings
in Business and Economics, https://doi.org/10.1007/978-3-030-04726-9_22

1 Introduction

Increasing customer satisfaction is the primary objective of companies to survive in the current competitive market. *Tourism* is called as one of the fastest growing industry over the past decades with no signs of slowing down in the 21st Century [1]. But hospitality and tourism industry has been reframed in the recent years. The presence of accommodation sharing, home-exchange, and hospitality exchange have overshadowed hotel services around the world. For instance, Airbnb is an accommodation sharing website that provides a platform to rent rooms or apartments offering over 3 million lodging listing, while the company does not own any lodging. As a consequence, although the profit of tourism industry has been soared, it does not necessarily increase the profit of hotels and hostels. In order to increase the attractiveness of hotels in comparison to other lodgings formats, the quality of service in hotels should be significantly better than their adversaries. High service quality is increasingly recognized as a critical factor in the success of travel and tourism industry. Service quality has shown to have influences on customer loyalty, satisfaction, and business performance [2].

Travelers' satisfaction in hotel service industry cannot be maximized unless managers understand travelers' expectations from a provided service and measure the quality of provided services accordingly. However, measuring service quality is difficult for many reasons. First, quality of service is evaluated voluntarily and is highly associated with respondent judgment or biases. Individuals usually have wide range of perceptions toward quality of service, depending on their own experience, preferences, and beliefs. Ignoring the variability in perception, many of previous research studies have utilized conventional cardinal or ordinal scales to measure service quality. In such scale-based measurement system, scores do not necessarily represent user preferences. This is because respondents have to internally convert preferences to ordinal scales, and the conversion may introduce distortion of the preference being captured [3]. Second, the service is comprised of both tangible and intangible/subjective attributes that complicate the accuracy of measurements. Examples of unmeasurable aspect of services are the safety and comfort of a service. Additionally, some services are perishable, inseparable, or heterogeneous that would increase the hardship of true measurement of service quality. For instance, the airline industry provides heterogeneous services to passengers in same flight, based on the class categories of passenger's ticket. While many studies have introduced toolboxes to quantify the quality of services, non-of-them are generalizable to all service industry. Choosing an appropriate tool that fits the needs and attributes of the hotel service industry is essential for accurate service assessment. For this purpose, this research modifies SERVQUAL method to fully adapt with services provided in hotels. Also, Fuzzy set theory is used to decrease the impact of judgmental preferences to express the perception ranges through the judgments of persons. Integrating Analytic Hierarchy Process (AHP) with Fuzzy set theory, his research seeks for the most important criteria in hotel service quality. Finally, using the weights of the criteria that are obtained by AHP, a ranking between two types of hotels in cities with historic sites are assessed using TOPSIS, a well-structured multi-criteria decision making method.

The rest of this study is structured as follows: Sect. 2 describes the important aspects for the assessment of service quality of hotel industry and presents the evaluation framework and methodology that are used in this study. Section 3 discusses the procedure and results of empirical study by evaluating and ranking of service quality in hotel industry in Yazd, Iran. The final results of the empirical study are presented and discussed in the Sect. 4 followed by implications and recommendations for future research.

2 Literature Review

In this section, we present a review of the relevant literature. In the first subsection, we explore the history and methods of service quality assessment. Then, two multi-criteria decision making methods, AHP and TOPSIS, that are used to build this research framework, are briefly discussed. In order to reduce the vagueness of the qualitative assessments, we used fuzzy numbers and fuzzy theory in designing our surveys. Last subsection of this Section surveys recent fuzzy numbers and fuzzy theory methods.

2.1 Service Quality Assessment

Service industry constitutes over 50% of GDP in developing countries and its importance grows as the economy of developing countries evolves [4]. The importance of service industry elevated the researchers to categorize, analyze, and evaluate the quality of service continuously. Quality of service is used in many disciplines that offer service to customers such as telecommunications [5, 6], healthcare [7, 8], web-services [9], bank and insurance services [10], and so forth. One of the first service quality classifications was developed during the 1980s where Gronroos [11] distinguished between technical and functional service quality. Technical quality refers to the delivery of the core service, while functional quality refers to the way in which the customer receives the service. Lehtinen and Lehtinen [12], discussed three distinct service quality dimensions: physical; interactive; and corporate quality. Physical quality includes the physical aspects associated with the service such as the reception area and equipment in hotel industry. Interactive quality involves the interaction between the customer and the service provider; and finally corporate quality includes the firm's image or reputation. A common notion in service quality assessment is a comparison of what the customers feel a service provider should offer (i.e., customers' expectations) versus the service provider's actual performance [13]. This notion was created and then validated by a research conducted in [14] using twelve groups of consumers in four different services including retail banking, telecommunications, securities brokerage, and product repair and maintenance. Their efforts has led to a new definition for service quality evaluation using the degree of discrepancy between customers' perceptions and expectations. Perception-expectation gap formed a service quality assessment method called SERVQUAL that consists of five main dimensions and total of 22 sub-dimensions: The main dimensions include tangibles (appearance of physical elements), reliability (ability to provide the promised service in a reliable and accurate manner), responsiveness (promptness and helpfulness), assurance (courtesy, credibility, competence) and empathy (easy access, good communications and customer understanding). SERVQUAL is mostly assessed with a seven-point scale surveys ranging from 'strongly disagree' to 'strongly agree'. Although SERVQUAL has been substantially successful for assessing service quality, it has been criticized on both conceptual and methodological grounds. Some studies argue that the instability of the SERVQUAL is probably due to the type of service sector under investigation [15]. Parasuraman et al. [16], concede that the generalization of the five dimensional structure of service quality remains in doubt and should be further investigated.

While SERVQUAL has been widely assessed in several service sectors, only a few studies have directly evaluated this method within the context of the hospitality industry [17, 18]. In an early research, Saleh and Ryan [18] applied SERVQUAL model to lodging services. They identified and used 33 attributes for hotel services rather than the 22 items existed in the original SERVQUAL model. Ramsaran-Fowdar [1] also found other sets of attributes for SERVQUAL analysis in hotel industry. Accordingly, a shortlist of attributes related to SERVQUAL sub-dimensions are adapted from the literature that are concise and compatible with hotel industry, as shown in Table 1.

2.2 Analytical Hierarchy Process (AHP)

Multiple Criteria Decision Making (MCDM) tackle the problems with more than one criterion to determine the best ranking among set of feasible alternatives. Analytical Hierarchy Process (AHP) [3] is one of the first MCDM methods that was introduced by Saaty [19]. AHP is a well-known technique for modeling subjective decision-making processes based on multiple attributes and can be used in both individual and group decision-making environments [20]. AHP has been used in social sciences, business administration, and service quality [2]. In AHP, multiple pair-wise comparisons are based on a standardized nine-point scale ranging from "equally important" to "extremely more important" [21].

In the AHP, the problem is decomposed into smaller independent criteria where each criterion may have its own sub-criteria. The hierarchical model is then solved by the evaluators who conduct pair-wise comparisons between criteria in each level. Furthermore, the relative importance derived from these pair-wise comparisons allows a certain degree of inconsistency within a domain. Saaty [19] used the principal eigenvector of the pair-wise comparison matrix derived from the scaling ratio to determine the comparative weight among the criteria [22]. The result of the pair-wise comparison on n criteria can be summarized in an $n \times n$ matrix A in which every element $a_{ij}(i, j = 1, 2, ..., n)$ is the quotient of weights of the criteria, as shown in (1).

$$A = \begin{bmatrix} a_{11} \; a_{12} \; \ldots \; a_{1n} \\ a_{21} \; a_{22} \; \ldots \; a_{2n} \\ \vdots \quad \vdots \quad \ddots \quad \vdots \\ a_{n1} \; a_{n2} \; \ldots \; a_{nn} \end{bmatrix} \qquad a_{ii} = 1, \;\; a_{ji} = \frac{1}{a_{ij}}, \; a_{ij} > 0 \tag{1}$$

Next, the mathematical process normalizes and finds the relative weights for each matrix. The relative weights are given by the right eigenvector (w) corresponding to the largest Eigen value (λ_{max}) [23], as shown in (2). If the pair-wise comparisons are completely consistent, weights can be obtained by normalizing any of the rows or columns of the matrix A.

$$Aw = \lambda_{\max} w \tag{2}$$

Table 1 Revised-SERVQUAL model representing hotel service quality attributes

Dimensions	Sub-dimensions (attributes)
Tangibility D_1	C1: convenient hotel location C2: hotel staff professional appearance C3: rooms attractiveness, comfort, spaciousness, and cleanliness C4: hotel lobby interior/exterior design C5: convenient restaurants with high quality diverse foods C6: image of the hotel
Reliability D_2	C1: providing the services within the promised time-frame C2: well-trained and experienced staff C3: staff with good communication skills C4: staff providing the right services at the first request C5: accuracy in room and hotel billing and food orders
Responsiveness D_3	C1: staff willingness to assist tourists promptly C2: availability of staff to provide requested service at anytime C3: quick and convenience check-in and check-out process
Assurance D_4	C1: staff friendliness C2: courteous employees C3: ability of staff to instill confidence in tourists
Empathy D_5	C1: providing special attention to the tourists C2: availability of room service in regular bases C3: understanding the tourists' needs C4: listening carefully to complaints and compensate accordingly C5: recognizing the tourists commitments and offering loyalty program

2.3 TOPSIS

Technique for Order Performance by Similarity to Ideal Solution (TOPSIS) is a ranking methodology that was first introduced by Tzeng and Huang [24]. TOPSIS ranks the alternatives based on their distance to the positive-ideal solution and the negative ideal solution. The positive-ideal solution is a solution that maximizes the benefit criteria and minimizes the cost criteria, whereas the negative ideal solution maximizes the cost criteria and minimizes the benefit criteria. The method uses a term called "relative closeness" that considers and correlates the distance to the positive and negative ideal solutions for each alternative [3]. For instance, Let A^{+} and A^{-} be the positive and negative ideal solutions, respectively. Let $x = \left(x_{ij}\right)$ denote the performance matrix where x_{ij} is the performance of alternative i to criterion j and $W = \{w_1, \ldots, w_n\}$ is the weight factor where $\sum w_i = 1$. TOPSIS can be defined with the following steps:

Step 1: Establish the normalized performance matrix $n = \left(n_{ij}\right)$ in order to unify matrix entries' units using:

$$n_{ij} = \frac{x_{ij}}{\sqrt{\sum_{i=1}^{m} x_{ij}^{2}}} \quad \forall_{i,j} \in \{1, \ldots, \mathrm{m}\} \tag{3}$$

Step 2: Calculate weighted normalized performance matrix using $v_{ij} = w_j \times n_{ij} \quad i \in \{1, \ldots, m\}, j \in \{1, \ldots, n\}$.

Step 3: Determine the ideal solution and negative ideal solution with the following equations:

$$A^{+} = \left(v_1^{+}, v_2^{+}, \ldots, v_n^{+}\right) = \left\{\left(\max v_{ij} \middle| j \in J\right), \left(\min v_{ij} \middle| j \in J'\right), \quad i = 1, \ldots, m\right\} \tag{4}$$

$$A^{-} = \left(v_1^{-}, v_2^{-}, \ldots, v_n^{-}\right) = \left\{\left(\min v_{ij} \middle| j \in J\right), \left(\max v_{ij} \middle| j \in J'\right), \quad i = 1, \ldots, m\right\} \tag{5}$$

where $j = \{1, \ldots, m\}$ belongs to the benefit criteria and $j' = \{1, \ldots, n\}$ belongs to the cost criteria.

Step 4: Calculate the Euclidean distance between positive and negative ideal solutions for each alternative:

$$S_i^{+} = \sqrt{\sum_{j=1}^{n} (v_{ij} - v_j^{+})^2} \;, \quad S_i^{-} = \sqrt{\sum_{j=1}^{n} (v_{ij} - v_j^{-})^2} \quad i = 1, \ldots, m \tag{6}$$

Step 5: Calculate the relative closeness $\left(C_i^{*}\right)$, $0 \le ci^{*} \le 1$, to the ideal solution of each alternative:

$$C_i^{*} = \frac{S_i^{-}}{S_i^{+} + S_i^{-}} \qquad i = 1, \ldots, m \tag{7}$$

Step 6: Rank the preference order of set of alternatives in corresponds to the descending order of C_i^*.

2.4 *Fuzzy Set Theory*

Zadeh [25] introduced fuzzy set theory to orient the rationality of uncertainty as a result of imprecision and vagueness. The contribution of fuzzy set theory into the science is in capability of representing and handling the conditions that data is unclear and out of formal binary values [26]. Fuzzy theory can be divided into two categories: (1) *Fuzzy systems* that are designed for uncertain and complex systems in which approximate reasoning of the system can be hardly represented by a mathematical model, and (2) *Fuzzy logics* that enable decision-making with estimated values where incomplete or uncertain information exist [26].

A simple representation of fuzzy set theory is as follows. Let X refer to a universal set. Then, a fuzzy subset of X is calculated by its membership function $\mu_{\bar{A}} : x \rightarrow [0, 1]$ which assigns a real number $\mu_{\bar{A}}(x)$ to each element in $x \in X$ in the interval $[0, 1]$. The value of $\mu_{\bar{A}}(x)$ represents the grade of membership of x in $\bar{A}$. As the value of $\mu_{\bar{A}}(x)$ gets closer to unity, the grade of membership of x in $\bar{A}$ becomes higher [27]. Triangular Fuzzy Numbers (TFN) is a special type of fuzzy number with three parameters, each representing the linguistic variable associated with a degree of membership of 0 or 1. Since it is shown to be very convenient and easily implemented in arithmetic operations, the TFN is also commonly used in practice. A triangular fuzzy number $\tilde{m}$ can be defined by a triplet (a, b, c). The membership function μ_m is given in Eq. (8) [28]:

$$\mu_{\overline{m}} = \begin{cases} \dfrac{x-a}{b-a} & a \leq x \leq b \\ \dfrac{c-x}{c-b} & b \leq x \leq c \end{cases} \tag{8}$$

Algebraic operations such as addition ($\oplus$), multiplication ($\otimes$), subtraction (Θ), and division ($/$) are practically conceivable for the triangular fuzzy numbers [22, 29]. For instance, addition of two TFN is calculated in Eq. (9).

$$(L_1, M_1, U_1) \oplus (L_2, M_2, U_2) = (L_1 + L_2, M_1 + M_2, U_1 + U_2) \tag{9}$$

Statistical operations are also computable for triangular fuzzy numbers. For instance, average of multiple TFNs is $A_{ave} = (A_1 + A_2 + \cdots + A_n)/n$, or equivalently $A_{ave} = [(L_1 + \cdots + L_n) + (M1 + \cdots + M_n) + (U_1 + \cdots + U_n)]/n$. Fuzzy sets have vague boundaries and are therefore well-suited for linguistic terms (such as "very" or "somewhat") or natural phenomena (e.g., temperatures) [30].

Linguistic variables represent linguistic terms such as words and sentences [31]. Each linguistic variable can be interpreted with a fuzzy number, a real number within a specific range. In this research, the weight for each revised-SERVQUAL criterion is captured by a fuzzy number that considers decision maker's vagueness in providing pair-wise comparisons. Afterwards, in order to rank the hotel alternatives with a nonfuzzy ranking method, a de-fuzzification method should be employed. Defuzzification is a technique to convert a fuzzy number into crisp numbers with locating the Best Non-fuzzy Performance (BNP) value [3]. There are several methods that serve this purpose such as Mean-of-Maximum, Center-of-Area, and a-cut Method. This study utilizes the Center-of-Area method since this method is simple while capturing critical information [29]. The defuzzified value of a TFN number ($TFN = (L, M, U)$) can be obtained using Eq. (10).

$$\text{BNF} = \frac{[(U - L) + (M - L)]}{3} + L \tag{10}$$

3 Research Framework

In order to assess and rank hotel service quality, a framework is proposed as shown in Fig. 1. First, we identify the evaluating criteria and attributes with respects to a revised-SERVQUAL that is tailored for hotel and tourism industry. After constructing the evaluation criteria hierarchy, their weights are calculated using Analytic Hierarchy Process (AHP) method. The measurement of service quality corresponding to each criterion is conducted under the setting of fuzzy numbers. Finally, Technique for Order Preference by Similarity to Ideal Solution (TOPSIS) is utilized to achieve the final ranking among hotels. A detailed descriptions of each step is discussed as follows.

Due to the variations in service standards, quantifying travelers' expectation has become more complicated than ever. Services provided in brand-new hotels may be different with the hotels located in the historic sites of the cities. In middle-east, there are many historic sites registered to UNESCO in which any change to the site and the buffer zone is restricted or prohibited. City of Yazd in Iran, is one of the first urban cities that is inscribed on the world heritage list in 2017. The city has 195 ha of world heritage property with 665 ha of buffer zone. In order to better maintain the historic sites and following the UNESCO restriction rules, a number of traditional houses have been recently converted into atmospheric hotels.

In this research, a comparison between modern and atmospheric hotels in middle-east is provided. For this purpose, two hotels that have similar share from the market and have 4-star ratings are selected in Yazd. Hotel (A) is a modern hotel that is 20 min away from city historic site but built in 1990s with 190 rooms that are designed with traditional city theme. Hotel (B) is an atmospheric hotel in the heart of the historic site which was a residential mansion build over 200 years ago. Before comparing the hotels, it is necessary to understand the importance or the weight for each of the dimensions and sub-dimensions of modified-SERVQUAL. Consequently, 50 surveys were distributed between different hotel managers and tourist guides in Yazd, Iran.

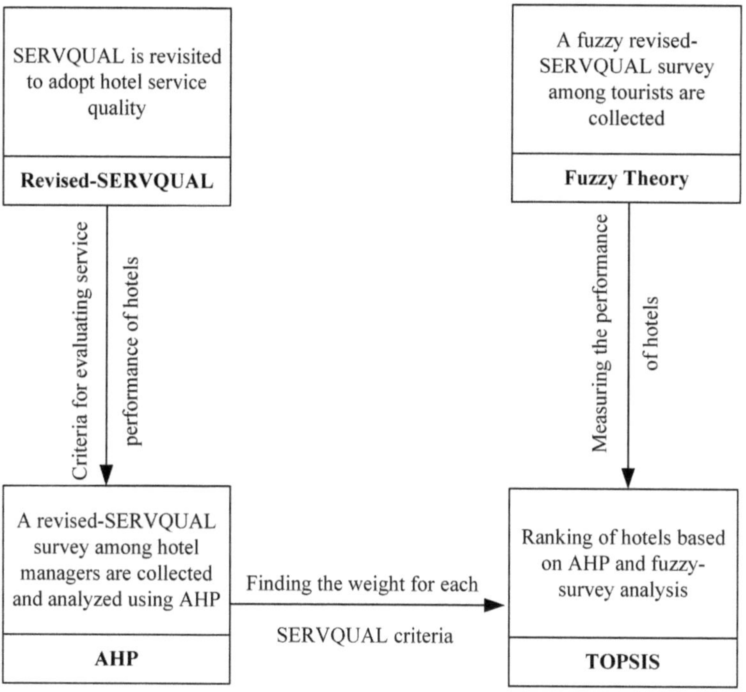

Fig. 1 Proposed framework for hotel service quality assessment

Out of the 50 surveys, 30 were returned with a return rate of 60%. We used AHP method based on criteria in Table 1 for evaluating hotel service quality. The results are shown in Fig. 2.

As shown in Fig. 2, responsiveness and reliability dimensions are considered the most valuable service quality dimensions among others in hotel industry. Correspondingly, the rest of dimensions in the order of importance are assurance, empathy, and tangibility. The sub-dimensions related to service quality in hotel industry are also prioritized both in comparison with other criteria in same dimension (local comparison) and with all criteria from every dimensions (global comparison) as shown in Fig. 2. "Staff willingness to assists tourists promptly" and "Staff providing the right service at the first request" are the most important sub-dimensions.

After finding the weights of the revised-SERVQUAL dimensions and sub-dimensions using AHP, another survey is developed for understanding travelers' expectations. For each question in the survey, travelers are asked to respond to the question in accordance with five linguistic terms (i.e. strongly disagree, disagree, fair, agree, and strongly agree). However, since tourists from different nations have significant cultural differences, there is no guarantee that a response such as "fair" from two tourists meant the same. Therefore, fuzzy set numbers are used in this survey. For this purpose, in order to determine the fuzzy range for each term, responders

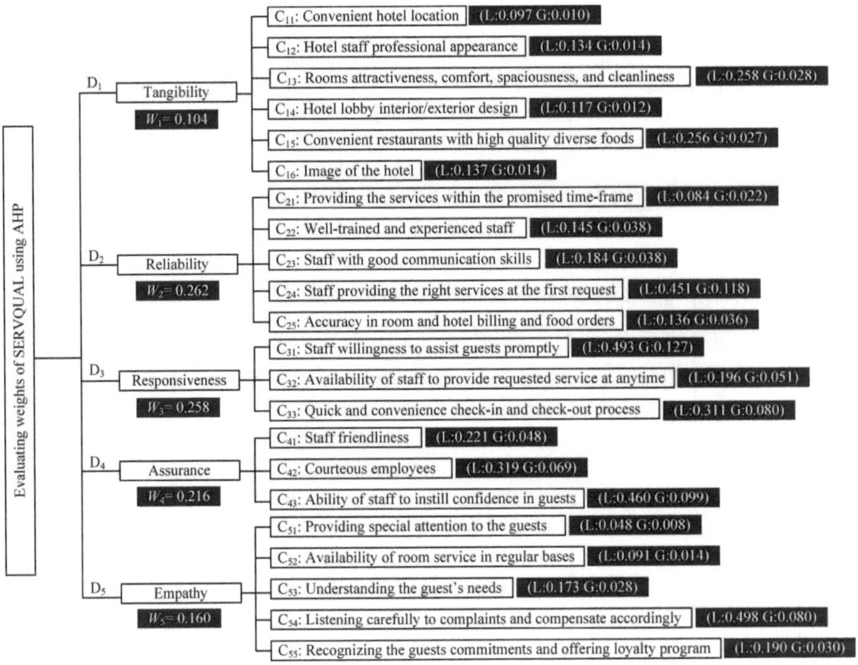

Fig. 2 Evaluating importance of revised-SERVQUAL dimensions using AHP. Note that for each sub-dimension L denotes local importance of sub-dimension compared with other attributes in same dimension and G denotes global importance of a sub dimension among all sub dimensions

were asked to specify a range from 1 to 100 corresponding to each linguistic term. For instance "fair" may mean a range between 50 and 60 for one respondent, and 50–75 for another respondent. Candidate tourists are selected from Hotel (A) and Hotel (B). Out of the 60 surveys, 50 were returned with a return rate of 83%. For each criteria, averages of fuzzy numbers were calculated using center area method, as shown in Table 2. In the next step, the fuzzy results for each criteria is de-fuzzified, using Eq. (10). The pair-wise comparison between each criterion is performed and shown with (*) in Table 2. The comparison reveals that Hotel A has superiority in physical aspects whereas Hotel B has advantageous in assurance and reliability aspects. Since de-fuzzified values are crisp, it can be easily used with TOPSIS to find the final rank of the two types of hotels by calculating the distance of hotels with ideal solution. The TOPSIS results (Similarity to ideal solution Hotel (A): 0.27658; Hotel (B): 0.72342) shows that atmospheric hotels are more in favor for the tourists.

Table 2 Analysis of fuzzy revised-SERVQUAL surveys for modern and atmospheric hotels

Revised-SERVQUAL		Fuzzy evaluation of service criteria		Defuzzified evaluation	
Dimension	Sub-dimension	Hotel A (Modern)	Hotel B (Atmospheric)	Hotel A (Modern)	Hotel B (Atmospheric)
Tangibility	C_{11}	(53.05, 68.83, 80.27)	(60.00, 65.80, 70.71)	67.38*	65.50
	C_{12}	(52.36, 70.26, 73.94)	(53.57, 63.71, 74.28)	65.52*	63.85
	C_{13}	(67.89, 80.58, 93.95)	(68.46, 80.00, 91.54)	80.81*	80.00
	C_{14}	(67.63, 81.71, 93.68)	(48.66, 59.30, 70.00)	81.01*	59.32
	C_{15}	(17.63, 34.45, 46.32)	(54.00, 65.16, 76.33)	32.80	65.16*
	C_{16}	(54.69, 64.84, 80.00)	(52.71, 63.14, 73.14)	66.51*	62.99
Reliability	C_{21}	(39.47, 44.79, 59.47)	(63.00, 71.63, 82.00)	47.91	72.21*
	C_{22}	(44.72, 57.64, 70.53)	(64.00, 74.50, 85.00)	57.63	74.50*
	C_{23}	(46.58, 57.55, 73.68)	(64.00, 71.50, 83.00)	59.27	72.84*
	C_{24}	(55.79, 70.00, 83.95)	(66.25, 75.78, 85.31)	69.91	75.78*
	C_{25}	(35.52, 47.29, 56.31)	(64.06, 71.25, 83.40)	46.37	72.90*
Responsiveness	C_{31}	(43.68, 57.10, 70.00)	(62.18, 71.56, 80.93)	56.93	71.55*
	C_{32}	(52.63, 65.05, 77.36)	(52.05, 60.73, 69.41)	65.01*	60.73
	C_{33}	(68.68, 80.45, 92.10)	(54.11, 62.94, 70.59)	80.41*	62.54
Assurance	C_{41}	(36.05, 50.92, 63.68)	(55.81, 60.15, 70.29)	50.22	62.10*
	C_{42}	(53.62, 65.58, 77.89)	(57.05, 63.03, 73.00)	65.37*	64.36
	C_{43}	(35.53, 46.84, 58.15)	(49.41, 58.82, 68.23)	46.84	58.82*
Empathy	C_{51}	(44.74, 58.18, 64.21)	(55.88, 65.29, 74.70)	55.71	65.29*
	C_{52}	(55.78, 65.81, 78.68)	(40.00, 56.76, 65.88)	66.76*	54.21
	C_{53}	(38.95, 52.24, 61.84)	(54.41, 63.68, 72.74)	51.01	63.68*
	C_{54}	(33.94, 48.42, 56.58)	(49.70, 58.09, 67.65)	46.31	58.48*
	C_{55}	(26.58, 33.42, 47.10)	(53.25, 62.20, 71.18)	35.70	62.21*

4 Conclusion and Implications

In this study, a new framework for understanding the service quality in hotel and tourism industry is proposed. Using a questionnaire to compare the expectation of customers from the service provided in a hotel versus their perception, gave an opportunity to rank the most important features that affect customers satisfaction. Two multi-criteria decision making namely Analytical Hierarchical Process (AHP) and Technique for Order Preference by Similarity to the Ideal Solution (TOPSIS) are used to identify, evaluate, and rank features affecting service quality in hotel industry. Moreover, the fuzzy logic is used with a membership function to measure the linguistic variables, decreasing the complications with interpreting linguistic terms. In general, AHP method is used to obtain service quality criteria weight. Then a survey with triangular fuzzy numbers are used to better capture travelers' expectations while considering differences in background and culture. With TOPSIS, the performance of one modern and one atmospheric hotel in a historic city is evaluated using the weight of criteria, obtained from AHP, and defuzzified matrix of performance, obtained from surveys.

Tourists are more cognizant to staff performance rather than physical layout of hotels. Investing on developing experienced staff is strongly proposed. The ability of staffs to instil confidence in tourists is also a key to success. As customers are the one who define the boundaries of quality of service, instilling confidence increases loyalty and trust, and has its own revenues for hotels. Quick and easy check-in/check-out procedure is still a crucial issue for many customers that need to be designed and operated efficiently. This study possesses a few limitations: firstly, our survey respondents were selected among limited hotel managers and hotel industry specialists. This may raise questions regarding representativeness of preference of hotels, especially in generalizing the results. It is strongly advised to consider the size and dimensions of this research before mapping it as a general description in hotel industry. For further research, we recommend to use fuzzy AHP and fuzzy TOPSIS for more accurate outcomes instead of using fuzzy numbers for linguistic terms.

References

Ramsaran-Fowdar RR. Developing a service quality questionnaire for the hotel industry in Mauritius. J Vac Market. 2007;13:19–27.

Moutinho L, Curry B. Modelling site location decisions in tourism. J Travel Tour Market. 1994;3:35–57.

Tsaur S-H, Chang T-Y, Yen C-H. The evaluation of airline service quality by fuzzy MCDM. Tour Manag. 2002;23:107–15.

Cali M, Ellis K, te Velde DW. The contribution of services to development: the role of regulation and trade liberalisation. Overseas Development Institute London, England;2008.

Esmailpour A, Salehi S, Safavi N. Quality of service differentiation measurements in 4G networks. Wirel Telecommun Symp (WTS). 2013;2013:1–5.

Salehi S, Li L, Shen C-C, Cimini L, Graybeal J. Traffic differentiation in dense WLANs with CSMA/ECA-DR MAC protocol. 2018. arXiv:1806.09582.
Mobin M, Li Z, Amiri M. Performance evaluation of tehran-qom highway emergency medical service system using hypercube queuing model. In: IIE annual conference. Proceedings;2015, p. 1175.
Vahdat V, Griffin J, Stahl JE. Decreasing patient length of stay via new flexible exam room allocation policies in ambulatory care clinics. Health Care Manage Sci. 2017;1–25.
Zada VV, Abbasi S, Barazesh F, Abdi R. E-service websites quality measurement through a revised ES-QUAL. Global J Technol. 1;2013.
Saeedpoor M, Vafadarnikjoo A, Mobin M, Rastegari A. A servqual model approach integrated with fuzzy AHP and fuzzy topsis methodologies to rank life insurance firms. In: Proceedings of the international annual conference of the American society for engineering management;2015, p. 1.
Gronroos C. Service management and marketing: customer management in service competition, vol. 3. Wiley Chichester;2007.
Lehtinen U, Lehtinen JR. Service quality: a study of quality dimensions. Service Management Institute;1982.
Zeithaml VA, Parasuraman A, Berry LL. Delivering quality service: balancing customer perceptions and expectations. Simon and Schuster;1990.
Parasuraman A, Zeithaml VA, Berry LL. Servqual: a multiple-item scale for measuring consumer perc. J Retail. 1988;64:12.
Babakus E, Mangold WG. Adapting the SERVQUAL scale to health care environment: an empirical assessment. Enhan Knowl Develop Market. 1989;9:67–8.
Parasuraman A, Berry LL, Zeithaml VA. Refinement and reassessment of the SERVQUAL scale. J Retail. 1991;67:420.
Bojanic DC, Drew Rosen L Measuring service quality in restaurants: an application of the SERVQUAL instrument. Hospit Res J. 1994;18:3–14.
Saleh F, Ryan C. Client perceptions of hotels: A multi-attribute approach. Tour Manag. 1992;13:163–8.
Saaty TL. Analytic hierarchy process. In: Encyclopedia of operations research and management science. Springer;2013, pp. 52–64.
Bolloju N. Aggregation of analytic hierarchy process models based on similarities in decision makers' preferences. Eur J Oper Res. 2001;128:499–508.
Chen S-H, Wang H-H, Yang K-J. Establishment and application of performance measure indicators for universities. The TQM J. 2009;21:220–35.
Chiu Y-C, Chen B, Shyu JZ, Tzeng G-H. An evaluation model of new product launch strategy. Technovation. 2006;26:1244–52.
Dağdeviren M, Yavuz S, Kılınç N. Weapon selection using the AHP and TOPSIS methods under fuzzy environment. Expert Syst Appl. 2009;36:8143–51.
Tzeng G-H, Huang J-J. Multiple attribute decision making: methods and applications. Chapman and Hall/CRC;2011.
Zadeh LA. Fuzzy sets. Inf Control. 1965;8:3.
Kahraman C, Cebeci U, Ulukan Z. Multi-criteria supplier selection using fuzzy AHP. Logist Inf Manage. 2003;16:382–94.
Sakawa M. "Fuzzy multi objective and multilevel optimization: multiple criteria optimization" state of the art annotated bibliographic surveys. Kluwer Academic Publishers;2002, pp. 172–226.
Chamodrakas I, Alexopoulou N, Martakos D. Customer evaluation for order acceptance using a novel class of fuzzy methods based on TOPSIS. Expert Syst Appl. 2009;36:7409–15.

Abdolvand M, Toloie A, Taghiouryan M. The evaluation of custom service quality by SERVQUAL fuzzy. In: Applied international business conference. Sarawak, Malaysia;2008, pp. 367–80.
Friedlob GT, Schleifer LL. Fuzzy logic: application for audit risk and uncertainty. Manag Audit J. 1999;14:127–37.
Chen C-T. A fuzzy approach to select the location of the distribution center. Fuzzy Sets Syst. 2001;118:65–73.

Estimating the Effect of Social Influence on Subsequent Reviews

Saram Han and Chris K. Anderson

Abstract This study proposes an effective way of using retailer-prompted review data from TripAdvisor to measure the social network effect in self-motivated online reviews by overcoming the reflection problem. After applying the network effect model, we find that self-motivated review ratings are positively associated with previous corresponding peer reviews. We further show that the size of this peer effect attenuates as the peer reviews are located further away from the first page. This study suggests that reviewer ratings are more strongly influenced by peer ratings located on the visible page.

Keywords Peer effect · eWOM on-line review

1 Introduction

Online user reviews have become a critical part of electronic word-of-mouth (e-WOM) research. Given the importance of opinion sharing in collective social processes, a wealth of studies have focused on the impact customer reviews have on corporate revenue [1, 8, 9] and on online reputation [6]. While several studies have concentrated on the dynamics of online opinion formation, only a few of these studies have explored the influence of online consumer reviews on other reviewers. Unlike traditional customer surveys, consumer reviewers in an online community can see what other members have written. In light of the potential for social interaction among these online reviewers, several researchers have recently endeavored to find evidence for this social influence effect [7, 10, 16]. In the context of online reviews, social influence refers to the tendency for one's opinion to be influenced by other reviews [14].

S. Han (✉) · C. K. Anderson
School of Hotel Administration, Cornell University, Ithaca, NY, USA
e-mail: sh2322@cornell.edu

H. Yang and R. Qiu (eds.), *Advances in Service Science*, Springer Proceedings in Business and Economics, https://doi.org/10.1007/978-3-030-04726-9_23

2 Related Work

Research by Li and Hitt [12] and by Godes and Silva [10] show that online reviews submitted by users later in a product's life cycle differ considerably from those submitted early in a products life cycle. Sikora and Chauhan [16] suggests using a Kalman filter to estimate the time variant sequential bias in online reviews. However, none of these studies provide an estimate of the exact amount of social influence derived from other reviewers.

The most relevant study from this group is the work of Askalidis et al. [2]. Askalidis et al. [2] measured the level of social influence of multiple product ratings by comparing self- motivated consumer reviews with retailer-prompted consumer reviews. The present study uses hotel reviews from TripAdvisor and makes the same distinction between self-motivated and retailer- prompted consumer reviews. Askalidis et al. [2] found that the trend of self-motivated reviews were negatively related to the trend of retailer-prompted reviews, thus concluding a negative social influence bias.

However, there are two limitations in the study by Askalidis et al. [2] that we overcome in the present study. First, while it might be possible to identify a negative temporal trend in self-motivated reviews, it was not possible for Askalidis et al. [2] to identify at which stage of in the transaction process initial peer reviewers influenced subsequent reviewer. Social influence may arise at two points: (a) reviews at the pre-purchase stage that affect expectation and thus influence satisfaction, and (b) reviews at the evaluation stage that affect the reviewer when evaluating the product or service [5]. We aim to identify the peer effect at the evaluation stage by comparing the peer effect levels of reviews on different pages (i.e., first, second, and third page) on self-motivated reviewers who have seen these previous reviews. Second, the research assumes that the products true performance is consistent across time. According to expectation disconfirmation theory, satisfaction is a function of customer's pre-purchase expectations [15]. Therefore, controlling the expectations of each reviewer is critical for measuring the antecedent's effect on online review ratings. This is especially important in the case of experience-based goods and services because, by virtue of their inconsistent nature, their true performance and expectation is in a constant state of flux. We therefore account for this by using a monthly average reviewer rating in our model.

Therefore, our study is distinct from that of Askalidis et al. [2] by virtue of our efforts to: (a) identify the peer effect of the reviewers on each previous page at the evaluation stage, and (b) control for the expected performance that each reviewer may have perceived at the pre-purchase stage.

The research questions guiding this study are: (a) when reviewers write reviews on TripAdvisor, are their ratings influenced by what other previous reviewers have written; and (b) does this effect reduce as the previous reviews are further away from the first page, which is less visible to the reviewer? We find that reviews on the first page have a strong positive effect on the next review; however, the effect size of this social influence bias reduces as the reviews are located further away from the first page, disappearing altogether by the third page.

3 Method

In order to identify the level of social influence bias, we follow the peer effect identification strategy of Brock and Durlauf [4], Bramoull´e et al. [3], and Tiwari and Richards [17], using a random intercept regression model with online hotel reviews from http://www.TripAdvisor.com. Since identifying the social influence bias induced from the recent reviewers is a similar research problem as identifying the amount of the network or peer effect, for convenience of discussion, we will refer to the previous reviewers as the peer reviewers.

3.1 Data Collection

We collected online reviews directly from http://www.TripAdvisor.com for 1,638 hotels in four US states (NY, CA, NV, and FL), written between January 1, 2015 and December 31, 2017. We restricted the data so that each hotel had at least 100 retailer-prompted and self-motivated reviews each. Each review provides an indicator of whether it was collected through retailer invitation, thus allowing us to distinguish between data collection channels. Note that retailer-prompted reviewers submit their ratings by replying an email invitation from the retailer. The reviewer subsequently writes their review on an isolated page in the absence of social signals (e.g., other reviews). In contrast, the review submission process allows self-motivated reviewers to see what previous consumer reviewers have written before evaluating their own hotel experience. As a result, we had a sample of 189,212 reviews from 282 hotels for the analysis. This data consists of reviewer identification numbers, ratings (number of stars on a scale ranging 1–5), travel year/month when reviewer stayed at the hotel, reviewed time, and an identifier indicating whether the review had been prompted by the retailer.

3.2 Estimation Model

According to Manski [13], members in a same network behave in similar ways for three reasons: (a) contextual effects, such as demographics or psychographics within a peer group; (b) correlation effects, or environmental factors that are common to a set of peers; and (c) endogenous effects, or true induced-behavior effects in which the choices made by one peer affect the decision making of others [17]. We are interested in identifying this endogenous effect in relation to subsequent reviewer ratings.

TripAdvisors default screen for hotel reviews is in reverse chronological order; therefore, unless the user changes the review order, we can assume that reviewers are more likely to be exposed to other recent reviews for the same hotel. As such, we assume that the content of the current reviewers review is more likely to be influenced by these other recent reviews. Furthermore, because TripAdvisor lists no more than five reviews per page, it is reasonable to assume that the reviews on the first page have the most influence. Therefore, we define consumer reviews in the order of $t-1$, $t-2$, $t-3$, $t-4$, and $t-5$ as the peer reviews on the first page for the reviewer t ($t=1, 2, \ldots, n$). Accordingly, as the reviews are ordered chronologically, $t-5 \cdot p+4$, $t-5 \cdot p+3$, $t-5 \cdot p+2$, $t-5 \cdot p+1$, and $t-5 \cdot p$ are the five reviews that appear on page p, where reviewer t is assumed to be influenced by the reviews they can see. Once reviewer t clicks on the Next page button, reviewer t is assumed to be influenced by the reviews she sees on the next page. We denote the satisfaction rating of reviewer t for hotel h as y_{ht}, which is numerically coded on a five-point Likert scale, where 5 represents the highest and 1 represents the lowest rating. Reviewer t and $t-5 \cdot p+4, \ldots, t-5 \cdot p$ are coded as being in the same peer group. We assume a recency effect in which the most recently presented review will most likely influence the reviewer. Using this data, we created three $N \times N$ adjacency matrices G_p ($p=1, 2, 3$), where each element in G_p represents a row-normalized coefficient that describes how reviewer t is close to each $t-5 \cdot p+4, \ldots, t-5 \cdot p$. For example, from G_1, reviewer t is closest to $t-1$; therefore, $G_{p,t,t-1} = 5/15$; in contrast, $t-5$ is the furthest, thus $G_{p,t,t-5} = 1/15$. Otherwise, assuming the new reviewer does not see the reviews on other pages, we assign zero. In this way, multiplying the social adjacency matrix G_p by the satisfaction rating creates a weighted-average rating value that reflects not only the positive or negative value of each peer rating, but also the strength of the social relationship to each other peer [17].

One problems inherent to social network analysis is what Manski [13] describes as a reflection problem. This occurs when the peer group affects the behavior of individuals within the group. As such, any econometric model of individual behavior produces biased results unless the problem is addressed econometrically [11]. Under our network formation, peers in the same page are very likely to experience similar influences; therefore, the rating of t and her peers $t-5 \cdot p+4, \ldots, t-5 \cdot p$ are highly correlated, thus making it difficult to identify an endogenous peer effect. We avoid this reflection problem by taking advantage of the natural experiment research setting of TripAdvisor, where it is random whether the reviewer t is self-motivated or retailer-prompted. This random design allows us to control for unobserved similarities and extract the effect which is solely due to the peers. In order to control for unobserved heterogeneity in each hotel, we use a random intercept model where the intercept is allowed to vary across different hotels.

Therefore, we specify our full model in a matrix notation as follows:

$$Y = \alpha + \beta_1 StayedAvgRating + \beta_2 SelfMotivated + \beta_3 rank + \beta_4 rank \cdot SelfMotivated + \sum_{p=1}^{3} \beta_{4+p} G_p Y + \sum_{p=1}^{3} \beta_{7+p} G_p Y \cdot SelfMotivated + e \quad (1)$$

where each Y, rank, StayedAvgRating, and SelfMotivated is a vector of length N. Note that we follow the model and the variable notification of Askalidis et al. [2], except for the inclusion of the adjacency matrices G_p. The coefficients of $G_p \cdot Y$ measure the unobserved prior similarities between the review t and peer reviews on the page p. These slopes test no more than the correlation between the peers and the retailer-prompted reviewer (i.e., SelfMotivatedt = 0), who does not see the peer reviews. Therefore, β_5, β_6, and β_7 estimate the correlations in the rating between peers, which is not a causal effect of peers. In contrast, β_8, β_9, and β_{10}, which are the statistic of interest of this study, estimate the effects of the peer reviewers in each page p on the reviewers rating at the evaluation stage. StayedAvgRating is the average rating of the reviews of hotel h for the month when the reviewer stayed at the hotel. The coefficient of StayedAvgRating variable will detect the satisfaction changes caused by hotel performance. Self Motivated is a binary variable indicating whether it is a self-motivated review. rank indicates the chronological order in which a review was submitted among all other reviews for the same hotel. Thus, β_3 measures the trend of retailer-prompted reviews and β_4 tests the trend of self-motivated reviews. Askalidis et al. [2] suggests the negative β_4 as evidence of negative social influence. However, as this temporal trend is likely sufficiently explained by the self-motivated reviewers exposure to other peer reviews, we expect that once we consider the effect of the peer ratings, the trend effect disappears.

3.3 Result

The four models in Table 1 test whether including the peer effect of each page improves the model fit across pages 1–3. Model 2 suggests that self-motivated reviewers ratings have a positive correlation with the ratings of the previous five reviews on the first page. By adding the interaction term between the weighted average of the peer ratings and self-motivation, the model fit becomes increasing statistically significantly according to the reduced AIC and BIC, and the increased log likelihood. Our analysis reveals that, on average, one unit increase of the weighted average of the peer ratings on the first page increases the rating of the subsequent self-motivated review by 0.05.

Note that this effect cannot be attributed to the correlation between the ratings submitted during the similar time period as this effect is controlled by β_5, β_6, and β_7 in our model. In Model 3, we test whether the peer ratings in the second page impact the new rating. Unlike Model 2, Model 3 has a minor fit improvement from the nested model, and the slope for the $G_2 \cdot Y$ is not statistically significantly different from

Table 1 Results for random intercept regression models to estimate the peer effects

	Model 1	Model 2	Model 3	Model 4
Intercept	0.205*** (0.022)	0.309*** (0.026)	0.327*** (0.027)	0.331*** (0.028)
StayedAvgRating	1.166*** (0.008)	1.168*** (0.008)	1.168*** (0.008)	1.168*** (0.008)
SelfMotivated	−0.151*** (0.007)	−0.378*** (0.030)	−0.417*** (0.036)	−0.424*** (0.039)
rank	−0.000 (0.000)	−0.000 (0.000)	−0.000 (0.000)	−0.000 (0.000)
rank · SelfMotivated	0.000 (0.000)	0.000 (0.000)	0.000 (0.000)	0.000 (0.000)
$G_1 \cdot Y$	−0.086*** (0.005)	−0.111*** (0.006)	−0.109*** (0.006)	−0.108*** (0.006)
$G_2 \cdot Y$	−0.076*** (0.004)	−0.077*** (0.004)	−0.083*** (0.006)	−0.083*** (0.006)
$G_3 \cdot Y$	−0.035*** (0.004)	−0.036*** (0.004)	−0.036*** (0.004)	−0.037*** (0.006)
SelfMotivated · $G_1 \cdot Y$		0.054*** (0.007)	0.048*** (0.008)	0.047*** (0.008)
SelfMotivated · $G_2 \cdot Y$			0.015* (0.008)	0.014* (0.008)
SelfMotivated · $G_3 \cdot Y$				0.004 (0.008)
AIC	519687.912	519639.449	519645.609	519655.247
BIC	519789.268	519750.940	519767.236	519787.010
Log likelihood	−259833.956	−259808.724	−259810.804	−259814.624
Num. obs.	186,392	186,392	186,392	186,392
Num. groups: hotel	282	282	282	282
Var: hotel (Intercept)	0.000	0.000	0.000	0.000
Var: residual	0.951	0.951	0.951	0.951

$^{***}p < 0.001$, $^{**}p < 0.01$, $^{*}p < 0.1$

zero within p<0.001. Finally, the result of Model 4 shows that the reviews on the 3rd page have no effect on the new ratings. This result implies that reviewers refer to the reviews on the first page but rarely to the reviews on the following pages before rating the hotel.

One noticeable result is that after accounting for the peer effects, neither the rank variable nor the interaction variable rank · Self Motivated, which measures the relative upward/downward trend of the retailer-prompted and self-motivated reviews respectively, are significant. This findings contradicts those of Askalidis et al. [2].

We argue that most of this effect trend can be explained by the influence of the peers and the perceived performance of the hotel during the month the reviewer stayed.

4 Discussion

We estimated the effect of existing peer ratings, which may influence new ratings in TripAdvisor hotel reviews. We showed that existing reviews strongly affect new reviews when they are on the first page. The peer effect remains in the second page, but decreases as the page is located further away from the first page. This effect completely disappears by the third page. This result implies that previously submitted peer ratings that reviewers see immediately before they evaluate their hotel stay influences how they rate the product or service.

There are several interesting future possibilities for this work. Future studies might review whether the textual content is influenced by the peer reviews. For example, the topics that reviewers write about might be influenced by what previous reviewers have written. It might also be interesting to compare the peer effects of different customer review platforms. Considering that not all online review platforms order reviews chronologically, it might be interesting to see whether the peer effect exist in other such platforms.

This study has implications both for marketing practitioners and academia in terms of providing a better understanding of how reviewers are influenced by other reviewers when evaluating a product or service. The findings of this study can benefit system managers and service providers looking to collect representative opinions on a minimal budget. Broadly speaking, our findings suggest that simply hiding or randomly displaying previous reviews can reduce the peer effect as much as conducting an expensive customer survey.

References

1. Anderson M, Magruder J. Learning from the crowd: regression discontinuity estimates of the effects of an online review database*. Econ J. 2012;122(563):957–89.
2. Askalidis G, Kim SJ, Malthouse EC. Understanding and overcoming biases in online review systems. Decis Support Syst. 2017;97:23–30.
3. Bramoull´e Y, Djebbari H, Fortin B. Identification of peer effects through social networks. J Econ. 2009; 150(1):41–55.
4. Brock WA, Durlauf SN. Identification of binary choice models with social interactions. J Econ. 2007;140(1):52–75.
5. Chen Y, Xie J. Online consumer review: word-of-mouth as a new element of marketing communication mix. Manage Sci. 2008;54(3):477–91.
6. Dellarocas C, Zhang X, Awad NF. Exploring the value of online product reviews in forecasting sales: the case of motion pictures. J Interact Market. 2007;21(4):23–45.
7. Eryarsoy E, Piramuthu S. Experimental evaluation of sequential bias in online customer reviews. Inf Manag. 2014;51(8):964–71.

8. Ghose A, Ipeirotis PG. Estimating the helpfulness and economic impact of product reviews: mining text and reviewer characteristics. IEEE Trans Knowl Data Eng. 2011;23(10):1498–512.
9. Glaeser EL, Kim H, Luca M. Nowcasting gentrification: using yelp data to quantify neighborhood change. AEA Papers Proc. 2018;108:77–82.
10. Godes D, Silva JC. Sequential and temporal dynamics of online opinion. Market Sci. 2012;31(3):448–73.
11. Goldsmith-Pinkham P, Imbens GW. Social networks and the identification of peer effects. J Bus Econ Stat. 2013;31(3):253–64.
12. Li X, Hitt LM. Self-selection and information role of online product reviews. Inf Syst Res. 2008;19(4):456–75.
13. Manski CF. Identification of endogenous social effects: the reflection problem. Rev Econ Stud. 1993;60(3):531.
14. Muchnik L, Aral S, Taylor SJ. Social influence bias: a randomized experiment. Science. 2013;341(6146):647–51.
15. Oliver RL. Effect of expectation and disconfirmation on postexposure product evaluations: an alternative interpretation. J Appl Psychol. 1976;62(4):480–6.
16. Sikora RT, Chauhan K. Estimating sequential bias in online reviews: a Kalman filtering approach. Knowl Based Syst. 2012;27:314–21.
17. Tiwari A, Richards TJ. Social networks and restaurant ratings. Agribusiness, 2016;32(2):153–74. https://doi.org/10.1002/agr.21449.

Product and Service Design for Remanufacturing, Uncertainty and the Environmental Impact in the Closed-Loop Supply Chain Network

Qiang (Patrick) Qiang, Rong Fu and Shaowei Chen

Abstract We propose a closed-loop supply chain network model, in which the manufacturers compete with each other to determine the production quantity and remanufactureability service level. The network equilibrium solution is presented by utilizing the theory of variational inequalities. The properties of the solution are discuss.

1 Introduction

Remanufacturing of used products at their end of life (EOL) reduces both the need for natural resources and the waste produced. Remanufacturing can keep used products out of the waste stream longer and as a result, saturation of landfill is slowed down and air pollution is reduced.

With the mounting evidence of pollution and its dire consequences, many government legislations have been put in place to emphasize the extended producer responsibility (EPR), which includes the European end-of-life Vehicle (ELV) Directive, the Waste Electrical and Electronic Equipment (WEEE) Directive within the European Union, and the Electronics Recycling laws in the U.S. However, even with these take-back legislations, the global e-waste generation is still growing at an alarming rate. According to Ref. [1], the total amount of e-waste in has grown from 33.8 million tons to 41.8 million tons from year 2010 to 2014. In year 2014, only around 6.5 million tons of e-waste was reported as formally treated by national take-

Q. (Patrick) Qiang
Management Division, Penn State Great Valley School of Graduate Professional Studies, Malvern, PA, USA

R. Fu (✉) · S. Chen
Xi'an University of Finance and Economics, Xi'an, Shaanxi, China
e-mail: fuyd84@163.com

H. Yang and R. Qiu (eds.), *Advances in Service Science*, Springer Proceedings in Business and Economics, https://doi.org/10.1007/978-3-030-04726-9_24

back systems. There were 1.15 billion mobile phones sold in 2009 versus 674 million in 2004 [2]. To make things worse, the life span for mobile phones has dropped to less than two years.

Due to the escalating environmental concerns from shareholders, many firms have voluntarily participated in the product remanufacturing practice, as reported by BMW, IBM, DEC, and Fuji Xerox [3]. Some firms take "first-mover advantage" to incorporate remanufactureability concept into the product design so that they can remanufacture the product easily and at a cost-effective manner. It is worth pointing out that remanufactureability can be considered as a product service level design, which will have an impact on the cost and yield rate of the future refurbished product. For example, it has been reported that the tire manufacturers can make the choice of material and production technology to influence the retreadability of tires [4, 5]. In addition, firms like Kodak, Xerox, and General Motors "begin to design products with eventual remanufacturing in mind" [6]. For example, in order to achieve longer life span, Kodak recently replaced a few plastic parts in its high-speed copiers with more expensive, but reusable, stainless steel [6]. Fuji Xerox archived zero landfill of used products in Japan by building the concepts of easy disassembly, durability, reuse, and recycling into their equipment design. It is reported that all the company's equipment is developed with remanufactureability components [7, 8].

It's been discussed in the literature that design of remanufactureability service level has impact on production cost and consumer demand. For example, "Quality" is used in Ref. [9] to imply the product design issue and its impact on the manufacturing cost and the demand. In addition, reference [10] discussed the government's role to incentivize the manufacturers to design product to have low impact on environment and cheaper to remanufacture. Furthermore, as pointed out by reference [11], product design, sales and other company strategies are not well aligned with the remanufacturing activities, which resulted in such problems as the uncertain quality and quantity in the returning products. Reference [12] proposed a model to study the supply chain effect on the product life-cycle design, which has been shown to be significant. For a comprehensive review of the research on closed-loop supply chain please see reference [13].

In this paper, we would like to analyze the following research questions:

1. How does the design of remanufactureability impact on the production cost?
2. In addition to the cost advantage, how does the design of remanufactureability impact on the uncertainty in yield rate in terms of converting percentage from collected used product to remanufactured product?

2 The CLSC Model with Competition and Remanufactureability Design

We assume that there are M manufacturers supplying a homogenous product to N demand markets. m is denoted as a typical manufacturer while i is denoted as a typical demand market. The network structure of the CLSC in the paper is depicted in Fig. 1. The solid line indicates the product flow in the forward chain and the dashed line represents the product flow in the reverse chain.

Now we present the CLSC network model. We assume that there are two periods in our model. In the first period, manufacturers determine the product remanufactureability level and production quantity. In the second period, when the product sold in the first period is at its EOL, manufacturers will collect the used product from the consumers, which will be, in turn, remanufactured. In the above model, q_{mi}^{1n} and q_{mi}^{2n} denote the new product quantities supplied from manufacturer m to demand market i in period 1 and 2, respectively. q_{mi}^{2r} represents the refurbished product quantity shipped from manufacturer m to demand market i. We group the above variables to form the vectors Q^{1n}, Q^{2n}, and Q^{2r}, respectively. Furthermore, manufacturer m determines the remanufactureability level, $s_m \in [0, 1]$, of his product in the first

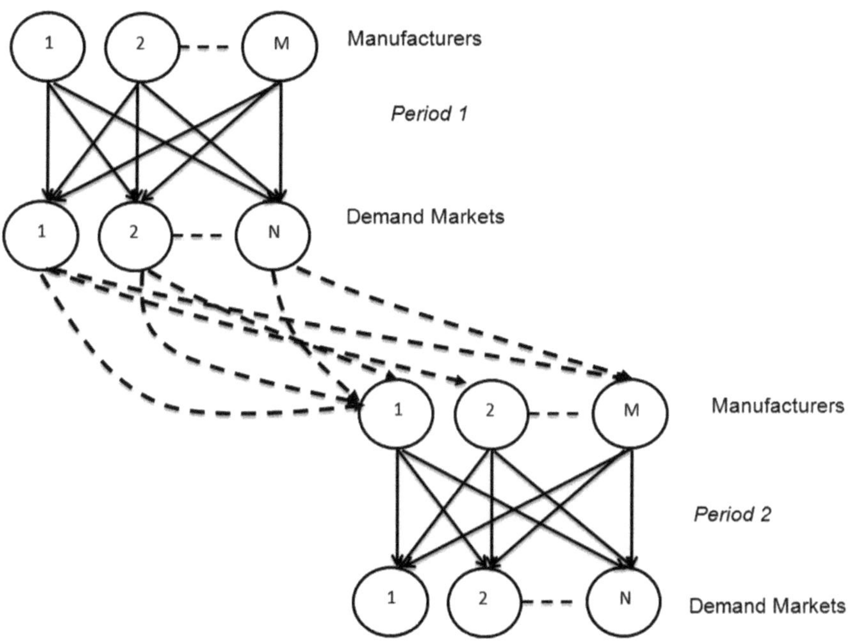

Fig. 1 Two-period CLSC network

period, which will impact the amount of the remanufactured product, manufacturing and remanufacturing costs. The higher the value of s_m, the more certain about the quantity of the collectable used product, which can be translated into a more certain manufacturing/remanufacturing cost. Furthermore, it is also cheaper to remanufacture the product in the second period. However, the above benefit is at the cost of the increasing marginal production cost of the new product. In this paper, we assume that the new and the remanufactured product share the same production facility. Hence, similar to the Kodak example discussed in Sect. 1, to increase the remanufactureability/reusability of the new product, a company will use more expensive technology and material and therefore, the manufacturing cost for the new product will be higher than that without the remanufacture-ability design. Similarly, s_m is grouped to form the vector s. In this paper, we use the term *remanufactured* and *refurbished* interchangeably and assume they have the same meaning.

In the first period, manufacturer m incurs a manufacturing cost of the new product, f_m^{1n} which is assumed dependent on the production volume Q^{1n} and the remanufactureability level s_m. c_m^1 denotes as the transaction cost associated with manufacturer m transacting with the demand markets. c_m^1 is assumed to depend on Q^{1n}. In the second period, besides the similar costs incurred in period one, manufacturers will have an additional remanufacturing cost f_m^{2r}, which is assumed dependent on the remanufacturing volume, Q^{2r} as well as s_m. δ is assumed to be the discount factor. For the sake of generality and to model the competition among the manufacturers, we assume that the manufacturing cost for one manufacturer depends on the production quantity by other manufacturers as well. Such an assumption is common in modeling the supply chain network competitions (see for example, reference [14]). In addition, we assume that manufacturers are multi-criteria decision makers. They want to minimize the risks associated with the uncertainties caused by different s_m, which is captured by the risk functions. Risk functions have been used in reference [15] to represent the general risk associated with uncertainties. For example, we can use variance of the cost functions to as a specific risk function. We know that the purpose of the remanufacturing is to reduce pollution to save the environment. Therefore, we assume that the manufacturers also want to minimize the emission induced by producing the new and the refurbished products. They are denoted by e_m^{1n}, e_m^{2n} and e_m^{2r}, which depend on remanufacturability s_m and quantity Q^{1n}, Q^{2n}, and Q^{2r}, respectively. ω_1 is the weight decision maker associated with the risk level and ω_2 is the weight associated with the emission level. The higher the value of ω_1, the more risk averse the manufacturer is. Similarly, the higher the value of ω_2, the more environmental awareness the manufacturer has.

At the demand side, the demand of the new/remanufactured product, at one demand market is affected by the following two factors: (1) its own price; (2) the price of the remanufactured/new product (to reflect the consumer's perception on the quality of the new and remanufactured product). As discussed in Sect. 1, the consumer's perception of quality is well documented in the CLSC literature and it is often used to describe the phenomenon of the remanufactured product cannibalizing the market share of the new product. At each demand market, there are two types of demand: the demand for the new product and the remanufactured product. In period one, since there is no EOL product, the supply and the demand for the remanufactured product is equal to zero. Denote ρ_i^{1n}, ρ_i^{2n} and ρ_i^{2r} as the inverse demand function at demand market i for the new product in period one, period two, and the remanufactured product in period two, respectively. Furthermore, we assume that Manufacturer m has a target recycle quantity z_m, which can be set by the government and is given in the model.

The objective of manufacturer m has several folds. First, he wants to maximize its profit (cf. (1) below), represented by the revenue minus the production and transaction costs in both periods with the expected profit in the second period discounted back to period one. Second, he wants to minimize the risks associated with remanufacturing uncertainties. Finally, he wants to minimize the associated emission. Thus, manufacturer m seeks to

$$\begin{aligned}\max_{q_{mi}^{1n}, q_{mi}^{2n}, q_{mi}^{2r}, s_m} U_m = & \left[\sum_{i=1}^{N} \rho_i^{1n}(d^{1n}) * q_{mi}^{1n} - f_m^{1n}(Q^{1n}, s_m) - c_m^1(Q^{1n})\right] \\ & + \delta * \left[\sum_{i=1}^{N} \rho_i^{2n}(d^{2n}, d^{2r}) * q_{mi}^{2n} + \rho_i^{2r}(d^{2n}, d^{2r}) * q_{mi}^{2r} - f_m^{2n}\left(Q^{2n}, s_m\right)\right. \\ & \left. - E\left[c_m^2(Q^{2n}, Q^{2r}, s_m)\right] - E\left[f_m^{2r}(Q^{2r}, s_m)\right]\right] \\ & - \omega_1 * R_m(Q^{1n}, Q^{2n}, Q^{2r}, s_m) \\ & - \omega_2 * \left(e_m^{1n}\left(Q^{1n}, s_m\right) + e_m^{2n}\left(Q^{2n}, s_m\right) + E\left[e_m^{2r}\left(Q^{2r}, s_m\right)\right]\right)\end{aligned} \quad (1)$$

At the demand markets, the following constraints need to be satisfied.

$$d_i^{1n} = \sum_{m=1}^{M} q_{mi}^{1n}, i = 1, \ldots, N, \quad (2)$$

$$d_i^{2n} = \sum_{m=1}^{M} q_{mi}^{2n}, i = 1, \ldots, N, \quad (3)$$

$$d_i^{2r} = \sum_{m=1}^{M} q_{mi}^{2r}, i = 1, \ldots, N, \quad (4)$$

$$\sum_{i=1}^{N} q_{mi}^{2r} = E\left[\beta_m^r(s_m)\right]\left(\sum_{i=1}^{N} q_{mi}^{1n}\right), \quad (5)$$

$$\sum_{i=1}^{N} q_{mi}^{2r} \geq z_m, \tag{6}$$

$$q_{mi}^{1n}, q_{mi}^{2n}, \text{ and } q_{mi}^{2r} \geq 0, \; i = 1, \ldots, N, \tag{7}$$

$$s_m \in [0, 1]. \tag{8}$$

In Eq. (1), the result from the first bracket is the profit in the first period and the profit from the second period is in the second bracket associated with the discount factor. Constraints (2)–(4) state that the demand for both types of the product at each period will be satisfied. Constraint (5) specifies that the amount of remanufactured product for manufacturer m in the second period is equal to the amount of the new product produced in the first period multiplied by an expected yield rate, β_m^r, which depends on s_m. Constraint (6) states that the remanufactured products of Manufacturer m cannot be less than the lower bound zm. Combining Constraints (5) and (6), we have the following

$$\frac{z_m}{E\left[\beta_m^r(s_m)\right]} - \sum_{i=1}^{N} q_{mi}^{1n} \leq 0 \tag{9}$$

We assume that $E[\beta^r(s_m)]$ is non-decreasing in s_m and concave, it is easy to verify the left side of Eq. (9) is convex.

We assume that the cost, risk, and emission functions in (1) are continuously differentiable and convex. Moreover, we assume that the manufacturers compete in a non-cooperative manner in the sense of Nash (1950 and 1951), with each trying to maximize his own profits. Hence, we define feasible set κ as $\kappa \equiv \{(Q^{1n}, Q^{2n}, Q^{2r}, s) | q_{mi}^{1n} \geq 0, q_{mi}^{2n} \geq 0, q_{mi}^{2r} \geq 0, s_m \in [0, 1]$ and (2)–(4), (9) are satisfied $\forall m = 1, \ldots, M, i = 1, \ldots, N\}$. We then verify that the above feasible set is closed and convex.

The optimality conditions for all firms simultaneously, under the above assumptions, can be expressed using the variational inequality formulation in Theorem 1.

Theorem 1 *(Variational Inequality Formulation of the CLSC with Competition, design for remanufactureability, and Uncertain Yield)*

The equilibrium conditions governing the CLSC with competition and design for remanufactureability coincide with the solution of the variational inequality given by determining $(Q^{1n}, Q^{2n}, Q^{2r}, s) \in \kappa$*, such that*

$$\sum_{m=1}^{M}\sum_{i=1}^{N}\left[\frac{\partial f_m^{1n}\left(Q^{1n^*},s_m^*\right)}{\partial q_{mi}^{1n}}+\frac{\partial c_m^1\left(Q^{1n^*}\right)}{\partial q_{mi}^{1n}}-\rho_i^{1n}\left(d^{1n^*}\right)-\left(\sum_{j=1}^{N}\frac{\partial \rho_j^{1n}\left(d^{1n^*}\right)}{\partial q_{mi}^{1n}}*q_{mj}^{1n^*}\right)\right.$$

$$\left.+\omega_1\frac{\partial R_m\left(Q^{1n^*},Q^{2n*},Q^{2r^*},s_m^*\right)}{\partial q_{mi}^{1n}}+\omega_2\frac{\partial e_m^{1n}\left(Q^{1n^*},s_m^*\right)}{\partial q_{mi}^{1n}}\right]\times\left(q_{mi}^{1n}-q_{mi}^{1n^*}\right)$$

$$+\sum_{m=1}^{M}\sum_{i=1}^{N}\left[\delta\frac{\partial f_m^{2n}\left(Q^{2n^*},s_m^*\right)}{\partial q_{mi}^{2n}}+\delta\frac{\partial E\left[c_m^2\left(Q^{2n^*},Q^{2r^*},s_m^*\right)\right]}{\partial q_{mi}^{2n}}+\omega_1\frac{\partial R_m\left(Q^{1n^*},Q^{2n^*},Q^{2r^*},s_m^*\right)}{\partial q_{mi}^{2n}}\right.$$

$$\left.+\omega_2\frac{\partial e_m^{2n}\left(Q^{2n^*},s_m^*\right)}{\partial q_{mi}^{2n}}-\delta\rho_i^{2n}\left(d^{2n^*},d^{2r^*}\right)-\delta\sum_{j=1}^{N}\left(\frac{\partial \rho_j^{2n}\left(d^{2n^*},d^{2r^*}\right)}{\partial q_{mi}^{2n}}q_{mj}^{2n^*}+\frac{\partial \rho_j^{2r}\left(d^{2n^*},d^{2r^*}\right)}{\partial q_{mi}^{2n}}*q_{mj}^{2r^*}\right)\right]$$

$$\times\left(q_{mi}^{2n}-q_{mi}^{2n*}\right)+\sum_{m=1}^{M}\sum_{i=1}^{N}\left[\delta\frac{\partial E\left[f_m^{2r}\left(Q^{2r^*},s_m^*\right)\right]}{\partial q_{mi}^{2r}}+\delta\frac{\partial E\left[c_m^2\left(Q^{2n^*},Q^{2r^*},s_m^*\right)\right]}{\partial q_{mi}^{2r}}\right.$$

$$+\omega_1*\frac{\partial R_m\left(Q^{1n^*},Q^{2n^*},Q^{2r^*},s_m^*\right)}{\partial q_{mi}^{2r}}-\delta\rho_i^{2n}\left(d^{2n^*},d^{2r^*}\right)$$

$$-\delta\left(\sum_{j=1}^{N}\frac{\partial \rho_j^{2n}\left(d^{2n^*},d^{2r^*}\right)}{\partial q_{mi}^{2r}}*q_{mj}^{2n^*}\right)-\delta\rho_i^{2r}\left(d^{2n^*},d^{2r^*}\right)$$

$$\left.-\delta\left(\sum_{j=1}^{N}\frac{\partial \rho_j^{2r}\left(d^{2n^*},d^{2r^*}\right)}{\partial q_{mi}^{2r}}*q_{mj}^{2r^*}\right)+\omega_2\frac{\partial E\left[e_m^{2r}\left(Q^{2r^*},s_m^*\right)\right]}{\partial q_{mi}^{2r}}\right]\times\left(q_{mi}^{2r}-q_{mi}^{2r^*}\right)$$

$$+\sum_{m=1}^{M}\left[\frac{\partial f_m^{1n}\left(Q^{1n^*},s_m^*\right)}{\partial s_m}+\delta\frac{\partial f_m^{2r}\left(Q^{2n^*},s_m^*\right)}{\partial s_m}+\delta\frac{\partial E\left[f_m^{2r}\left(Q^{2r^*},s_m^*\right)\right]}{\partial s_m}+\delta\frac{\partial E\left[c_m^2\left(Q^{2n^*},Q^{2r^*},s_m^*\right)\right]}{\partial s_m}\right.$$

$$\left.+\omega_1\frac{\partial R_m\left(Q^{1n^*},Q^{2n^\&},Q^{2r^\wedge},s_m^*\right)}{\partial s_m}+\omega_2\left(\frac{\partial e_m^{1n}\left(Q^{1n^*},s_m^*\right)}{\partial s_m}+\frac{\partial e_m^{2n}\left(Q^{1n^*},s_m^*\right)}{\partial s_m}+\frac{\partial E\left[e_m^{2r}\left(Q^{2r^*},s_m^*\right)\right]}{\partial s_m}\right)\right]$$

$$\times\left(s_m-s_m^*\right)\geq 0,\ \forall\left(Q^{1n},Q^{2n},Q^{2r},s\right)\in\kappa. \tag{10}$$

Proof The formulation is developed using the standard variational inequality theory (cf. in reference [16, 17]). Note that we have substituted the demand functions (2)–(4) into the objective function. ■

Remark 1 In the above, we utilize the theory of variational inequality (VI) to obtain the equilibrium solutions for the CLSC network because the VI has been shown to be a powerful tool to study the Nash equilibrium in a network game as shown in this paper. Further- more, there are existing algorithms related to VI can be used to calculate the network equilibrium to further study the behavior of various decision makers.

Remark 2 The variational inequality problem (10) can be rewritten in standard form as follows: determine $X^* \in \kappa^1$ satisfying

$$< F(X^*)^T, X - X^* > \geq 0, \forall X \in \kappa^1, \tag{11}$$

where $X \equiv (Q^{1n}, Q^{2n}, Q^{2r}, s)^T, \kappa^1 \equiv \kappa$ and

$$F(X) \equiv (F_{mi}^{1n}, F_{mi}^{2n}, F_m^1, F_{mi}^{2r}), \tag{12}$$

with indices $m = 1, \ldots, M$, $i = 1, \ldots, I$, and the functional terms preceding the multiplication signs in (10), respectively. Here $\langle \cdot, \cdot \rangle$, denotes the inner product in Θ-dimensional Euclidian space where $\Theta = M\,I + M\,I + M + M\,I$.

3 Qualitative Properties

In this section, we present qualitative properties of the solution and the function that enters the variational inequality (8).

As the proposed CLSC model studies the behavior of the decision-makers under network equilibrium, it is important to know under what conditions such an equilibrium exists. In particular, if we are interested in knowing the new and the remanufactured product flow as well as the remanufactureability level at the equilibrium, it is crucial to know if such an equilibrium is unique in the first place. Theorem 2 discusses the conditions where at least one CLSC network equilibrium exists. Theorems 3 and 4 lead to the conditions in Theorem 5 which discusses the conditions where the unique network equilibrium can be identified.

We now provide conditions for existence of a solution to variational inequality (10).

Theorem 2 *(Existence) Assume that the market sizes for the demand markets are bounded from up above and denoted as D_i^n and D_i^r for $i = 1, \ldots, N$ for the new and remanufactured product, respectively. There exists a solution to the variational inequality (10).*

Proof Since all variables are nonnegative, the lower bounds of the variables are zero. In addition, since the market size is fixed, according to constraints (2)–(4), the production volume of both the new and the remanufactured product is bounded by the market size. Furthermore, constraint (9) indicates Q^{1n} is also bounded. In addition, the maximum remanufactureability level is 1. Therefore, the feasible set of the variational inequality (10) is compact.

By the assumptions, the functions that comprise $F(X)$ in (11) are continuous. According to the standard theory, variational inequality (11) admits a solution [18, 16]. ■

Theorem 3 *(Monotonicity) Assume that the manufacturers' production costs, f_m^{1n}, f_m^{2n}, f_m^{2r} the transaction cost with the demand markets, c_m, the risk function R_m, and the emission functions e_m^{1n}, e_m^{2n}, e_m^{2r} are convex. In addition, assume that the inverse demand function is monotone decreasing. Thus, the vector function F that enters the variational inequality (10) is monotone, that is, for any X' and $X'' \in \kappa^1$ with $X' \neq X''$, $\langle (F(X') - F(X'')^T, (X' - X'') \rangle \geq 0$.*

Proof The expression $(F(X') - F(X''))^T \cdot (X' - X'')$ is equal to the expression (after some algebraic simplification)

$$\sum_{m=1}^{M}\sum_{i=1}^{N}\left[\frac{\partial f_m^{1n}\left(Q^{1n'}, s_m^{'}\right)}{\partial q_{mi}^{1n}} - \frac{\partial f_m^{1n}\left(Q^{1n''}, s_m^{''}\right)}{\partial q_{mi}^{1n}} + \frac{\partial c_m^{1}\left(Q^{1n'}\right)}{\partial q_{mi}^{1n}} - \frac{\partial c_m^{1}\left(Q^{1n''}\right)}{\partial q_{mi}^{1n}} - \sum_{j=1}^{N}\left(\rho_j^{1n}\left(d^{1n'}\right) - \rho_j^{1n}\left(d^{1n''}\right)\right.\right.$$

$$\left. + \frac{\partial \rho_j^{1n}\left(d^{1n'}\right)}{\partial q_{mi}^{1n}} * q_{mj}^{1n'} - \frac{\partial \rho_j^{1n}\left(d^{1n''}\right)}{\partial q_{mi}^{1n}} * q_{mj}^{1n''}\right) + \omega_1\left(\frac{\partial R_m\left(Q^{1n'}, Q^{2n'}, Q^{2r'}, s_m^{'}\right)}{\partial q_{mi}^{1n}} - \frac{\partial R_m\left(Q^{1n''}, Q^{2n''}, Q^{2r''}, s_m^{''}\right)}{\partial q_{mi}^{1n}}\right)$$

$$\left. + \omega_2\left(\frac{\partial e_m^{1n}\left(Q^{1n'}, s_m^{'}\right)}{\partial q_{mi}^{1n}} - \frac{\partial e_m^{1n}\left(Q^{1n''}, s_m^{''}\right)}{\partial q_{mi}^{1n}}\right)\right] \times \left(q_{mi}^{1n'} - q_{mi}^{1n''}\right)$$

$$+ \sum_{m=1}^{M}\sum_{i=1}^{N}\left[\delta\left(\frac{\partial f_m^{2n}\left(Q^{2n'}, s_m^{'}\right)}{\partial q_{mi}^{2n}} - \frac{\partial e_m^{2n}\left(Q^{2n''}, s_m^{''}\right)}{\partial q_{mi}^{2n}}\right) + \delta\left(\frac{\partial E\left[c_m^{2}\left(Q^{2n'}, Q^{2r'}\right)\right]}{\partial q_{mi}^{2n}} - \frac{\partial E\left[c_m^{2}\left(Q^{2n''}, Q^{2r''}\right)\right]}{\partial q_{mi}^{2n}}\right)\right.$$

$$+ \omega_1\left(\frac{\partial R_m\left(Q^{1n'}, Q^{2n'}, Q^{2r'}, s_m^{'}\right)}{\partial q_{mi}^{2n}} - \frac{\partial R_m\left(Q^{1n''}, Q^{2n''}, Q^{2r''}, s_m^{''}\right)}{\partial q_{mi}^{2n}}\right)$$

$$+ \omega_2\left(\frac{\partial e_m^{2n}\left(Q^{2n'}, s_m^{'}\right)}{\partial q_{mi}^{2n}} - \frac{\partial e_m^{2n}\left(Q^{2n''}, s_m^{''}\right)}{\partial q_{mi}^{2n}}\right) - \sum_{j=1}^{N}\left(\rho_j^{2n}\left(d^{2n'}, d^{2r'}\right) - \rho_j^{2n}\left(d^{2n''}, d^{2r''}\right)\right.$$

$$\left.\left. + \frac{\partial \rho_j^{2n}\left(d^{2n'}, d^{2r'}\right)}{\partial q_{mi}^{2n}} * q_{mi}^{2n'} - \frac{\partial \rho_j^{2n}\left(d^{2n''}, d^{2r''}\right)}{\partial q_{mi}^{2n}} * q_{mi}^{2n''} + \frac{\partial \rho_j^{2r}\left(d^{2n'}, d^{2r'}\right)}{\partial q_{mi}^{2n}} * q_{mj}^{2r'} - \frac{\partial \rho_j^{2r}\left(d^{2n''}, d^{2r''}\right)}{\partial q_{mi}^{2n}} * q_{mj}^{2r''}\right)\right]$$

$$\times\left(q_{mi}^{2n'} - q_{mi}^{2n''}\right) + \sum_{m=1}^{M}\left[\frac{\partial f_m^{1n}\left(Q^{1n'}, s_m^{'}\right)}{\partial s_m} - \frac{\partial f_m^{1n}\left(Q^{1n''}, s_m^{''}\right)}{\partial s_m} + \delta\left(\frac{\partial f_m^{2n}\left(Q^{2n'}, s_m^{'}\right)}{\partial s_m} - \frac{\partial f_m^{2n}\left(Q^{2n''}, s_m^{''}\right)}{\partial s_m}\right)\right.$$

$$+ \delta\left(\frac{\partial E\left[f_m^{2r}\left(Q^{2r'}, s_m^{'}\right)\right]}{\partial s_m} - \frac{\partial E\left[f_m^{2r}\left(Q^{2r''}, s_m^{''}\right)\right]}{\partial s_m}\right)$$

$$+\omega_1\left(\frac{\partial R_m\left(Q^{1n'},Q^{2n'},Q^{2r'},s_m'\right)}{\partial s_m}-\frac{\partial R_m\left(Q^{1n''},Q^{2n''},Q^{2r''},s_m''\right)}{\partial s_m}\right)$$

$$+\omega_2\left(\frac{\partial e_m^{1n}\left(Q^{1n'},s_m'\right)}{\partial s_m}-\frac{\partial e_m^{1n}\left(Q^{1n''},s_m''\right)}{\partial s_m}+\frac{\partial e_m^{1n}\left(Q^{1n'},s_m'\right)}{\partial s_m}-\frac{\partial e_m^{1n}\left(Q^{1n''},s_m''\right)}{\partial s_m}\right.$$

$$\left.+\frac{\partial e_m^{2n}\left(Q^{1n'},s_m'\right)}{\partial s_m}-\frac{\partial e_m^{2n}\left(Q^{1n''},s_m''\right)}{\partial s_m}+\frac{\partial E\left[e_m^{2r}\left(Q^{2r'},s_m'\right)\right]}{\partial s_m}-\frac{\partial E\left[e_m^{2r}\left(Q^{2r''},s_m''\right)\right]}{\partial s_m}\right]\times\left(s_m'-s_m''\right)$$

$$+\sum_{m=1}^{M}\sum_{i=1}^{N}\left[\delta\left(\frac{\partial f_m^{2n}\left(Q^{2n'},s_m'\right)}{\partial q_{mi}^{2r}}-\frac{\partial f_m^{2r}\left(Q^{2r''},s_m''\right)}{\partial q_{mi}^{2r}}\right)+\delta\left(\frac{\partial E\left[c_m^{2}\left(Q^{2n'},Q^{2r'}\right)\right]}{\partial q_{mi}^{2r}}-\frac{\partial E\left[c_m^{2}\left(Q^{2n''},Q^{2r''}\right)\right]}{\partial q_{mi}^{2r}}\right)\right.$$

$$+\omega_1\left(\frac{\partial R_m\left(Q^{1n'},Q^{2n'},Q^{2r'},s_m'\right)}{\partial q_{mi}^{2r}}-\frac{\partial R_m\left(Q^{1n''},Q^{2n''},Q^{2r''},s_m''\right)}{\partial q_{mi}^{2r}}\right)$$

$$-\sum_{j=1}^{N}\left(\frac{\partial\rho_j^{2n}\left(d^{2n'},d^{2r'}\right)}{\partial q_{mi}^{2r}}*q_{mi}^{2r'}-\frac{\partial\rho_j^{2n}\left(d^{2n''},d^{2r''}\right)}{\partial q_{mi}^{2r}}*q_{mi}^{2r''}+\rho_j^{2n}\left(d^{2n'},d^{2r'}\right)-\rho_j^{2n}\left(d^{2n''},d^{2r''}\right)\right)$$

$$-\left(\rho_j^{2n}\left(d^{2n'},d^{2r'}\right)-\rho_j^{2n}\left(d^{2n''},d^{2r''}\right)\right)$$

$$-\sum_{j=1}^{N}\left(\frac{\partial\rho_j^{2n}\left(d^{2n'},d^{2r'}\right)}{\partial q_{mi}^{2r}}*q_{mi}^{2r'}-\frac{\partial\rho_j^{2n}\left(d^{2n''},d^{2r''}\right)}{\partial q_{mi}^{2r}}*q_{mi}^{2r''}+\rho_j^{2r}\left(d^{2n'},d^{2r'}\right)-\rho_j^{2r}\left(d^{2n''},d^{2r''}\right)\right)$$

$$\left.+\omega_2\left(\frac{\partial E\left[e_m^{2r}\left(Q^{2r'},s_m'\right)\right]}{\partial q_{mi}^{2r}}-\frac{\partial E\left[e_m^{2r}\left(Q^{2r''},s_m''\right)\right]}{\partial q_{mi}^{2r}}\right)\right]\times\left(q_{mi}^{2r'}-q_{mi}^{2r''}\right)=(I)+(II)+(III)+(IV),$$

(13)

Based on the assumptions on the cost, inverse demand, risk, and emission functions, one can have $(I) \geq 0$, $(II) \geq 0$, $(III) \geq 0$, $(IV) \geq 0$, and $(V) \geq 0$. Therefore (11) must be greater than or equal to zero, under the above assumptions, and, hence, $F(X)$ is monotone. ■

Next, in Theorem 4, we require additional properties on various functions in variational inequality (9) to obtain strict monotonicity of $F(X)$. The proof of Theorem 4 is similar to that of Theorem 3, which is skipped here.

Theorem 4 *(Strict Monotonicity) Assume that the manufacturers' production costs, f_m^{1n}, f_m^{2n}, f_m^{2r}, the transaction cost with the demand markets, c_m, the risk function R_m, the emission functions e_m^{1n}, e_m^{2n}, e_m^{2r} are strictly convex. In addition, assume that the inverse demand function is strictly monotone decreasing. Then the vector function F that enters variational inequality (9) is strictly monotone, that is, for all X' and $X'' \in \kappa^1$, $\left\langle F\left(X'\right)-F\left(X''\right)^T,\left(X'-X''\right)\right\rangle > 0$.*

With Theorems 2 and 4, we can study the uniqueness of the solution to the variational inequality problem (10).

Theorem 5 *(Existence and Uniqueness of a Solution to the Variational Inequality Problem) The conditions of Theorem 4 are assumed to hold. Then, the function that enters the variational inequality (10) has a unique solution in* κ.

Proof Follows from reference [16]. ■

4 Conclusion

In this paper, we proposed a closed-loop supply chain network model that captures the impact of remanufactureability level on the production cost and uncertainty in the yield rate. We analyze the model from the network perspective, in which manufacturers compete with each other in terms of production quantity of the new and remanufactured product as well as the remanufactureability level. We used variational inequalities to obtain solutions the network equilibrium. In addition, properties of the solutions are discussed. For the future study, we would like to obtain empirical data to validate our model to generate managerial insights.

References

1. https://i.unu.edu/media/unu.edu/news/52624/UNU-1stGlobal-E-Waste-Monitor-2014-small.pdf.
2. http://www.theinquirer.net/inquirer/news/1603334/mobile-phone-industry-pulls-recession.
3. Ayres RU, Ferrer G, Van Leynseele T. Eco-efficiency, asset recovery and remanufacturing. Eur Manag J. 1997;15:557–74.
4. BIPAVER. Difficult tyre retail future ahead. Eur Rubber J May. 1998;46–52.
5. Bozarth M. Radial truck tire retreadability survey results. Tire Retreading/Repair J. 2000;44:3–8.
6. Deutsch CH. Second time around, and around; remanufacturing is gaining ground in corporate America. New York Times, 14 July 1998.
7. Benn S, Dunphy D. A case of strategic sustainability: the Fuji Xerox eco manufacturing centre. Innov Manage Policy Pract. 2004;6:258–68.
8. Fuji Xerox. Technical Report, Corporate societal responsibility: knowledge learning through sustainable global supply chain management, Company Report, No.15. 2005.
9. Orsdemir A, Kemahlioglu-Ziya E, Parlakturk A. Competitive quality choice and remanufacturing. Prod Oper Manage. 2014;23:48–64.
10. Atasu A, Van Wassenhove L, Sarvary M. Efficient take? back legislation. Pro-Duction Oper Manage. 2009;18:243–58.
11. Guide VDR, Harrison TP, Van Wassenhove LN. The challenge of closed-loop supply chains. Interfaces. 2003;33:3–7.
12. Chung WH, Kremer GE, Wysk RA. Life cycle implications of product modular architectures in closed-loop supply chains. Int J Adv Manufact Technol. 2014;70:2013–2028.
13. Govindan K, Soleimani H, Kannan D. Reverse logistics and closed-loop supply chain: a comprehensive review to explore the future. Eur J Oper Res. 2015;240:603–26.
14. https://www.researchgate.net/publication/222535504_A_supply_chain_network_equilibrium_model_with_random_demands.

15. Nagurney A, Cruz J, Dong J, Zhang D. Supply chain networks, electronic commerce, and supply side and demand side risk. Eur J Oper Res. 2005;164:120–42.
16. Nagurney A. Network economics: a variational inequality approach, Second and Revised Edition. Boston, Massachusetts: Kluwer Academic Publishers; 1999.
17. Gabay D, Moulin H. On the uniqueness and stability of Nash-equilibria in noncooperative games, in Applied Stochastic Control in Econometrics and Management Science. In: Bensoussan A, Kleindorfer P, Tapiero CS, editors. North-Holland, Amsterdam, The Netherlands. 1980. p. 271–94.
18. Kinderlehrer D, Stampacchia G. An introduction to variational inequalities and applications. New York: Academic Press; 1980.

Towards the Determinants of Successful Public-Private Partnership Projects in Jamaica: A Proposed Methodology

Kenisha Iton and Delroy Chevers

Abstract The Caribbean countries have a relatively large infrastructure deficit that has affected their economic growth. Infrastructure is essential for growth by providing critical services and facilities. Infrastructure is a major determinant that drives competitiveness, and as such is vital if these economies are to become competitive, grow and developed. Undoubtedly, investing in infrastructure is beneficial for developing countries, but such initiative is usually accompanied by high costs. Mobilizing financial resources needed for infrastructure investment can be challenging for governments in developing countries, with Jamaica being no exception. Public-private partnerships (PPPs) have become alternative ways of raising needed funds for capital intensive public projects, thereby providing a unique solution to speed up infrastructure development. However, the success of these initiatives is inconclusive. Hence, this study seeks to propose a research methodology to assess the major determinants of successful implementation of PPPs in Jamaica. It is hoped that the study will provide useful insights which can assist decision makers in their desire to implement successful PPPs and by extension promote economic and social development in Jamaica.

Keywords Economic development · Infrastructure · Jamaica
Public-private partnerships · Nominal group technique

1 Introduction

Most governments in the Caribbean are struggling to improve their infrastructure due to huge debts, lagging economies, and tight budgets [1]. Jamaica, a small island in the Caribbean is no exception. The country's debt to gross domestic product (GDP)

K. Iton
SALISES, University of the West Indies, Kingston, Jamaica

D. Chevers (✉)
Mona School of Business, University of the West Indies, Kingston, Jamaica
e-mail: delroy.chevers@uwimona.edu.jm

H. Yang and R. Qiu (eds.), *Advances in Service Science*, Springer Proceedings in Business and Economics, https://doi.org/10.1007/978-3-030-04726-9_25

ratio represents one of the highest in the world [2]. Its foreign exchange rate against the United States (US) dollar is very low at 129:1 and unemployment rate stands at 13.7%. In essence, the country faces tremendous constraints in human and financial resources.

But investment in infrastructure is essential for a nation's growth and development [3]. However, such investments usually require huge financial outlay [4]. This condition can be challenging for a country like Jamaica, which faces severe constraints [5]. Public-private partnerships (PPPs) have become alternative ways of raising needed funds for capital intensive public projects [6, 7], thereby providing a unique solution to speed up infrastructure development [8]. They have helped improve the infrastructure and service delivery in both developed and developing countries [9, 10]. PPPs are expected to bring a variety of benefits to the government by making use of the private finance for the development of public infrastructure [11]. These partnerships have led to considerable poverty reduction and sustainable development [12].

Public-private partnership is defined as an agreement between the government and one or more private partners (which may include the operators and the financers) according to which the private partners deliver the service in such a manner that the service delivery objectives of the government are aligned with the profit objectives of the private partners and where the effectiveness of the alignment depends on a sufficient transfer of risk to the private partners [13]. Although the literature makes reference to the benefits of PPPs [12], their success is believed to depend on the management of the contract and the inherent risks of such projects [14].

Poor governance, lack of policy framework and lack of transparency hinder the success of PPPs in developing countries [15]. In addition, political interference in the tendering process is a major hindrance to the success of PPPs. In general, governments need to build political support for PPPs to reduce interference by political players [15]. In terms of the lack of transparent legal frameworks to govern the management of PPPs, a study found that most developing countries lack policy and legal frameworks to guide the successful implementation of PPP projects [16]. There is the notion that institutions in developing countries usually have weak governance structures [17, 18]. As a result, it has been found that corruption and mistrust among shareholders can affect the successful implementation of PPP projects in developing countries [16].

The findings in the literature regarding the benefits of PPPs is inconclusive [19]. In addition, there is relatively little research in this domain in Jamaica. As a result, this study seeks to propose a research methodology to assess the major determinants of successful implementation of PPPs in Jamaica. Hence the following research questions:

- What policies are in place to govern the contracts issued under PPPs in Jamaica?
- What are the major determinants of successful PPPs implementation in Jamaica?

It is hoped that the study will provide useful insights which can assist decision makers in their desire to implement successful PPPs and by extension promote economic and social development in Jamaica.

2 Literature Review

The Caribbean countries, which are classified as developing economies, have a relatively large infrastructure deficit that has affected their economic growth [1]. Infrastructure is essential for growth by providing critical services and facilities [3]. It is defined as the basic physical and organizational structures and facilities like buildings, roads and power supplies needed for the operation of a society or enterprise. Infrastructure is listed in the global competitiveness report as a major determinant that drives competitiveness [20]. Interestingly, Jamaica is ranked 86th in the global competitiveness report, with infrastructure scoring the third lowest of the basic requirements at 3.8 on a 1–7 scale [20]. This score, which is just above the mid-point, indicates that Jamaica's infrastructure is not well established and advanced.

In the midst of this challenge in Jamaica there is the notion as stated by [3] that, "economic infrastructure keeps the country running. The nation needs power plants to fuel its homes, offices, industries, and support services, such as street lighting, and security systems. It needs roads, railways, airports and ports to move people and commodities and it must have good communications. The availability of infrastructure is a key factor for companies when making decisions on where to invest" (p. 5).

Undoubtedly, investing in infrastructure is beneficial for developing countries due to their characteristics of huge debts and tight budgets, but such initiative is usually accompanied by high costs [4]. Mobilizing financial resources needed for infrastructure investment can be challenging for governments in developing countries, with Jamaica being no exception. Public-private partnerships provide a unique solution to speed up infrastructure development [8]. But proper management and efficient utilization of the infrastructure is needed if economies in the Caribbean want to become competitive, grow and attain sustainable development [1, 21].

This is even more relevant in the case of Jamaica which suffers from scarce resources and low economic performance. The economic measures in Jamaica as distilled by the [22] are:

- Currency = US $1 = J $128
- GDP per capita (US $) = 5,003.8
- Unemployment (% of labour force) = 13.7
- Labour force participation (female/male %) = 57.6/72.1
- Balance of payments (million US $) = −1,110
- Tourist arrivals at national borders (000) = 2,080.

There is a widely held view that economic infrastructure is a strategic issue because it leads to competitiveness [3]. As a result, efforts were made to speed up the implementation of infrastructural development projects and by extension increase the economic performance of the country. So in August 2011, Jamaica's Investments Division of the Ministry of Finance and the Economy established a PPP Unit as part of an initiative to promote public–private partnerships in Jamaica. In 2012, the Government of Jamaica (GOJ), established a Policy and Institutional Framework for the implementation of a PPPs programme in the country. PPP as distilled by [23] is “a long-term procurement contract between the public and private sectors, in which the proficiency of each party is focused in designing, financing, building and operating an infrastructure project or providing a service, through the appropriate sharing of resources, risks and rewards” (p. 5). The literature makes reference to four types of PPPs. These are: design, build and finance (DBF); design, build, finance, operate and maintain (DBFM); design, build and operate (DBO); and concession PPPs.

In September 2012, the Cabinet approved the Jamaican PPP Policy, which sets out the principles that should guide decision-making by Ministries, Departments and Agencies which are considering utilising PPPs to improve infrastructure and the delivery of public services. In summary, the Jamaican government turned to PPPs as a means to overcome the deficits in infrastructure needs. Private companies have undertaken development projects in the ailing infrastructures in the energy, transport, and water sectors [1]. These sectors are important for economic growth and their improvement can have a significant impact on the Jamaican economy. The country requires transport infrastructure to hasten economic growth while facing tight budgetary constraints. By leveraging PPPs, Jamaica has embarked on projects to improve the transport infrastructure as a way of improving tourism and economic growth. Some examples include the Sangster’s International Airport, the Port Authority and the Highway Facility.

This decision to embrace PPPs in 2011 gave Jamaica the platform to institutionalize the required policies, roles, dedicated units and staff in place to properly manage and monitor the implementation of PPPs. Table 1 outlines the strategic moves that has been made by the Jamaica government to this end. Both Jamaica and Trinidad and Tobago are somewhat advanced in setting up established policies, procedures and structures to effectively manage PPPs.

Although tremendous gains can be achieved through the implementation of PPPs, the success of these projects are inconclusive [19]. Poor governance, lack of policy framework and lack of transparency hinder the success of PPPs in developing countries. In addition, corruption and mistrust among shareholders can affect the successful implementation of PPP projects in developing countries [16]. In the Corruption Perception Index Report 2016, Jamaica was ranked 83 out of 176 countries behind the Bahamas, Barbados, St. Lucia, St. Vincent and the Grenadines, Dominica, Grenada and Cuba [25]. Due to the high likelihood of corruption and political interference in Jamaica, it is quite possible for PPPs to fail than succeed in the country.

One of the main reasons for engaging in PPPs in developing countries is to raise capital required for infrastructure projects [12]. The North Luzon toll way in Philippines raised approximately US $371 million through PPPs. The Colombo Port project in Sri Lanka raised $175 million through a build-operate-transfer PPP contract [15]. The bulk terminal project raised US $7.5 million through a build-operate-transfer PPP contract in Ethiopia [15]. By 2013, 15 transport infrastructure projects in Turkey had been financed through PPPs [26]. According to [10], over US $6 billion had been spent to finance transport projects in South Asia through PPPs by 2007. As indicated in the above studies, governments in developing countries can offload the financing burden to the private sector.

Table 1 Public-private partnership architecture in the Caribbean

Figure 0.2: Limited PPP Policy and Institutional Architecture in the Caribbean

	Policy	Law	Detailed Guidelines	Defined Roles	Dedicated Units(s)	Staff with PPP Experience	Dedicated Project Prep Funding
Jamaica	✓ (2012)	✗	Underway	✓	✓ (DBJ & MOF)	Moderate	✗
Trinidad & Tobago	✓ (2012)	✗	Underway	✓	✓ (MOF)	Low	✗
Dominican Republic	✗	✗	✗	✗	✗	Moderate	✗
Haiti	✗	✗	✗	✗	✓ (MOF)	Low	✗
Suriname	✗	✗	✗	✗	✗	Low	✗
OEC States	✗	✗	✗	✗	✗	Low	✗

✓ In place (date) ✗ Absent Low Moderate High; Development Bank of Jamaica (DBJ); Ministry of Finance (MOF)

Source: Authors, based on Castalia and World Bank research

Source [24]

It is important to note that the outcome of a given PPP project depends on the strengths and weaknesses encountered. Project teams have the responsibility of identifying inherent challenges and successes at the beginning of the project to avoid negative results upon completion. It is posited that the determinants of successful PPP projects are contractual arrangements, project characteristics, project participants and interactive processes [27]. Contractual arrangements refers to contract type, award methods and risk allocation. Project characteristics refers to issues such as political and economic risks. Project participants are concern about inter-organizational conflicts, while interactive processes refers to the facilitation of effective coordination throughout the project life [27].

It is believed that high quality service should be the basis of all exchanges [28]. However, the common problems among stakeholders are high costs of tendering, cost restraints, complex negotiation, and conflicting project objectives [14]. Participation in core public activities and sufficient rewards are the main causes of conflicting project objectives. Identifying these issues early enough helps avoid conflicts that may arise when commissioning the project. The impact of the project may also be affected by the cost of maintaining and operating the facility as high operation and maintenance costs can affect the efficiency of the facility eliminating its intended impact.

A case study by [29] on telecommunication PPPs in Lebanon illustrated the importance of efficient management of PPPs. According to the study, the government had to maintain its involvement in the partnership and a transparent regulatory framework had to be established. In Lebanon, poor management made telecommunication companies exceed subscriber limits, evade fees and taxes, and provide poor network coverage.

An analysis of PPPs by [30] indicated that PPP projects can fail when the private contractors fail to transfer risk. This forces the government to bear the major cost of the project leading to failure of the project. As the economy recuperates from the effects of the financial meltdown, the Jamaican government is willing to absorb some of these risks and costs, but the government would like these occurrences to be as minimal and as seldom as possible.

3 Methodology

This is an exploratory study in which both primary and secondary data will be used. The primary data will be obtained from focus group sessions while the secondary data will be gathered from the Contractor General Report, the Auditor General Report the Statistical Institute of Jamaica Report and the literature review. The purpose of the secondary data is to assess the management, performance and outcome of the respective PPP projects, as well as findings from the literature regarding the determinants of successful PPP projects. These determinants—contractual arrangements, project characteristics, project participants and interactive processes—will form the basis of the initial discussion in the focus group sessions.

The purpose of the focus groups is to capture the views of key personnel involved in the implementation of the relevant PPP projects. The focus groups, supported by the nominal group technique, will be conducted with project sponsors, project managers and project administrators in Jamaica. The targeted participants will be the Contractor General, project sponsors and project managers who are involved and knowledgeable about a recently implemented PPP in Jamaica. The main objective of these focus groups is to ascertain from participants what they believe are the major determinants to deliver successful PPP projects. In doing so, information will be gathered regarding the management and outcomes of projects before the PPP policy and after the policy. A comparative analysis will be conducted with similar type and size PPP projects before 2012 and after 2012. It is hoped that the performance and outcomes of these projects before 2012 and after 2012 will be explored. Ethical approval will be sought to conduct this study and privacy will be maintained.

The initial questions to stimulate the discussion in the focus group will be pre-tested by about three project managers at the University of the West Indies. The selection of the University of the West Indies is based on convenience. However, effort will be made to ensure that the selected individuals are experts and quite knowledgeable about the implementation of PPPs. Relevant adjustments will be made to the questions regarding the wording, compound questions, ambiguous questions, number of questions, and similar concerns.

The scope of the study will be infrastructural projects throughout Jamaica. Letters of consent to participate in the study will be sent to potential participants. Based on the feedback, those persons who are willing to participant will be invited to a focus group session. Permission will be sought to record he sessions in an effort to capture all pertinent information and possible quotations.

Focus groups will be used to arrive at consensus on the determinants of successful PPPs implementation. A focus group approach will be taken because it is considered a cost-effective and efficient way to collect insights from experts in a domain of interest [31].

There are two commonly used group decision techniques to enable and improve group decision-making processes, namely the Delphi technique and the nominal group technique [32, 33]. The nominal group technique (NGT) is selected for this study because it is considered more superior in relation to the other group decision techniques [34]. In comparison to the Delphi technique, the NGT usually contribute to greater objectivity, yield more creative ideas, minimizes opinion differences and consensus is generally agreed on quicker [32, 35]. NGT also minimizes many of the problems that freely interacting groups may experience like group think, free loading and destructive dominance [34, 36].

The NGT process consists firstly of a creative thinking phase, then idea generation phase, followed by evaluation and finally a decision making phase [37]. The detailed procedure is set out below as distilled by [37]:

1. Participants independently and silently generate ideas regarding the problem and its solution in writing on index cards. On 3" × 5" index cards, participants will be asked to silently identify what they consider the major determinants of successful PPPs implementation.
2. Cards are collected by the independent facilitator, shuffled and randomly returned to the participants. In a round-robin format, participants will be asked to verbally state one idea per card until all participants have completed their list of ideas. The independent facilitator will write each idea on a slip chart or white board for visual display to all participants. The purpose of shuffling the cards will be to hide the identity of the person who generated the idea. In addition, the purpose of using an independent facilitator will be to control for researcher biases.
3. Each idea will be discussed for clarification, deeper insights and subsequent evaluation, without lobbying or identifying the originator of the idea.
4. Participants will be asked by the independent facilitator to silently and independently rank and select their major determinants. These major determinants will be written on newly distributed index cards.
5. The independent facilitator will collect the newly completed index cards with the ranked determinants. The cards will be shuffled and the independent facilitator will write the list of ranked determinants per index card on the flip chart or white board. The determinants are tallied and the most frequent determinants identified.
6. Participants will be asked to silently and independently rank these determinants on index cards and return the cards to the facilitator. Again, the facilitator will shuffle the cards and write the top ranked determinants on the flip chart or white board.
7. The top ranked determinants will be tallied. The determinant with the highest score will be presented to the participants as the top ranked determinant, with the determinant with the second highest score being presented as number two, and so forth.

8. Participants will be shown the resulting list of top ranked determinants and their acceptance sought. If there is no consensus, the ranking of the determinants will be repeated until there is consensus.

A pilot study of the procedure will be conducted with a small group of PPP project managers in Kingston. The pilot study will provide the opportunity to assess and refine the NGT process if necessary.

In summary, the NGT has been found to be an effective way to achieve consensus among participants in a group decision setting [37].

4 Conclusion

It is expected that the focus group sessions using the nominal group technique will not only be able to identify the major determinants of successful PPP projects but also to arrive at quick consensus on the determinants. Subsequent to the discovery of the major determinants from the focus group sessions, a research model will be formulated. This model will be presented to the research community for further refinement and validation in an attempt to ascertain a reasonable R^2 between the determinants and successful PPP projects. It is hoped that the derived research model will be able to explain a large majority of the variance in successful PPP implementation. Such knowledge can increase the likelihood of implementing successful PPPs, which by extension can lead to the growth and development of Jamaica.

References

1. Caribbean Development Bank. Public-private partnership in the Caribbean: building on early lessons. Caribbean Development Bank. 2014.
2. Williams D, Jones O. Factors associated with longevity of small, family-owned firms. Int J Entrepreneuship. 2010;14:37–56.
3. Morse A. Planning for economic infrastructure. National Audit Office; 2013. p. 1–44.
4. Camdessus M. Financing water for all. Report of the world panel on financing water infrastructure. Marseilles, France: World Water Council; 2003.
5. Chevers DA. The effectiveness of internal audit in Jamaican commercial banks. J Account Manage Inf Syst. 2016;15(3):522–41.
6. Murray S. Value for money? Cautionary lessons about P3 s from British Columbia. Canadian Centre for Policy Alternatives: Ontario; 2006.
7. Richter J. Public-private partnerships for health: a trend without alternatives? Development. 2004;37(2):43–8.
8. Lubis H, Majid NN. Developing a standardized assessment for PPP infrastructure project. J East Asia Soc Transp Stud. 2013;10:1–20.
9. Engel E, Fischer R, Galetovic A. The basic public finance of public-private partnerships. J Europ Econ Assoc. 2011.
10. Nataraj G. Infrastructure challenges in South Asia: the role of public-private partnerships. Asian Development Bank Institute; 2007.

11. Roehrich JK, Lewis MA, George G. Are public-private partnerships a healthy option? A systematic literature review. Soc Sci Med. 2014;113:110–9.
12. Hearne R. Origins, development and outcomes of public private partnerships in Ireland: the case of PPPs in social housing regeneration; 2009.
13. Organization for Economic Co-operation and Development. OECD—Annual Report 2008. Organization for Economic Co-operation and Development; 2008. p. 1–118.
14. Stanley M. Infrastructure financing and public-private partnerships. J Appl Corp Fin. 2011;23(3):30–8.
15. Nyagwachi JN. South African public private partnership projects. Port Elizabeth: Nelson Mandela Metropolitan University; 2008.
16. Reich R. Public-private partnerships for public health. Harvard Series on Population and International Health; 2002. p. 1–216.
17. Nicholson L. Jamaican family-owned businesses: homogeneous or non-homogeneous? Soc Econ Stud. 2010;59(3):7–29.
18. Zaidi SMS. Instituting corporate governance in developing, emerging and transitional economies. The Institute of Chartered Accountants of Pakistan; 2006. p. 1–38.
19. Fernandez RN, Carraro A, Hillbrecht RO. Efficiency, cost and benefits in contracts of public-private partnerships. Nova Econ. 2016;26(2):369–92.
20. Schwab K. The Global Competitiveness Report 2014–2015. World Economic Forum; 2014. p. 1–565.
21. Government of the Republic of Trinidad and Tobago. State enterprise investment programme (SEIP); 2014.
22. Department of Economic and Social Affairs. World statistics pocketbook. 2016th ed. New York: United Nations; 2016.
23. Monroe-Ellis P. Kingston container terminal public private partnership: analysis of contingency liability exposure. Auditor General's Department; 2017. p. 1–21.
24. Martin H. Caribbean infrastructure PPP roadmap. World Bank Group; 2014. p. 1–48.
25. Ugaz J. Corruption perceptions index 2016. Transparency International; 2017. p. 1–12.
26. Ermela K. Role of public-private partnership in infrastructure development: focus on Albania. Adv Res Sci Areas. 2013; 209–14.
27. Zhang X (ASCE). Critical success factors for public-private partnerships. J Constr Eng Manage. 2005;131(1):3014.
28. Vargo SL, Ajaka MA. Service-dominant logic as a foundation for service science: clarifications. Serv Sci. 2018;1(1):32–41.
29. Jamali D. Success and failure mechanisms of public private partnerships (PPPs) in developing countries: insights from the Lebanese context. Int J Public Sector Manage. 2004;17(5):414–30.
30. Edwards P, Shaoul J. Partnerships: For better or worst? Account Audit Account J. 2003;16(3):397–421.
31. Kontio J, Bragge J, Lehtola L. The focus group method as an empirical tool in software engineering. In: Shull F editor. Guide to advanced empirical software engineering (Springer); 2008.
32. Fretheim A, Schunemann HJ, Oxman AD. Improving the use of research evidence in guideline development: 5 group processes. Health Res Policy Syst. 2006;4(17).
33. Hsu C, Sandford BA. The delphi technique: making sense of consensus. Pract Assess Res Eval. 2007;12(10):1–11.
34. Tseng K, Lou S, Diez CR, Yang H. Using online nominal group technique to implement knowledge transfer. J Eng Educ. 2006;335–45.
35. Delbecq AL, Van de Ven AH, Gustafson DH. Group techniques for program planning: a guide to nominal group and delphi processes. Middleton, WI: Greenbriar; 1986.
36. Duggan EW, Thachenkary CS. Integrating nominal group technique and joint application development for improved systems requirements determination. Inf Manage. 2004;41(4):399–411.
37. Delbecq AL, Van de Ven AH, Gustafson DH. Group techniques for program planning: a guide to nominal group and Delphi processes. Glenview, Illinois: Scott, Foresman & Company; 1975.

Cognitive Solutioning of Highly-Valued IT Service Contracts

Shubhi Asthana, Aly Megahed and Ahmed Nazeem

Abstract Different service providers compete to win high valued IT service contracts in a tender kind of process, by providing comprehensive solutions that would fulfil the requirements of their client. Client requirements include services such as account management, storage systems, databases, and migrating the client infrastructure to the cloud. Preparing a solution is a step-wise process where a client shares the Request for Proposals (RFP) which documents the details of services required, and then service providers prepare solutions to fulfil these RFPs. The latter, usually referred to as "solutioning", can be a lengthy, time-consuming process. Therefore, solutioning automation could result in efficiency increases and cost reductions for the providers. In this paper, we propose an automated, cognitive, end-to-end solutioning methodology that is comprised of 3 steps. The first step involves a textual analytics approach for mining the RFP documents, and extracting the client requirements and constraints. Second, we formulate an optimization model that chooses the optimal set of offerings (that the provider can offer) and their attribute values that cover the client requirements at a minimum cost. Finally, we require market benchmarks to compute pricing of the chosen offerings. Market benchmarks are sometimes unknown for some offerings or for some attributes of these offerings. Thus, in that third step, we show an iterative method to estimate the missing benchmarks along with a confidence score. We validate our methodology by applying it to a real-world application with a comprehensive dataset. We show that it takes minutes to run our method, compared to the days and sometimes weeks for the case of manual solutioning. We also illustrate that our methodology provides more accurate solutions compared to manual solutioning.

Keywords IT services · Text analytics · Optimization · End-to-end solutioning
Estimating prices · Service benchmarking

S. Asthana (✉) · A. Megahed · A. Nazeem
IBM Research - Almaden, 650 Harry Road, San Jose, CA 95120, USA
e-mail: sasthan@us.ibm.com

H. Yang and R. Qiu (eds.), *Advances in Service Science*, Springer Proceedings in Business and Economics, https://doi.org/10.1007/978-3-030-04726-9_26

1 Introduction

Service Providers compete to win high-valued outsourcing contracts, in a tender kind of process, by designing and proposing solutions to potential clients [9, 17, 18], as a response to the clients' request for proposals (RFP). The kind of services included in these contracts are often complex [8, 24], worth hundreds of millions of dollars and very hard to quote [3, 6, 7, 15]. Traditionally, service providers prepare such solutions manually and it could take them a few days or weeks to do so. Solution designers use their expertise to come up with such complex solutions. The process of preparing these solutions is typically referred to as "solutioning".

In addition to the significant time and resources needed for manual solution preparation, there are also multiple challenges solution designers struggle with, resulting in preparing solutions that might not be optimized for increasing the chances of the provider to win the contracts. We present three main challenges in that regard. First, the RFP documents are usually long and require very experienced solution designers to manually read and extract the client requirements. Obviously, automating this could result in significant time and cost reduction for providers. Second, there is usually a large set of offerings that can be used to fulfil each customer's requirement. Finding the optimal set of offerings that fulfil all requirements at a minimum possible cost is almost impossible to guarantee in manual solutioning.

Additionally, during the negotiation process with the client, requirements do change and thus, solution designers would need to come up with a new set of offerings to respond to such changes. This might take as long as it took them the first time and cannot be optimized either. Third, the service provider may not have the market benchmarks for every offering included in the solution and hence they may need to be computed or obtained from a third-party vendor. Benchmarks differ from one offering to another, from one delivery location to another, from one client geography to another, from one client industry type to another, and even for one configuration to another for the same offering. Without the benchmarks, the pricing computed by the service provider may be inaccurate and have less competitive pricing and lower win rates.

In this paper, we provide an automated, cognitive, end-to-end solutioning methodology that overcomes these challenges and provides a competitive solution. Our approach is composed of three steps. In the first step, we show a novel method that uses textual analytics to come up with list of services and offerings from the clients RFP document. In the second step, we present an optimization model that chooses the optimal set of offerings and their attribute values, with a guarantee of fulfilling the client requirements at a minimum cost. In the last step, we present a method based on machine learning and analytical techniques to accurately infer the missing benchmarks through an iterative process, and then price the offerings included in the solution. We then apply it to real data at one of the world's largest IT service providers. Our results show that our method is both more efficient as it prepares a solution in a matter of minutes rather than days or weeks, as well as effective because of it yields a higher accuracy compared to manual solutioning.

The rest of this paper is organized as follow: In Sect. 2, we review the related literature. We then discuss our methodology in Sect. 3, and provide a proof-of-concept implementation and some results in Sect. 4. We end with providing the conclusions and directions for future work in Sect. 5.

2 Literature Review

In this section, we briefly review the related literature of optimizing offerings in the solutions as well as estimating benchmarks since there is limited literature on cognitive text analytics of RFP documents. Current state of the art includes having detailed cost structure of historic and market deals as well as meta-data for such deals like geography, client, industry, etc. [4, 5, 14, 16].

The "bottom-up" approach [1] is used for preparing solutions with some attributes. This approach involves estimating the cost of individual activities at a granular level that together form the total cost for each individual IT service. While this bottom up approach is effective in estimating the detailed cost structure, it is very time consuming and requires a detailed solution, i.e. knowing a lot of details about the structure of the services to be added to the solution.

Also, some works include using questionnaire design tool to collect anonymous data as feedback. Another prior art includes using fixed benchmarks when data is unavailable and uses performance metrics to better them [11]. In Saavedra and Smith [21], have shown that machine and program characterization help identify features that dominate benchmark results and hence can be used to reference the benchmark by identifying their characteristics. Other methods of benchmarking, which do not include any steps in our method, but help compare information quality across organizations and provide a baseline include [12, 13, 19, 20, 22]. All these methods are not only time-taking but also cannot show accuracy of the benchmarks that are computed.

Thus, as seen from the above literature, there has been a lot of work in solutioning and bench-marking contracts using different techniques, but none of them provides an effective method which comes with a solution within some accuracy bounds in a short period of time, after cognitively extracting the client requirements from the RFPs and choosing the optimal set of offerings to include in the solution. This is the objective of this work.

3 Methodology

Our methodology is a three-step approach that aims at solving the problem of preparing an efficient solution for the clients RFP. The inputs to our model are the RFP as well as some client attributes, such as the geography and industry of the client, and a historical database of benchmarks used by the service provider for assessing and pricing contracts appropriately. In addition, we assume that we are given a minimum expected confidence score for standardizing a service benchmark. The output of our approach is an optimized choice of offerings which cover the client's requirements at a competitive price.

3.1 Map Offerings and Services Based on User Input

In this step, we use the user inputs such as the RFP, and deal attributes like geography, industry, and time of deal, etc. The RFP document contains the textual description of the client's environment and the client's requirements. The objective is to use text analytics to extract the client requirements and constraints automatically from these documents, and map them to the service offerings that the provider can offer.

The rest challenge is to identify whether each statement in the document is a requirement or not. Usually, a big portion of the document is devoted to describing the client environment, the bidding procedure, etc. Hence, the first task in our text analysis is to build a set of rules and patterns that can be used to decide whether a given text should be identified as a requirement or not. These rules and patterns include dictionary-based keywords, enclosing sections headers, context, and linguistic features. The output of this step is the set of statements that represent the client's textual requirements. The second challenge is to categorize the extracted client's textual requirements into the service provider's taxonomy of services and constraints. The output of this step is to assign a service or a constraint to each textual requirement. To this end, we execute 3 steps:

- Build a Lexicon of keywords for each service constraint category. This was done using both automated words frequency analysis for the service catalogue and manual expert validation.
- Match each requirement text and its context (e.g. enclosing section header, table label, and paragraph header) against the services/constraints Lexicon and obtain a score for each matching.
- Build a set of rules that takes the matching scores for each requirement with the services/constraints Lexicon and assign a service/constraint to each requirement.

The final step in this stage is to aggregate the services and constraints that were identified from the individual textual requirements at the contract level. We apply a straightforward mapping from the cognitively identified services to standard offerings. Our output of this step is the set of offerings that can provide the deal services, while adhering to the deal constraints.

3.2 Cost Optimization

In this step, we formulate an optimization model that determines the optimal set of offerings to include in the solution, at a minimum total cost. We define some notation then present our mathematical formulation for this step.

Let F be the set of offerings, R be the set of requirements extracted from the RFP (See the first step above), $|R|$ be the cardinality of set R, and I_f be the set of attributes for offering $f \in F$. We also define c_{iv} as the cost of attribute value $v \in V_{if}$, if that value is chosen for attribute $i \in I_f$ of offering $f \in F$. Further, we define e_{fr} as a given indicator (extracted in step 1), that is 1, if offering $f \in F$ can be used to fulfill requirement $r \in R$, and zero otherwise. Next, we define our decision variables as follows: X_{fr} is 1, if offering $f \in F$ is chosen in the solution to fulfill requirement $r \in R$, and zero otherwise. Y_{iv} is 1, if value $v \in V_{if}$ is chosen for attribute $i \in I_f$ of offering $f \in F$, and zero otherwise. We now present the formulation of our optimization model:

$$Min \sum_{i \in I_f} \sum_{f \in F} \sum_{v \in V_{if}} c_{iv} \cdot Y_{iv} \tag{1}$$

$$s.t. \sum_{f \in F} e_{fr} \cdot X_{fr} = 1 \quad \forall r \in R \tag{2}$$

$$A_{rf}.Y = b_{rf} \cdot X_{rf} \quad \forall r \in R, \forall f \in F \tag{3}$$

$$\sum_{v \in I_{if}} Y_{iv} \leq 1 \quad \forall f \in F, \forall i \in I_f \tag{4}$$

$$\sum_{v \in V_{if}} Y_{iv} \geq \frac{X_{fr}}{|R|} \quad \forall r \in R, \forall f \in F, \forall i \in I_f \tag{5}$$

$$Y_{iv} \in \{0, 1\} \quad \forall v \in V_{if}, \forall i \in I_f, \forall f \in F \tag{6}$$

$$X_{fr} \in \{0, 1\} \quad \forall f \in F, \forall r \in R \tag{7}$$

where, objective function (1) minimizes the total cost of the solution, which is the cost of all chosen values for all attributes across all offerings. Constraint (2) ensures that exactly one offering is used to fulfill each requirement. The constraints imposed by each requirement, on the attribute values of each the offering used to fulfill it, are introduced in constraint (3). Note that in constraint (3), A_{rf}, Y, and b_{rf} are in matrix form and are meant to illustrate, in linear programming standard notation, the constraints imposed by a requirement on the attribute values of the offering covering it. Constraint (4) makes sure that at most one categorical value for each attribute is chosen. This is meant to put an upper limit on the number of values to be chosen for each attribute. That upper limit is obviously 1. Constraint (5) puts the lower limit for attribute value choice. That lower limit is choosing one value for an attribute if its offering is used to fulfill one requirement at least. Lastly, constraints (6) and (7) are the binary restrictions of our variables.

3.3 *Estimation for Missing Benchmarks*

We use a historical database of benchmarks from the contracts that have been executed in the past. However, sometimes these benchmarks maybe missing for a few offerings and hence service providers may not be able to calculate prices accurately. We treat that issue in this step as follows.

Since we have the pricing for some offerings but not for others, we use machine learning recommendation systems to estimate the missing values. That is, we train the recommenders on the known data and then use them to amend the missing values. The features used in this training are meta-data of the contract, such as the geography of the client, the delivery location of the offerings, and the services included in the solution.

Each feature is given a weight w depending on how indicative they are to the pricing of the service, and a weighting function is used to come up with a confidence score for the benchmark values that we amend. The confidence score indicates an estimate on how accurate the benchmark data would be. We then assess the contracts that use these draft benchmarks to calibrate and come up with pricing solutions that can be used in the solutioning process. Assessing the performance of the draft benchmarks on winning contracts, we can re-iterate the confidence scores on contracts based on whether the contracts were won or lost. If the aggregate confidence score of all the draft benchmarks for all services under an offering go higher than a given threshold value and multiple corresponding contracts are won, we standardize such set of benchmarks. By standardizing the benchmarks, the service provider can use them with high confidence for pricing future contracts.

4 Implementation and Numerical Results

In this section, we show a proof-of-concept implementation of our method in Sects. 4.1 and 4.2, and present some numerical results.

4.1 *Textual Analytics to Map to a Set of Standard Offerings and Cost Optimization*

For our first experiment, we gathered a set of 15 RFPs and 15 standard offerings. The RFPs text was annotated by a handful of domain experts to categorize each relevant statement in the documents to one of the provider's services or constraints. 10 of these documents were used to construct the rules, while 5 were used for testing. A Lexicon of about 3000 keywords for 20 services and constraints was built by running the TF-IDF algorithm on the service catalogue documentation. IBM BigInsights SystemT [2] was used to extract the linguistic features of the text. The set of rules for requirement identification and categorization were built by statistically studying the most common patterns and matchings scores to the Lexicon for each category in the training documents. The average accuracy of the requirement identification sub-model is about 92%, and the average accuracy of requirement classification sub-model is about 83%.

For the service/constraints to standard offerings mapping, we collected the necessary information for each of the 15 standard offerings to understand the services they provide and their associate constraints. The service to offerings mapping and the constraints filtering takes usually a few seconds. Lastly, for the cost optimization model, we implemented it on the Python programming language and used CPLEX [10] to solve it. The model is an integer programming model (See, for example, [23]). For realistic instances, it takes an average of 3 min to run.

4.2 Computing the Confidence Scores for the Draft Benchmarks

For our second experiment, we selected a set of 300 requirements from the aforementioned repository. We evaluated our method on a set of 7 offerings, 246 geographies, and 3 delivery locations. We then analyzed the computation of the confidence score based on the features used to estimate the draft benchmarks. Four features were used and were given random weights in order to proceed with our method. The used features are the client geography, service delivery-from location, scope (which services exactly were included), and the configuration of the services.

Table 1 gives the requirements which have a collection of offerings and the confidence scores computed based on feature similarity. The results are the total confidence score for every service under an offering.

Table 1 Excerpt of the dataset of the offerings and confidence scores of draft benchmarks

Offering	Geography	Services	Feature similarity on which confidence scores are based	Total confidence score
X	Asia, Philippines	Service A	Geography 0.1 Delivery location 0.2	0.3
		Service B	Scope 0.3 Delivery location 0.2	0.5
		Service C	Scope 0.3 Configuration 0.4	0.7
Y	Europe, Croatia	Service D	Configuration 0.4 Geography 0.4	0.8
		Service E	Scope 0.3	0.3
		Service F	Scope 0.3 Delivery location 0.2	0.5

5 Conclusions and Directions for Future Work

In this paper, we presented a novel, automated, cognitive end-to-end solutioning methodology for highly-valued IT service contracts. We showed a method for mapping the text in RFP documents to a set of requirements and offerings that the IT service provider can offer. We also formulated an optimization model that determines the optimal set of offerings to include at a minimum cost. Lastly, we illustrated how we can amend missing data from the benchmark prices so that we can have complete data to use for pricing the offerings in the solution.

We applied our methodology to real-world data and showed how it brings both efficiency and effectiveness to the solution. The solution is efficient as it is provided in a matter of minutes compared to the traditional time taken which can be days or weeks to prepare the solution manually. We also illustrated the effectiveness of our method by showing its high accuracy which results in higher win rates for the provider, when implemented.

There are multiple directions for future research to our work. We can perform a more comprehensive numerical study for the effect of confidence scores in refining the draft benchmarks. That is, one would evaluate how well the draft benchmarks are performing and use it directly in pricing contracts by standardizing them. Numerical comparisons would then illustrate a more quantitative analysis on the efficiency of our method of computing confidence scores. Another possible extension of this work is to come up with natural language processing ways to treat typographical errors in the RFP documents. This can help identify the requirements more efficiently. Lastly, our optimization model is an integer programming model which might take a long time to solve for large instances. Thus, a direction for future research is to come up with heuristics for calculating the optimal set of offerings and attribute values.

References

1. Akkiraju R, Smith M, Greenia D, Jiang S, Nakamura T, Mukherjee D, Pusapaty S. On pricing complex IT service solutions. In: Service research and innovation institute global conference; 2014. p. 55–64.
2. Chiticariu L, Krishnamurthy R, Li Y, Raghavan S, Reiss F, Vaithyanathan S. SystemT: an algebraic approach to declarative information extraction. In: Proceedings of the 48th annual meeting of the association for computational linguistics (ACL'10). Association for Computational Linguistics: Stroudsburg, PA, USA; 2010. p. 128–37.
3. Fukuda MA, Gajananan K, Jiang S, Megahed A, Nakamura T, Smith MA. Effectiveness of service complexity configurations in top-down complex services design. U.S. Patent Application 14/977,383, filed 22 June 2017.
4. Fukuda MA, Gajananan K, Jiang S, Megahed A, Nakamura T, Smith MA. Method and system for determining an optimized service package model for market participation. U.S. Patent Application 15/050,986, filed 24 August 2017.
5. Fukuda MA, Gajananan K, Jiang S, Megahed A, Nakamura T, Smith MA. Augmenting missing values in historical or market data for deals. U.S. Patent Application 15/192,875, filed 28 December 2017.

6. Fukuda MA, Gajananan K, Jiang S, Megahed A, Nakamura T, Smith MA. Top-down pricing of a complex service deal. U.S. Patent Application 15/192,884, filed 28 December 2017.
7. Gajananan K, Megahed A, Abe M, Nakamura T, Smith M. A top-down pricing algorithm for IT service contracts using lower level service data. In: 2016 IEEE international conference on services computing (SCC); 2016. p. 720–7.
8. Gamma N, Do Mar Rosa M, Da Silva M. IT services reference catalog. In: IFIP/IEEE international symposium on integrated network management (IM); 2013. p. 764–7.
9. Greenia D, Qiao M, Akkiraju R. A win prediction model for IT outsourcing bids. In: Service research and innovation institute global conference; 2014. p. 39–42.
10. IBM ILOG CPLEX CPLEX, V12.1: User's Manual for CPLEX. International Business Machines Corporation; 2009.
11. Iosup A, Ostermann S, Yigitbasi MN, Prodan R, Fahringer T, Epema D. Performance analysis of cloud computing services for many-tasks scientific computing. IEEE Trans Parallel Distrib Syst. 2011;22(6):931–45.
12. Kahn BK, Strong DM, Wang RY. Information quality benchmarks: product and service performance. Commun ACM. 2002;45(4):184–92.
13. Lueger RJ, Barkham M. Using benchmarks and benchmarking to improve quality of practice and services. In: Developing and delivering practice-based evidence: a guide for the psychological therapies; 2010. p. 223–56.
14. Megahed A, Asthana S, Becker V, Nakamura T, Gajananan K. A method for selecting peer deals in IT service contracts. In: 2017 IEEE international conference on AI mobile services (AIMS); 2017. p. 1–7.
15. Megahed A, Gajananan K, Abe M, Jiang S, Smith M, Nakamura T. Pricing IT services deals: a more agile top-down approach. In: International conference on service-oriented computing (ICSOC). Berlin: Springer; 2015. p. 461–73.
16. Megahed A, Gajananan K, Asthana S, Becker V, Smith M, Nakamura T. Top-down pricing of IT services deals with recommendation for missing values of historical and market data. In: International conference on service-oriented computing (ICSOC). Springer International Publishing; 2016. p. 745–60.
17. Megahed A, Ren GJ, Firth M. Modeling business insights into predictive analytics for the outcome of IT service contracts. In: 2015 IEEE international conference on services computing (SCC); 2015. p. 515–21.
18. Megahed A, Yin P, Nezhad HRM. An optimization approach to services sales forecasting in a multi-staged sales pipeline. In: 2016 IEEE international conference on services computing (SCC). IEEE; 2016. p. 713–9.
19. Nissinen A, Grnroos J, Heiskanen E, Honkanen A, Katajajuuri JM, Kurppa S, Mkinen T, Menp I, Seppl J, Timonen P, Usva K. Developing benchmarks for consumer-oriented life cycle assessment-based environmental information on products, services and consumption patterns. J Clean Prod. 2007;15(6):538–49.
20. O'Mahony M, Oulton N, Vass J. Market services: productivity benchmarks for the UK. Oxford Bull Econ Stat. 1998;60(4):529–51.
21. Saavedra RH, Smith AJ. Analysis of benchmark characteristics and benchmark performance prediction. ACM Trans Comput Syst (TOCS). 1996;14(4):344–84.
22. Wang L, Zhan J, Luo C, Zhu Y, Yang Q, He Y, Gao W, Jia Z, Shi Y, Zhang S, Zheng C. Bigdatabench: a big data benchmark suite from internet services. In: 2014 IEEE 20th international symposium on high performance computer architecture (HPCA); 2014. p. 488–99.
23. Wolsey LA, Nemhauser GL. Integer and combinatorial optimization. Wiley; 2014.
24. Yin P, Nezhad HRM, Megahed A, Nakamura T. A progress advisor for IT service engagements. In: 2015 IEEE international conference on services computing (SCC). IEEE; 2015. p. 592–9.

A Predictive Approach for Monitoring Services in the Internet of Things

Shubhi Asthana, Aly Megahed and Mohamed Mohamed

Abstract In Internet of Things (IoT) environments, devices offer monitoring services that would allow tenants to collect real-time data of different metrics through sensors. Values of monitored metrics can go above (or below) certain predefined thresholds, triggering the need to monitor these metrics at a higher or lower frequency since there are limited monitoring resources on the IoT devices. Also, such triggers might require additional metrics to be included or excluded from the monitoring service. An example for this is in a healthcare application, where if the blood pressure increases beyond a certain threshold, it might be necessary to start monitoring the heart beat at a higher frequency. Similarly, the change of the environmental context might necessitate the need to change/update the monitored metrics. For instance, in a smart car application, if an accident is observed on the monitored route, another route might need to be monitored. Whenever a trigger happens, there are optimization-based methods in the literature that calculate the optimal set of metrics to keep/start measuring and their frequencies. However, running these methods takes a considerable amount of time, making the approach, of waiting until the trigger happens and executing the optimization models, impractical. In this paper, we propose a novel system that predicts the next trigger to happen, run the optimization-based methods beforehand, and thus have the results ready before the triggers happen. The prediction is built as a tree structure of the state of the system followed with its predicted child nodes/states, and the children states of these children… etc. Whenever part of that predicted tree actually occurs, one can remove the calculations of the part that did not occur to save storage resources.

Keywords Internet of Things (IoT) · Monitoring services · Resource constraints
Optimization · Predictive analytics

S. Asthana (✉) · A. Megahed · M. Mohamed
IBM Research, Almaden, 650 Harry Road, San Jose, CA 95120, USA
e-mail: sasthan@us.ibm.com

H. Yang and R. Qiu (eds.), *Advances in Service Science*, Springer Proceedings in Business and Economics, https://doi.org/10.1007/978-3-030-04726-9_27

1 Introduction

The Internet of Things (IoT) refers to the networked interconnection of devices that can include sensors as well as software to run it [1]. There are various kind of sensors such as GPS sensors, security cameras, temperature sensors, ... etc. The IoT devices are equipped with such sensors that can allow them to collect information real-time to measure different metrics like location co-ordinates, accelerometer, heart rate, blood pressure, ... etc. [2, 3]. Sensors are typically either built-in on the devices or remotely connected to them via the internet. Examples of devices for the latter case are modems, Raspberry Pi, and smart phones.

In an IoT environment, different tenants are allowed to access their data through a set of services [4–6]. These services are referred to as monitoring services [7, 8]. A solution bundles these distinctive services to cater to customer needs [9]. Tenants need to set the parameters of the monitoring services, i.e., decide which metrics to monitor, the allocation of metrics/sensors to devices, and the frequency of monitoring such metrics. Note that deciding these optimal decisions and policies is not trivial because of the presence of computing resource constraints at the IoT devices, such as the CPU, memory, disk storage, and network bandwidth. There is prior literature on how to determine the optimal decisions for the system in order to set the monitoring services. See, for example, our prior work references in [10–13], where optimization-based approaches are discussed and implemented.

Additionally, there are triggers that happen in IoT environments that necessitate the need to change the state of the system. That is, the metrics to be monitored and their frequencies, as well as the allocation of metrics/sensors to devices, might need to be changed. The two main types of these triggers are metric value-based triggers and environmental triggers [10]. An example of the former type is in an IoT healthcare application (see, for example [14, 15]), where whenever the heartbeat of a patient increases above a certain threshold, the blood pressure might need to be monitored at a higher frequency. Another example but for the latter type of triggers is in a smart car application, where the observance of an accident in the route being monitored might require monitoring alternative routes.

Therefore, whenever a trigger happens, the aforementioned optimization-based approaches are being executed to determine the new optimal decisions for the system. However, these optimization models take a relatively long time to run (sometimes tens of minutes) and thus, simply waiting for the triggers to happen and then execute them is not practical.

In this paper, we propose a new proactive approach in which we predict the next triggers to happen and execute the optimization models beforehand. Then, whenever the trigger happens, we are ready with the optimal solutions. We model our system as a tree structure, where the current trigger (referred to as the state of the system) is a node and the possible triggers that could happen after it are child nodes. Then, each of these triggers would have child nodes of its own, and so on. We keep predicting the order of occurrence of the nodes and running the corresponding optimization models, until one of the triggers actually occurs, in which case, we delete the part of the predicted tree that did not occur to save memory resources, and keep predicting deeper nodes for the part of the tree that has the children of the trigger that just occurred as well as execute of the corresponding optimization models and save their solutions, and so on. We could also stop the predictions and proactive execution of the optimization models, even before a trigger happens, if the available computing resources are exhausted. In that latter case, we delete the least likely triggers and their solutions from the memory and wait until the next trigger to happen before reiterating on our approach.

Since this is a work-in-progress paper, we present our approach and ideas for its implementation featuring how effective it could be. In the future, we plan on carrying out actual experiments to validate our proposed method. The rest of this paper is outlined as follows. In Sect. 2, we present our proposed approach and in Sect. 3, we end the paper with our conclusions and directions for future work.

2 Methodology

Given resource constraints of the monitoring system, it is impossible to store the next states for each metric that can get triggered in an IoT environment. Therefore, the objective of our research is to develop an interactive approach for monitoring services that is able to predict the likely set of metrics that would get triggered, and run the optimization model that determines the optimal frequencies of these metrics [10–13]; so that the solution for next states for these metrics is ready. That way, if a metric gets triggered, its optimal solution is ready and can be implemented immediately.

Our methodology is a two-step approach where we obtain the current state of the system and predict the next states. It continuously monitors the resource capacities while processing the solution for every state. Section 2.1 defines the notation and inputs to our methodology and Sect. 2.2 illustrates its overall underlying algorithm.

2.1 Notation and Inputs to Our Method

We first define some notations and then present our mathematical formulation that handles the predictive monitoring problem. Let N be the number of metrics monitored, I be the set of metrics where $I=\{1, \ldots, N\}$ and any $i \in I$ is an integer. So, $|I| = N$, where N is the number of metrics and $|I|$ is the cardinality of set I. Let S be the one-dimensional ordered set of metric states of size N, $\forall s \in S$, s is either 0 or 1. $|S| = N$, where $|S|$ is the cardinality of set S.

We also define t as the trigger with a flag 0 or 1 and a value Vt which represents the ID of the metric that is triggered. This value is an integer between 1 and N inclusive. We also define Y as a set of 2^N ordered sets, each with a size N. Therefore, $\forall Yi \in Y$, there are N elements where all of them are zeros except for element number i, where $i \in \{1, 2, \ldots, N\}$. Next, we define w_{if} as the weight of metric $i \in I$ when it is in status $f \in F$. That is, $F = \{0, 1\}$, where when $f = 0$, it means that the metric is not triggered, and $f = 1$ means that the metric is triggered.

We further define sol_i as the solution corresponding to state Y_i and $size_s$ is the solution space occupied by solution sol_i. Z is the maximum storage space of the system, and L are the number of levels for which solutions are calculated. Lastly, we define *Saved* as the set of ordered pairs of metrics $i \in I$ and their solutions sol_i that are saved so far. In the next section, we define our methodology in detail.

2.2 Algorithm of Our Methodology

In this step, we first get the current state of the system S; obtaining the list of metrics N being monitored and then we predict the priorities of the next states to the current state of system S. This priority list Y is an array of size N whose values are a rank for each metric out of N metrics. The rank is anywhere between 0 and $N - 1$, and no two metrics have the same ranking. To predict these priorities, we build classification model for whether a state will occur or not. We train the classification model on historical data and then apply it to the potential children states of the current state to get the probability score of occurrences for each of them. Then, we rank them in a descending manner according to that probability.

For each metric $i \in I$ in the priority list Y, we check if the solution sol_i exists for the corresponding state to that metric, i.e., whether we have calculated its solution before or not. If it does exist, we go on to prioritizing the remaining children of the current state and then calculating their solutions (via the optimization method of the prior art). Then, we go to the top metric in the list and do the same for its children, then to the next and do the same to its children, and so on. We stop when either a trigger happens, when we had calculated a maximum given depth of L levels of the current state, or if we run out of storage resources for the stored solutions. If it is the former, then we restart as before. If it is the latter, then we delete the least likely solutions and those away from the current state, from our solution list, to save some storage.

Also, whenever any trigger actually occurs, we delete the part of the tree solutions stored corresponding to the part of the triggers' sub-tree that did not happen. Additionally, we might terminate our predictions and optimization-method execution if they reach a certain depth, even if neither the new trigger occurred nor we ran out of storage space. Algorithms 1 and 2 illustrate our overall methodology, where Algorithm 1 overviews our overall method except for the part of handling storage overflow which is captured in Algorithm 2.

Algorithm1: Algorithm for Metric State Monitoring

Given State i to begin the procedure at and current Saved
if trigger t == 1 then
  for all Yi ∈ Y do
    S(i) ← state of the system with respect to metric i
    if (i, sol_i) ∈ *Saved* then
      continue
    else
      check storage() procedure
    end if
    repeat trigger procedure for child nodes of S(i) upto L levels
  end for
end if

Algorithm2: Algorithm for Resource Optimization Modeling.

Procedure for check storage() procedure
Given current Saved, state x to run the optimization model for.
if $\sum_{i \in I:(i,sol_i) \in Saved} size_i < Z$ then
  sol_x ← run the optimization model
  Saved ← Saved ∪ {(x, sol_x)}
else
  Delete all ordered pairs (k; solk) from Saved, where k is not a child node of state x in the immediate L levels of it.
  sol_x ← run the optimization model
  Saved ← Saved ∪ {(x, solx)}
end if

3 Conclusions and Directions for Future Work

In this work-in-progress paper, we presented a novel methodology that incorporates a predictive monitoring model for determining the next trigger in an internet of things (IoT) monitoring system. Our approach is proactive; it predicts the priority of the next triggers that would happen in the system and runs the corresponding metrics choice and monitoring frequency optimization models, so that when the trigger actually happens, the solution is ready to be implemented right away. We also presented a method to deal with storage limitations via both checking the storage capacity and deleting less important solutions, and through deleting the solution parts that are not relevant to the triggers that end up occurring. Our method treats the issue in the related literature, where systems have to wait for running the optimization models, after the trigger happens.

There are several directions for the in-progress and future research to this work. First, we are implementing this new method and incorporating it in a simulated test bed for an actual IoT system. Deploying it in this system and comparing the performance to the prior method (of waiting for the trigger to happen before running the optimization models) would be vital to validate our contribution. Second, one can apply different predictive analytics techniques and compare the results, for the priority prediction part of our methodology. Third, another direction for future research is extending our approach to the case of IoT systems with multi-tenants, where different triggers occur for different tenants while there are computing resources shared among the tenants.

References

1. Yang SH. Internet of things. Wireless Sens Netw. 2014: 247–61 (Springer).
2. Gubbi J, Buyya R, Marusic S, Palaniswami M. Internet of things (IoT): a vision, architectural elements, and future directions. Future Gener Comput Syst. 2013;29(7):1645–60.
3. Megahed A, Tata S, Nazeem A. Cognitive determination of policies for data management in IoT systems. In: International conference on service-oriented computing. 2017: 188–97 (Springer).
4. Mohamed M. Generic monitoring and reconfiguration for service-based applications in the cloud. Ph.D. thesis, Institut National des Telecommunications, 2014.
5. Gajananan K, Megahed A, Nakamura T, Abe M, Smith M. A top-down pricing algorithm for IT service contracts using lower level service data. In: Proceedings of the IEEE international conference on services computing (SCC), 2016: 720–7.
6. Megahed A, Gajananan K, Abe M, Jiang S, Smith M, NakamuraT. Pricing IT services deals: a more agile top-down approach. In: International conference on service-oriented computing. Berlin: Springer; 2015: 461–473.
7. Gil D, Ferńandez A, Mora-Mora H, Peral J. Internet of things: a review of surveys based on context aware intelligent services. Sensors 2016; 16(7): 1069.
8. Issarny V, Bouloukakis G, Georgantas N, Billet B. Revisiting service-oriented architecture for the IoT: a middleware perspective. In: International conference on service-oriented computing, 2016: 3–17 (Springer).
9. Megahed A, Asthana S, Becker V, Nakamura T, Gajananan K. A method for selecting peer deals in IT service contracts. In: Proceedings of the IEEE international conference on artificial intelligence and mobile services (AIMS), 2017: 1–7.
10. Tata S, Mohamed M, Megahed A. An optimization approach for adaptive monitoring in IoT environments. In: 2017 IEEE international conference on services computing (SCC), 2017: 378–385. IEEE.
11. Megahed A, Pazour J, Nazeem A, Tata S, Mohamed M. Monitoring services in the internet of things: an optimization approach. Computing. 2018 (To Appear).
12. Megahed A, Yin P, Nezhad HRM. An optimization approach to services sales forecasting in a multi-staged sales pipeline. In: Proceedings of the IEEE international conference on services computing (SCC), 2016: 713–719.
13. Fukuda MA, Gajananan K, Jiang S, Megahed A, Nakamura T, Smith MA. Method and system for determining an optimized service package model for market participation. U.S. Patent Application 15/050,986, filed 24 Aug 2017.
14. Asthana S, Megahed A, Strong R. A recommendation system for proactive health monitoring using IoT and wearable technologies. In: 2017 IEEE International Conference on AI & Mobile Services (AIMS), 2017: 14–21. IEEE.
15. Asthana S, Strong R, Megahed A. Healthadvisor: recommendation system for wearable technologies enabling proactive health monitoring. arXiv:1612.00800, 2016.

Managing Clinical Appointments in an Academic Medical Center

Chester Chambers, Maqbool Dada, Marlís González Fernández and Kayode Williams

Abstract According to the US Department of Health and Human Services, roughly 40% of all outpatient visits in the US are made to teaching hospitals. The educational mission of these settings adds stages to the care delivery process and, in some instances leads to a system which is a hybrid between a single and two-server queue. We consider the problem of determining appointment schedules for these settings which explicitly account for the teaching mission, high no-show rate, a blending of patient types, highly variable processing times, processing in multiple stages, processing times that are not independent across stages, and a dynamic policy regarding the involvement of a medical trainee. We formulate the resulting problem and develop structural results. We then use properties of the optimization problem to develop an intuitive Cyclic scheduling approach, which bundles multiple patients into each Cycle. The application of our modeling framework is illustrated for the AMC clinic that motivated this work.

1 Introduction

Roughly 40 million appointments are made each year for patients visiting outpatient clinics in Academic Medical Centers (AMC's) [1]. Our direct observation and data collection at several of these clinics reveals that attending physicians routinely alter process flow for clinic visits based upon system status. When waiting times are realized due to the clinic falling behind schedule attending physicians have options regarding activity times and process flow. These options include removing the resident or fellow from the process flow for selected patients. These state-dependent alterations imply that steady state results are not valid and models that ignore these

C. Chambers (✉) · M. Dada
Carey Business School, Johns Hopkins University, Baltimore, MD 21202, USA
e-mail: cchamber@jhu.edu

M. González Fernández · K. Williams
School of Medicine, Johns Hopkins University, Baltimore, MD 21205, USA

H. Yang and R. Qiu (eds.), *Advances in Service Science*, Springer Proceedings in Business and Economics, https://doi.org/10.1007/978-3-030-04726-9_28

actions are incomplete. Our objective is to develop and optimize a schedule that fully incorporates these elements.

Within the focal clinic the patient encounter routinely requires three stages with physicians. In Stage 1 the patient is examined by a medical trainee, typically a resident or fellow (Trainee). In Stage 2 the Trainee discusses the case with the attending physician (Attending). Stage 3 is a joint examination by the Trainee/Attending dyad. Let us refer to this as Process 1. The times associated with these stages are not independent. A more complex case tends to have longer times at each of the three stages.

The particular unit discussed here is a Physical Medicine and Rehabilitation Clinic with a no-show rate of 25%. In a typical 4-h session, the clinic used block scheduling to book 2–3 patients coming to the clinic for the first time with a particular problem (New) and 4–6 patients for follow-up visits (Return). Patients are seen by a team of one Attending and one Trainee. Patients were scheduled in 20-min time slots with New patients given two concurrent slots.

Most patients were treated using Process 1. However, congestion caused frequent patient waits and overtime was needed to complete the clinic schedule in roughly 50% of the observed sessions. In periods of high congestion the Attending sometimes chooses to treat some patients without the involvement of the Trainee. We label, this 1-Stage approach as Process 2. This ad hoc adaptation was used for roughly 30% of the observed patient visits. This compromised the clinic's educational mission. In addition, clinic management anticipated an increase in demand. Therefore, they sought a schedule and policy that would accommodate 3 New cases and 6 Return cases in each session with consistent but less frequent use of this practice without a marked increase in total patient waiting times and clinic overtime. For educational reasons, the Trainee had to be involved with all New cases. Given these needs it seemed natural to schedule 3 cycles having each cycle include 1 New and 2 Return patients.

We begin by outlining the basic model that assumes that all patients arrive at their appointment time. This develops problem structure and leads to two structural results. We then incorporate unreliable patients—meaning that we include no-shows within the problem structure. We focus on the case with three cycles, fitting the clinic that motivated this work, and uncover an additional structural result. We then turn to the application of our approach in the focal clinic. We uncover the cost minimizing schedule, and also develop a set of heuristics that are easier to explain and implement. We then close with a few summarizing comments.

2 Outline of the Basic Model

Consider a system with one server that needs to process L scheduled, single-stage jobs in the order of their arrival, over a targeted session time T. Each job arrives at the appointment time A_ℓ, where $1 \leq \ell \leq L$. (Notation is summarized in Table 1.)

Table 1 Summary of notation used in order of appearance

Symbol	Definition
L	Number of jobs to be scheduled
T	Targeted end of clinic session
A_ℓ	Appointment time for job ℓ
τ_ℓ	Processing time for job ℓ
$f_\ell, F_\ell, \bar{F}_\ell$	PDF, CDF, and complementary CDF of τ_ℓ
t_ℓ	Completion time for job ℓ
$g_\ell, G_\ell, \bar{G}_\ell$	PDF, CDF, and complementary CDF of G_ℓ
A_ℓ^*	Optimal appointment time for job ℓ
TC	Expected total of waiting and overtime costs
D_ℓ	Delay for job ℓ
$E[\cdot]$	Expectation operator
w_ℓ	Waiting cost per minute for job ℓ
k, ℓ, m	Indexes for jobs $1 \leq k, \ell, m \leq L+1$
$\tau_{k,m}$	Duration of continuous busy period starting with arrival of job k, including work from jobs $k - m$
$b_k, B_k, \bar{B}_k$	PDF, CDF, and complementary CDF of $t_{k,m}$

We model the total service duration for each job by the independently distributed random variable which takes realized values τ_ℓ with corresponding probability density function (PDF), cumulative distribution function (CDF), and complementary cumulative distribution function (CCDF) f_ℓ, F_ℓ, and $\bar{F}_\ell$ respectively. Let t_ℓ, represent the completion time for job ℓ measured from the start of the problem horizon at $t = 0$. We define g_ℓ, G_ℓ and $\bar{G}_\ell$ as the PDF, CDF, and CCDF of t_ℓ respectively.

The goal is to determine a set of optimal appointment times $\left(A_1^*, \ldots, A_{L+1}^*\right)$ for $L+1$ jobs. This schedule includes the virtual job, $L+1$ scheduled for time T. Thus delay for job $L+1$ is synonymous with overtime. Letting E refer to the expectation operator, the objective may be formally stated as,

$$\min E[TC(A_1, \ldots, A_{L+1})] = \left\{ \sum_{\ell=1}^{L+1} w_\ell E[D_\ell] \right\}$$
$$\text{s.t. } A_1 = 0 \leq A_2 \leq A_3 \ldots \leq A_{L+1} = T \tag{1}$$

Expected Total Cost is the sum of expected delays for each job $E[D_\ell]$ multiplied by a per minute waiting cost (w_ℓ). We allow w_ℓ to be job specific so that operating costs incurred during overtime operations are embedded in w_{L+1}. The expected delay for job ℓ is,

$$E[D_\ell] = \int_{t_{\ell-1}=A_\ell}^{\infty} (t_{\ell-1} - A_\ell) g_{\ell-1}(t_{\ell-1}) dt_{\ell-1}$$
$$= \int_{t_{\ell-1}=A_\ell}^{\infty} \left[1 - G_{\ell-1}(t_{\ell-1})\right] dt_{\ell-1} = \int_{t_{\ell-1}=A_\ell}^{\infty} \bar{G}_{\ell-1} dt_{\ell-1} \tag{2}$$

Our problem statement hinges on the ability to uncover G_ℓ for all ℓ. Given all F_ℓ, we can develop G_ℓ recursively. Since the system is presumed to be empty at time $t = 0$, we set:

$$G_0 \equiv \begin{Bmatrix} 1 \text{ for } t \geq 0, \\ 0 \text{ elsewhere} \end{Bmatrix} \tag{3}$$

The evolution of the system now depends on the service time of job 1. This leads to,

$$G_1(t) = \int_{\tau_1=0}^{t_1 - A_1} G_0(t - \tau_1) f_1(\tau_1) d\tau_1 \tag{4}$$

By induction, for $t \geq A_\ell$ we can recursively write,

$$G_\ell(t) = \int_{\tau_\ell=0}^{t_\ell - A_\ell} G_{\ell-1}(t_\ell - \tau_\ell) f_\ell(\tau_\ell) d\tau \tag{5}$$

Since total cost is a linear function of expected delays, and the only decisions are appointment times, we have particular interest in $\partial E[D_\ell]/\partial A_k$ where $k \leq \ell$. Consider the linkage between the delay for job ℓ and a small change in the appointment time for an earlier job k. There is no connection between A_k and D_ℓ if job k was itself delayed by an earlier job, or if job k starts on time, but the system is idle at any time between A_k and A_ℓ. However, if job k starts on time and the system is busy from that time until the arrival of job ℓ then there is a one to one correspondence between a change in A_k and D_ℓ. Thus the derivative of interest is meaningful iff the system is empty upon the arrival of job k and; a busy period that began at A_k extends beyond A_ℓ.

The probability that the server is free at A_k is simply $G_{k-1}(A_k)$ which is the CDF of the completion time of job $k-1$ evaluated at A_k. Let us define an auxiliary random variable $\tau_{k,m}$ with $m \geq k$ as the duration of an uninterrupted busy period caused by the work generated by jobs k through m. We label the PDF, CDF, and CCDF of this variable as $b_{k,m}$, $B_{k,m}$, and $\bar{B}_{k,m}$ respectively. Note from (5) that since the server is free when job 1 arrives at A_1, it initiates a busy period that is no less than τ_1. If τ_1 exceeds $A_2 - A_1$ then the arrival at A_2 will find the server busy, its service time will extend the busy period by τ_2, and we write $\tau_{1,2} > A_2 - A_1$. If, in addition we have $\tau_1 + \tau_2 > A_3 - A_1$ then the arrival at A_3 will also find the server busy, the busy period will be extended by τ_3, and we write $\tau_{1,3} > A_3 - A_1$, and so on. If we define terms such that $\tau_{k,k} \equiv \tau_k$ then $b_{k,k}\left(\tau_{k,k}\right) \equiv f(\tau_k)$ and recursively we can write,

$$b_{k,m}\left(\tau_{k,m}\right) = \int_{\tau=0}^{t_{k,m}-A_m} b_{k,m-1}\left(t_{k,m-1} = t_{k,m} - \tau\right) f(\tau) d\tau \tag{6}$$

To proceed with our analysis we make use of the following property.

Proposition 1 *When $0 < k < \ell$, the relationship between the expected delay of job ℓ and the appointment time A_k is described by,*

$$\begin{aligned}
\frac{\partial E[D_\ell]}{\partial A_k} &= G_{k-1}(A_k)\bar{B}_{k,\ell-1}(A_\ell - A_k), \\
\frac{\partial^2 E[D_\ell]}{\partial A_k^2} &= G_{k-1}(A_k) b_{k,\ell-1}(A_\ell - A_k) + g_{k-1}(A_k)\bar{B}_{k,\ell-1}(A_\ell - A_k) \geq 0, \textit{ and} \\
\frac{\partial^2 E[D_\ell]}{\partial A_k \partial A_\ell} &= -G_{k-1}(A_\ell) b_{k,\ell-1}(A_\ell - A_k) \leq 0
\end{aligned} \tag{7}$$

Thus $E[D_\ell]$ is decreasing and convex in A_ℓ and it is increasing and convex in A_k for all $k < \ell$. Since the expected delay for job $\ell > 1$ is increasing in A_1 and the expected delay for job 1 is 0, we conclude without loss of generality (WLOG) that the optimal value $A_1^* = 0$. We are also able to write the nested First Order Conditions (FOC) for the problem as,

$$\begin{aligned}
&\text{For all } 2 \leq \ell \leq L, \\
&\frac{dTC}{dA_\ell} = -w_\ell \bar{G}_{\ell-1}(A_\ell) + G_{\ell-1}(A_\ell) \sum_{m=\ell+1}^{L+1} w_m \bar{B}_\ell(A_m) = 0
\end{aligned} \tag{8}$$

The first term of each FOC represents the decrease in marginal expected delay cost for job ℓ when A_ℓ is increased. This marginal benefit is weighed by the probability that the server is busy when job ℓ arrives. The second set of terms represents the increase in delay costs for all subsequent jobs and is weighed by the complementary probability that the server is free when job ℓ arrives. Given the negative values of the cross-partials, it follows from [2] that *TC* is submodular, which leads to the following finding.

Proposition 2 *For a fixed value T, let $\mathbf{a_n^*} = (x_1, x_2 \dots x_n)$ be the set of optimal appointment times. If we add an additional job {n + 1} to be served after job n and find $\mathbf{a_{n+1}^*} = (y_1, y_2 \dots y_n, y_{n+1})$, then $y_i \leq x_i$ for all i from 1 to n.*

In other words, given an optimal schedule for *n* jobs, adding job *n + 1* implies that all optimal appointment times for jobs 2 through *n* are no later than in the case with only *n* jobs. This simplifies the search for an optimal schedule.

2.1 Incorporating Unreliable Patients

We extend the basic model to accommodate no-shows. One common response to the problems caused by no-shows is the use of double-booking. This complicates our problem in two respects. First, some random variables become mixtures. Second, it is possible to occupy the Trainee using Process 1 for one patient while the Attending uses Process 2 simultaneously for another. Consequently, we need to consider groups of patients. To this end, we schedule a group of patients as one "Cycle," which includes 1 New patient and 2 Return patients. We label the gap between consecutive cycles using Z and the three appointment times within the cycle as a_1, a_2, and a_3. When Process 2 is used in parallel to Process 1 it is easy to show that $a_1^* = 0$, and that any sample path with $a_2 > a_1$ is sub-optimal. Consequently, we have $a_1^* = a_2^* = 0$ and the search for the optimal policy (within each cycle) reduces to a search for a_3^*.

When a New patient is served using Process 1 the 3-tuple (n_1, n_2, n_3) represents the duration of the three stages and has joint PDF $\tilde{n}(n_1, n_2, n_3)$. The 3-tuple (r_1, r_2, r_3) represents the parallel values for a Return patient using Process 1 with joint PDF $\tilde{r}(r_1, r_2, r_3)$. If Process 2 is used for a Return patient we have a processing time of p with density $\tilde{p}$. Note that simultaneous use of Processes 1 and 2 may delay the start of stage 2 for the New patient.

At the start of a cycle at time A_ℓ there are four possibilities. With probability $F_{0,0}$ no patient arrives. With probability $F_{0,N}$ only the New patient arrives with total processing time $(n_1 + n_2 + n_3)$. With probability $F_{R,0}$ only a Return patient arrives with total processing time $(r_1 + r_2 + r_3)$. Finally, with probability $F_{R,N}$ two jobs arrive with total processing time $(\text{Max}(p, n_1) + n_2 + n_3)$. The third patient in the cycle as has the appointment time $A_\ell + a_\ell$.

The problem of determining the appointment times of C cycles can be represented as a problem with $2 \cdot C + 1$ jobs where job $2 \cdot C + 1$ represents the end of the session. The other odd numbered jobs correspond to the scheduled arrival of a "Composite" job at the start of each cycle caused by double booking, and the even numbered jobs each refer to a simple job corresponding to a Return patient arriving later in the cycle.

2.2 Properties of Optimal Appointment Times for the Problem of Three Cycles

We seek to schedule 3 cycles in which each cycle is comprised of two unreliable jobs. Three times are set for 3 (odd-numbered) composite jobs constituted of one New and one Return patient. Three additional times are set for 3 (even-numbered) simple jobs consisting of a single Return patient. For ease of exposition we often label the time between the odd-numbered jobs as the Z-gap and the time between an odd numbered job and the following even numbered job as the z-gap. For example, if the z-gap is fixed across all cycles we have $A_1 = 0$, $A_2 = z$, $A_4 = A_3 + z$, $A_6 = A_5 + z$, and $A_7 = T$. Thus, the problem involves only 3 controls, z, A_3, and A_5. Since odd numbered jobs represent the beginning of a cycle, we write G_1, G_3, and G_5 as,

$$G_\ell(t) = F_{0,0}G_{\ell-1}(t) + F_{0,N} \int_{\tau=0}^{t-A_\ell} G_{\ell-1}(t-\tau)d\tau_{0,N} + F_{R,0} \int_{\tau=0}^{t-A_\ell} G_{\ell-1}(t-\tau)d\tau_{R,0} + F_{N,R} \int_{\tau=0}^{t-A_\ell} G_{\ell-1}(t-\tau)d\tau_{N,R} \quad (9)$$

If we assume that the arrival probability of all Return jobs is ϕ, then for jobs 2, 4, and 6 we have,

$$G_\ell(t) = \bar{\phi}_R G_{\ell-1}(t) + \phi_R \int_{\tau=0}^{t-A_\ell} G_{\ell-1}(t-\tau)d\tau_{0,R} \quad (10)$$

Given the linkages between A_3 and Job 4, as well as between A_5 and Job 6 we have,

$$\frac{\partial TC_5}{\partial A_5} = -w_5\bar{G}_4(A_5) + w_6 \cdot G_4(A_5)\bar{B}_5(A_6) + w_7 \cdot G_4(A_5)\bar{B}_5(A_7) - w_6\bar{G}_5(A_6) + w_7 \cdot G_5(A_6)\bar{B}_6(A_7) \quad (11)$$

For the special case in which all z values are equal, analysis of 1 leads to the following result.

Proposition 3 *When $A_{\ell+1} = A_\ell + z$, $E[D_{\ell+1}]$ is decreasing in A_ℓ.*

Since a delay is only costly when the delayed job arrives, we include the arrival probabilities to write,

$$TC = \sum_{x=1}^{3} w_{2x}\phi_R E[D_{2x}] + \sum_{x=1}^{2} w_{2x+1}(\phi_N + \phi_R)E[D_{2x+1}] + w_7 E[D_7] \quad (12)$$

For this case TC is convex in each of A_3 and A_5 and the problem can be reduced to a line search in z because each z value implies optimal values of A_3 and A_5 which are uniquely determined by the FOCs.

3 Application in the AMC

Stated results assume a No-show rate of 25%, and combine 1 New patient using Process 1 and 1 Return patient using Process 2 at the start of each cycle. The second Return patient arrives at time z and uses Process 1. Our data set includes processing times for 23 New patients using Process 1, 45 Return patients using Process 1, and 43 Return patients using Process 2. We developed empirical distributions based on the observed data. Since the range of processing times is finite and measured in

Table 2 Expected delays and costs when OT costs $10 per minute

Policy	A2	A3	A4	A5	A6	PR OT	Exp Wait	Exp OT	Tot Cost	Err Min
Optimal	36	58	117	139	211	42.6	125.5	17.2	297.3	0.00
Fixed-Z	44	68	117	136	211	45.4	122.8	17.9	301.9	0.46
Fixed-*z*	50	67	117	144	194	44.1	124.1	18.4	307.7	1.04
3-Block	48	72	120	144	192	44.8	122.2	18.7	309.7	1.24
3-Block +	46	69	115	138	211	44.0	121.1	18.1	302.4	0.51
Fixed-*zZ*	50	72	122	144	194	44.9	120.4	18.9	309.2	1.19
Fixed-*zZ*+	45	68	113	136	211	42.9	124.0	17.8	302.3	0.50

Table 3 Expected delays and costs when OT costs $20 per minute

Policy	A2	A3	A4	A5	A6	PR OT	Exp Wait	Exp OT	Tot Cost	Err Min
Optimal	27	49	105	128	204	36.5	149.5	15.4	458.2	0.00
Fixed-Z	36	57	104	114	204	36.5	149.9	15.9	468.7	0.53
Fixed-*z*	42	56	98	131	173	37.9	158.8	16.0	478.1	1.00
3-Block	42	63	105	126	168	37.8	159.0	16.2	483.0	1.24
3-Block +	40	60	100	120	204	37.8	150.9	16.0	470.6	0.63
Fixed-*zZ*	42	64	106	128	170	38.2	155.2	16.4	482.6	1.23
Fixed-*zZ*+	38	61	99	122	204	37.1	149.0	16.0	469.7	0.58

one-minute increments, we are able to use a simple iterative scheme implemented in Mathematica® to compute G_ℓ functions given any vector of appointment times. These functions are used as the key module in a basic dynamic programming algorithm to find the optimal policy. For Tables 2 and 3, we set the cost per minute of patient delay at $1 while the overtime cost is set to $10 and $20 respectively. The first row of each table presents the optimal policy. When overtime cost rises the amount of expected overtime drops at the expense of increasing expected patient delay.

The optimal schedule yields cycles of uneven durations. In some settings, implementation can be simplified using heuristics that are more intuitive. The second policy listed on the table is labeled the Fixed-Z policy. This policy fixes the cycle length (values for $A_i - A_{i-1}$) but allows the duration from the start of the cycle to the arrival of the second Return patient (the z-gaps) to be determined independently for each cycle. In this case the number of decision variables is $C + 1$, which is 4 for the problem with 3 cycles. When comparing the costs of these policies it is convenient to divide Total Cost by the OT cost so that we can describe the cost added by deviating from the optimal policy using minutes of overtime operations as a unit of measurement (Err Min). Using this metric, we can say that the penalty associated with using the Fixed-Z policy, as opposed to the optimal policy is equivalent to less than 0.5 min of overtime operations. The Fixed-z policy selects a single value as the best z-gap but allows the time between cycles (the Z-gaps), to be determined independently. The cost under this policy is within 1 min of OT Cost of the optimal policy.

Several policies exist that have a fixed number of decision variables regardless of problem size. For example, in the 3-Block policy each cycle is the same length and the second Return patient arrives at a point in time, which corresponds to 2/3 of the cycle length. Note that the sum of these three blocks may be less than T and this policy is defined by a single parameter. The cost of the best 3-Block policy is within 1.2 min of OT Cost.

The 3-Block policy can be improved by making the time for the last appointment on the schedule a separate decision variable. We label this the 3-Block+policy. The resulting policy is fully defined by two variables and is computationally inexpensive since, given A_1 through A_5, TC is convex in A_6. This yields a policy that is within 0.63 min of the optimal. Another way to modify the 3-Block policy is to treat the Z-gap and z-Gap as two independent decision variables to yield the best Fixed-Zz policy. This is also a policy with only two controls, regardless of problem size. The optimal Fixed-Zz policy is within 1.2 min of OT Cost of the optimal cost. Finally, by making A_6 independent we define the three decision variable Fixed-Zz+policy. Making A_6 independent is virtually costless since TC is convex in A_6 given A_2 through A_5. The optimal Fixed-Zz+policy is within 0.6 min of OT Cost of the optimal policy. This scalable policy outperforms the Fixed-z policy, but is slightly inferior to the Fixed-Z policy.

4 Closing Comments

Specialty clinics in AMCs have process flows that differ from private practice settings. In particular the presence of a trainee creates a tradeoff between adding process steps while facilitating parallel processing. We model the strategic use of this capability and generate a number of results. This work incorporates the planned use of parallel processing, accounts for possible no-shows, accommodates arbitrary and job-specific service time distributions, as well as job specific no-show rates and waiting costs.

We use data collected in one AMC specialty clinic to illustrate how our model can be used to develop schedules for groups of patients organized as cycles. Because a pair of patients was treated as a composite job that can be served in parallel by the Attending/Trainee dyad we could represent cycles as alternating sets of composite and simple jobs, and we could treat the dyad as one single-server team. Using this construct, we were able to find the optimal schedule by using a dynamic programming scheme. We were also able to identify a series of simple heuristics that provided performance close to that of the optimal appointment schedule.

References

1. Hing E, Hall MJ, Ashman JJ, Xu J. National hospital ambulatory medical care survey: 2007 outpatient department summary. Natl Health Stat Rep. 2010;28:1–3.
2. Topkis DM. Minimizing a submodular function on a lattice. Oper Res. 1978;26:305–21.

Factors Influencing E-procurement Adoption in the Transportation Industry

Arim Park, Soohyun Cho, Seongtae Kim and Yao Zhao

Abstract On-demand matching services for commercial transportation needs have only recently entered the market in South Korea, with early players like Uber Freight and Convoy seeking new ways to curtail inefficiencies in the transportation sector. However, despite the availability of these services and the fact that advanced information technology can be readily applied in transportation-related fields, multiple factors, including negotiation power and inefficient matching rates between shippers and truckers, continue to influence transportation rates. With this in mind, we seek to conduct an empirical study by using logistic regression to identify the critical factors that might increase successful matching rates through the e-platform. Truckers consider travel time, fuel prices, truck and cargo types, load factor and O-D pairs to select a job when using the platform. They can use their search results to develop segmented operation strategies that vary according to the unique characteristics of individual truckers and shippers, the features of their services and other factors. By leveraging these and other aspects of the e-markets, truckers can expect to remain busy by filling any gaps in employment with the help of e-procurement.

1 Introduction

Freight transport is a critical activity in the supply chain, allowing for the storage and movement of products within the supply chain. Indeed, the downstream logistics of these services, moving from distributors to retailers, significantly affects consumers [1]. In South Korea, for example, over 90% of freight is transported by road (via

A. Park (✉) · Y. Zhao
Department of Supply Chain Management, Rutgers University, Newark, NJ 07102, USA
e-mail: arim.park@rutgers.edu

S. Cho
Department of Accounting Information System, Rutgers University, Newark, NJ 07102, USA

S. Kim
Chair of Logistic Management, Swiss Federal Institute of Technology Zurich, Weinbergstrasse 56-58, 8092 Zurich, Switzerland

H. Yang and R. Qiu (eds.), *Advances in Service Science*, Springer Proceedings in Business and Economics, https://doi.org/10.1007/978-3-030-04726-9_29

trucks). Yet despite the vital role that the trucking industry plays in the nation's economic development, analysts generally view the trucking market itself to be riddled with inefficiencies. The inefficient market structure of the Korean trucking sector renders it less competitive than other freight modes due to (1) the extraneous layers of brokers that wield inordinate market power through information asymmetry, and (2) a general lack of information transparency for both end users (customers) and actual service providers (truckers) [2, 3]. These have led many companies to focus on e-commerce shipping and logistics as a central part of their growth strategy. For example, Uber Freight, the on-demand trucking service created by Uber and introduced in 2017, is expanding its operations by looking for new and innovative ways to improve the day-to-day lives of truckers [4]. On-demand matching services for transportation, however, have only just begun in Korea, with early players like Uber Freight and Convoy still looking to solve inefficiencies in the transportation market. Thus, even though advanced information technology can be applied in the transportation sector, the hindrances to efficacy posed by factors such as popular routes, congestion rates, inter alia, have yet to be resolved. Driven by the pressures of cost and customer service, the transportation industry appears to be seeking to improve operational efficiency through research allocation and planning with e-commerce, leading analysts to project strong and continued growth of such services in the sector [5].

Our inquiry examines whether certain enabling factors can advance win-win situations between shippers and truckers through the scrupulous use of matching systems, allowing shippers to select the most appropriate truckers for their requests, and truckers to search for jobs at a lower cost through the e-platform. To this end, we use company data, who is a major enterprise that provides e-commerce matching service in South Korea. In the matching system, shippers seek to buy transportation capacities that are best matched with candidate truckers, i.e., the sellers of the transportation capacity. Using web-based tools, we are able to eliminate the intermediaries by providing direct contact between parties. In this way, shippers and truckers can avoid paying brokerage commissions. This means that web-based tools not only make the entire process faster and easier, but also significantly reduce clients' expenses. Thus, the main purpose of this study is to identify the main factors driving the phenomena described above by comparing transaction completions and cancellations in e-procurement within the trucking industry. Based on the data analysis results, short- and long-term risks and projected prices in the sector can be analyzed in order to suggest effective decision-making tools for the purposes of e-procurement in the transportation business. From a practical point of view, this might allow companies to re-evaluate their transportation services using the electronic marketplace to maximize their productivity.

2 Literature Review

In the past, procurement purchases were traditionally conducted via direct communication between the players involved in the supply chain. Recent developments in advanced technology allow for more productivity in the procurement process, enabling improved purchasing functions in an organization by culling redundant activities. Because many transportation companies also recognize that e-commerce can induce numerous benefits, they have begun adopting the e-marketplace to facilitate transportation operation services [6]. One merit of the e-marketplace is that it enables all players to easily exchange information regarding order entry, financial records, capacity status and other data, thus improving the efficiency of the matching mechanism between supply and demand [7, 8]. In calling for an e-commerce logistics trading system that contains a matching mechanism, previous study outlines several benefits of such a system [5]. The first advantage is facilitated calculation of the order-vehicle assignment results, with quick response after inputting the necessary information. Another merit is of such a system is the way it optimizes efficient allocation of transportation resources. Additionally, it contributes to analyzing the key factors from the dynamic environment. Numerous studies have examined matching systems in logistics, and many of those works have reported increased benefits, especially in terms of price considerations. However, very little research exists on the effectiveness of matching systems in transportation services with e-procurement by considering critical aspects such as load factor, among others.

3 Data

Generally, shippers register specific jobs using multiple websites; if they obtain a great deal with one specific information system, they cancel the transaction on the other sites. The high rate of rejections implies that shippers must have alternate truckers in place, along with appropriate mechanisms to access them [9]. Most times, truckers conduct the search, select the jobs, and then contact the shipper. Interested parties usually end up conducting negotiations via phone or fax.

The characteristics of on-demand matching services are summarized below.

- First come, first served system
- Shippers use spot contracts and consider full truckload
- There is no relationship between shippers and truckers
- Truckload transportation firms (truckers) generally handle shipments that are picked up at a location and driven directly to a single destination with no intermediate stops.

In this study, we use the dataset that was obtained from a major enterprise that provides logistics services using e-commerce matching in South Korea. The company collaborates with many truckers, creating business opportunities that allow for the continuous flow and employment of truckers and offer advantageous cost and time savings. We obtained data for shipments from January to August 2017 including origin and destination cities (15 cities in Korea), trucker and shipper names, requested tonnage from shippers, accepted tonnage from truckers and the total line haul price paid to the trucker. While the total transaction sample originally comprised 14,878 transactions, the cleaned dataset contained 14,421 transactions. However, the number of matching completions totaled only 139 cases, which resulted in a 0.9% successful matching rate. This low matching rate indicates that the current e-platform falls short of the expectations of shippers and truckers, underlining the need to identify significant factors such as freight rate that may contribute to higher matching rates in the e-marketplace [10]. An improved platform would directly benefit all players in the supply chain, including shippers, truckers and end consumers, via time and cost savings [11, 12]. The variables used in the study were selected based on recommendations by experts in the transportation market, as well as those indicated in the previous literature [13, 14, 15, 16].

4 Result

The following model uses logistic regression to identify the abovementioned factors by comparing successful matching cases to failed cases.

$$\begin{aligned} \mathrm{y(x)} = & \beta_0 + \beta_1(a) + \beta_2(b) + \beta_3(c) + \beta_4(d) + \beta_5(e) \\ & + \beta_6(f) + \beta_7(g) + \sum_{k=1}^{24} \beta_{8,k}(h) + \sum_{k=1}^{6} \beta_{9,k}(i) \end{aligned} \quad (1)$$

where, x = matching, a = window time, b = traveling time, c = window time × revision, d = freight rate per ton per km, e = fuel price, f = load factor, g = truck type (normal), h = origin-destination pairs and i = cargo types (chemistry). In addition, β indicates coefficient and equation shows the investigation of dependent variables (matching or fail) as dichotomous to test the predictive power of the independent variables and estimate their relative contribution to the dependent variables.

Table 1 displays the empirical findings for the multiple logistic regression model in which we examine the probability of increasing the successful matching rate in the Korean trucking industry's e-marketplace. The successful matching cases with different magnitudes were affected by different independent variables that consider the truckers' perspective. In terms of critical factors associated with successful matching rates in the platform, the results were significant for travel time, window time × revision, freight rate per ton/per km, fuel price, truck type, load factor, product type 2 (chemistry product) and some O-D pairs.

The results show the following implications: when using e-procurement, successful matching cases demonstrated less willingness (shorter time) on the part of shippers to wait and longer requested time between the loading and unloading date and time

Table 1 Logistic regression analysis

	Coefficient	Wald	Significance (Sig.)	Odds ratio
Travel time	0.373***	11.139	0.001	1.452
Window time × revision	−0.153**	6.451	0.011	0.858
Fuel price	−0.199*	2.884	0.089	0.819
Freight rate	0.162***	12.428	0.000	1.176
Normal trucks	0.588**	5.171	0.023	1.800
Load factor	2.623***	106.058	0.000	13.776
Chemistry	−0.615*	2.931	0.087	0.541
SCA-DC	0.732*	2.732	0.098	2.079
DC-SCA	0.776*	3.755	0.053	2.174
DC-DC	0.849*	2.771	0.096	2.337
DC-GJ	2.225***	24.493	0.000	9.250
DC-BS	1.551***	11.361	0.001	4.716
DN-GJ	3.136***	7.754	0.005	23.011
GJ-BS	1.529**	5.194	0.023	4.614
BS-SCA	1.293*	3.587	0.058	3.645
BS-DC	1.930**	9.546	0.002	6.892
BS-DN	2.772***	38.679	0.000	15.992
BS-GJ	1.418*	3.149	0.076	4.129
Number of total observations			14,421	
Likelihood ratio, (*p*-value)	Chi: 323.236	Df: 37	Sig.: 0.000	Nagelkerke R square: 0.215
Hosmer-Lemeshow test	Chi: 13.170	df: 8		Sig.: 0.106

*p value < 0.100; **p value < 0.05; ***p value < 0.01

(travel time) as compared to failed cases. Successful matching cases also featured higher transportation costs and lower fuel prices than in failed cases. Furthermore, shippers that own general purpose trucks (as opposed to specialized purpose trucks) are likely to find well-matching jobs due to high demand for such trucks. However, our analysis reveals that only some O-D pairs are relevant to the matching rates due to considerations of highway networks and traffic congestion related to truckers' preference for O-D pairs.

5 Conclusions and Future Research Directions

E-procurement in transportation services helps to create more transparent, effective and efficient freight markets, especially when considering the divergent considerations of shippers and truckers [2, 3]. According to our case study, truckers considered travel time, fuel price, truck type and cargo types, load factor and O-D pairs when selecting a job using an e-platform. While both players expect to match their operational requirements quickly in the e-market, they weigh considerations of time differently. To improve the current trading system, the matching process should be changed dynamically through negotiation, communication and tracking. The matching site should encompass an integrated system that can provide brokerage services to multiple independent truckers and the shipper to allow parties to reach a satisfactory deal.

Our paper is not without its limitations, which mainly arise from a relative lack of data. For example, the dataset relies on cross-sectional information comprising data collected at a single point in time. Thus, a future study might incorporate a longitudinal design in which the firm might improve its business prospects by reconstructing its resources based on our suggestions and in line with the evolution of capability and competence examined in this paper. Another limitation of this study stems from the characteristics of the on-demand matching freight service, which is still in its nascent stages in South Korea. We hope that this study will generate further interest and engagement with the emerging literature on e-platforms for freight services.

References

1. McKinnon A. Life without trucks: the impact of a temporary disruption of road freight transport on a national economy. J Bus Logist. 2006;27(2):227–50.
2. Hahn J, Sung HM, Park MC, Kho S, Kim D. Empirical evaluation on the efficiency of the trucking industry in Korea. KSCE J Civ Eng. 2015;19(4):1088–96.
3. Shin DS. Study on the efficiency of consignment contracts between goods carriers and drivers. J Transp Res. 1995;2:51–63.
4. https://www.supplychain247.com/article/uber_freight_introduces_personalized_load_matching.
5. Zhang M, Huang GQ, Xu SX, Zhao Z. Optimization based transportation service trading in B2B e-commerce logistics. J Intell Manuf. 2016:1–17.
6. Nandiraju S, Regan A. Freight transportation electronic marketplaces: a survey of the industry and exploration of important research issues; 2008.
7. Wang J, Liu D, Ip WH, Zhang W, Deters R. Integration of system-dynamics, aspect-programming, and object-orientation in system information modeling. IEEE Trans Ind Inf. 2014;10(2):847–53.
8. Wang J, Wang H, Zhang W, Ip WH, Furuta K. On a unified definition of the service system: What is its identity? IEEE Syst J. 2014;8(3):821–6.
9. Caplice C. Electronic markets for truckload transportation. Prod Oper Manag. 2007;16(4):423–36.
10. http://www.supplychainquarterly.com/topics/Technology/20180227-10-technologies-that-will-reshape-scm-software/.

11. Basole RC, Karla J. Value transformation in the mobile service ecosystem: a study of app store emergence and growth. Serv Sci. 2012;4(1):24–41.
12. Scott A, Parker C, Craighead CW. Service refusals in supply chains: Drivers and deterrents of freight rejection. Transp Sci. 2017;51(4):1086–101.
13. Budak A, Ustundag A, Guloglu B. A forecasting approach for truckload spot market pricing. Transp Res Part A: Policy Pract. 2017;97:55–68.
14. Joo S, Min H, Smith C. Benchmarking freight rates and procuring cost-attractive transportation services. Int J Logist Manag. 2017;28(1):194–205.
15. Özkaya E, Keskinocak P, Joseph VR, Weight R. Estimating and benchmarking less-than-truckload market rates. Transp Res Part E: Logist Transp Rev. 2010;46(5):667–82.
16. Smith LD, Campbell JF, Mundy R. Modeling net rates for expedited freight services. Transp Res Part E Logist Transp Rev. 2007;43(2):192–207.

MIX
Papier aus verantwortungsvollen Quellen
Paper from responsible sources
FSC® C105338

If you have any concerns about our products,
you can contact us on
ProductSafety@springernature.com

In case Publisher is established outside the EU,
the EU authorized representative is:
Springer Nature Customer Service Center GmbH
Europaplatz 3, 69115 Heidelberg, Germany

Printed by Libri Plureos GmbH
in Hamburg, Germany